21世纪高等学校数字媒体专业规划教材

数字媒体专业英语

（第2版）

◎ 周灵 薛雁丹 编著

清华大学出版社
北京

内 容 简 介

本书旨在让学生掌握较多相关专业英语词汇和数字媒体的基本概念，为阅读专业文献和书籍打下坚实的基础，同时为在以后工作中解决与专业英语相关的问题提供必要的知识保证。

本书结构编排完整、信息材料新颖。首先简单介绍了专业英语的翻译方法；第一部分阐述数字媒体相关概念，包括多媒体综述、数字图像处理、数字音频、媒体资产管理、虚拟现实技术等内容；第二部分对常用的数字媒体应用软件进行介绍，如Photoshop、Flash、Dreamweaver、3ds Max、Premiere、After Effects等；第三部分介绍包括数据传输技术、光纤技术、无源光网络技术、电视原理、电视接收机在内的电信与通信技术；第四部分阐述计算机系统构成；第五部分列举常用的编程语言，如Java、ActionScript、C++等；第六部分介绍计算机网络的概念、应用及网络安全相关知识。每篇课文都配有习题及参考译文，供相关专业的师生参考。

本书封面贴有清华大学出版社防伪标签，无标签者不得销售。
版权所有，侵权必究。举报：010-62782989，beiqinquan@tup.tsinghua.edu.cn。

图书在版编目（CIP）数据

数字媒体专业英语/周灵，薛雁丹编著. —2版. —北京：清华大学出版社，2019（2023.6重印）
（21世纪高等学校数字媒体专业规划教材）
ISBN 978-7-302-50749-9

Ⅰ. ①数… Ⅱ. ①周… ②薛… Ⅲ. ①数字技术－多媒体技术－英语 Ⅳ. ①TP37

中国版本图书馆CIP数据核字（2018）第172074号

责任编辑：魏江江　李　晔
封面设计：刘　键
责任校对：李建庄
责任印制：丛怀宇

出版发行：清华大学出版社
　网　　址：http://www.tup.com.cn, http://www.wqbook.com
　地　　址：北京清华大学学研大厦A座　　邮　　编：100084
　社 总 机：010-83470000　　邮　　购：010-62786544
　投稿与读者服务：010-62776969，c-service@tup.tsinghua.edu.cn
　质 量 反 馈：010-62772015，zhiliang@tup.tsinghua.edu.cn
印 装 者：涿州市般润文化传播有限公司
经　　销：全国新华书店
开　　本：185mm×260mm　　印　张：21.5　　字　　数：520千字
版　　次：2011年8月第1版　　2019年1月第2版　　印　　次：2023年6月第5次印刷
印　　数：10901～11400
定　　价：49.50元

产品编号：076089-01

出版说明

数字媒体专业作为一个朝阳专业，其当前和未来快速发展的主要原因是数字媒体产业对人才的需求增长。当前数字媒体产业中发展最快的是影视动画、网络动漫、网络游戏、数字视音频、远程教育资源、数字图书馆、数字博物馆等行业，它们的共同点之一是以数字媒体技术为支撑，为社会提供数字内容产品和服务，这些行业发展所遇到的最大瓶颈就是数字媒体专门人才的短缺。随着数字媒体产业的飞速发展，对数字媒体技术人才的需求将成倍增长，而且这一需求是长远的、不断增长的。

正是基于对国家、社会、人才的需求分析和对数字媒体人才的能力结构分析，国内高校掀起了建设数字媒体专业的热潮，以承担为数字媒体产业培养合格人才的重任。教育部在 2004 年将数字媒体技术专业批准设置在目录外新专业中（专业代码：080628S），其培养目标是“培养德智体美全面发展的、面向当今信息化时代的、从事数字媒体开发与数字传播的专业人才。毕业生将兼具信息传播理论、数字媒体技术和设计管理能力，可在党政机关、新闻媒体、出版、商贸、教育、信息咨询及 IT 相关等领域，从事数字媒体开发、音视频数字化、网页设计与网站维护、多媒体设计制作、信息服务及数字媒体管理等工作”。

数字媒体专业是个跨学科的学术领域，在教学实践方面需要多学科的综合，需要在理论教学和实践教学模式与方法上进行探索。为了使数字媒体专业能够达到专业培养目标，为社会培养所急需的合格人才，我们和全国各高等院校的专家共同研讨数字媒体专业的教学方法和课程体系，并在进行大量研究工作的基础上，精心挖掘和遴选了一批在教学方面潜心研究并取得了富有特色、值得推广的教学成果的作者，把他们多年积累的教学经验编写成教材，旨在为数字媒体专业的课程建设及教学起到抛砖引玉的示范作用。

本系列教材注重学生的艺术素养的培养，以及理论与实践的结合。为了保证出版质量，本系列教材中的每本书都经过编委会委员的精心筛选和严格评审，坚持宁缺毋滥的原则，力争把每本书都做成精品。同时，为了能够让更多、更好的教学成果应用于社会和各高等院校，我们热切期望在这方面有经验和成果的教师能够加入本套丛书的编写队伍，为数字媒体专业的发展和人才培养做出贡献。

21 世纪高等学校数字媒体专业规划教材

联系人：魏江江　weijj@tup.tsinghua.edu.cn

前　言

专业英语教学是大学英语教学及专业课教学的重要组成部分，是帮助和促进学生继续学习英语并学以致用的有效途径。而数字媒体代表着一个文理融合的全新领域，体现了技术与人文艺术的融合。

提起数字媒体，人们往往想到媒体行业，其实数字媒体的应用绝不仅仅局限于媒体。随着计算机技术、网络技术和数字通信技术的高速发展与融合，传统的广播、电影快速地向数字音频、数字视频、数字电影方向发展，与日益普及的计算机动画、虚拟现实等构成了新一代的数字传播媒体。因此，本教材适合数字媒体技术，计算机、通信电子，以及相关专业的大学本科、独立学院、高职高专的学生使用，也可供广大工程技术人员学习和参考。

通过“数字媒体专业英语”课程的学习，学生能掌握较多相关专业英语词汇和数字媒体的基本概念，为阅读专业文献和书籍打下坚实的基础，同时为在以后工作中解决与专业英语相关的问题提供必要的知识保证。本课程的目的是使学生不仅能学到专业英语词汇，扩大知识面，同时能掌握用英语表达专业知识的方法，提高阅读及理解专业英文资料的能力，掌握专业文献翻译的方法和技巧。

本教程的课文包括专业英语翻译方法、数字媒体、数字媒体应用软件、平面设计理论、通信和电信技术、计算机系统、编程语言、计算机网络等专业知识，取材丰富，所选篇目具有代表性。首先简单介绍了专业英语的翻译方法；第一部分阐述数字媒体相关概念，包括多媒体综述、数字图像处理、数字音频、媒体资产管理、虚拟现实技术等内容；第二部分介绍常用的数字媒体应用软件，如 Photoshop、Flash、Dreamweaver、3ds Max、Premiere、After Effects 等；第三部分介绍包括数据传输技术、光纤技术、无源光网络技术、电视原理、电视接收机在内的电信与通信技术；第四部分阐述计算机系统构成；第五部分列举常用的编程语言，如 Java、ActionScript、C++等；第六部分介绍计算机网络的概念、应用以及网络安全相关知识。

本书的主要特点如下：

（1）知识介绍系统。根据数字媒体技术的跨学科特点，系统介绍其基本理论与知识。

（2）结构编排完整。每篇课文都附有词汇、习题、译文等。

（3）信息材料新颖。所选素材具有很强的代表性、时代性、实用性、参考性。

（4）配有电子教案。为教师教学、学生自学提供方便。

本书 Part 0～Part 3 由周灵编写，Part 4 由周灵、薛雁丹共同编写，Part 5、Part 6 由薛雁丹编写，周灵、薛雁丹共同完成附录的整理工作，单园园完成全书的校对工作。

本书在编写过程中，得到了苑文彪老师的热情指导与帮助，在此表示感谢！

本书在选材时，参考了国内外有关书籍和资料，在此对这些文献的作者表示诚挚的敬意和谢意！

由于信息技术发展日新月异，新的知识在不断扩展更新，加上编者的学识和水平有限，书中疏漏、错误之处在所难免，敬请广大读者不吝批评指正。

编　者

2018 年 10 月于南京

目录

Part 0 专业英语翻译方法

0.1 专业英语翻译概述

0.1.1 专业英语的翻译标准

专业英语的翻译标准是准确规范、通顺易懂、简洁明晰。

1. 准确规范

所谓“准确”，就是忠实地传达原文的全部信息；所谓“规范”，就是译文要符合所涉及的某个专业领域的表达规范。要做到这一点，必须充分地理解原文所表述的内容，其中包括对词汇、语法、逻辑关系和科学内容的深入理解。例如：

Velocity changes if either the speed or the direction changes.

分析：在物理学中，速度和速率是两个不同的概念。速度（velocity）是矢量，有大小和方向；而速率（speed）是标量，有大小而没有方向。所以这句话应译成：

如果（物体运动的）速率和方向有一个发生变化，则（物体的）速度也会随之变化。

2. 通顺易懂

所谓“通顺易懂”，就是指译文的语言符合语法结构及表达习惯，容易为读者理解和接受。例如：

When a person sees, smells, hears or touches something, then he is perceiving.

当一个人看到某种东西，闻到某种气味，听到某个声音或触到某物时，他是在**察觉**。（不好）

当一个人看到某种东西，闻到某种气味，听到某个声音或触到某物时，他是在运用感官感受。（较好）

3. 简洁明晰

所谓“简洁明晰”，就是译文要简洁清楚，尽量避免烦琐、冗赘和不必要的重复。例如：

It should be realized that magnetic forces and electric forces are not the same.

磁力和电力的不一样是应该被认识到的。（生硬、啰唆）

应该认识到，磁力和电力是不同的。（简洁、较好）

0.1.2 专业英语文献的特点

1. 无人称句多

可以说，大多数的专业英语文章很少使用人称，这是由于专业文体的主要目的在于阐述科学事实、科学发现、实验结果等，故以客观陈述为主，因此专业英语大量使用被动语态。例如：

1s and 0 s can't be sent as such across network links.

1信号和0信号不能通过网络链路直接传送。

2．专业名词、术语多

专业英语专业性强，文体正式，使用大量的专业名词和术语。例如：

Third is **transmission mode**. The transmission mode includes **asynchronous** and **synchronous**. In **asynchronous transmission mode**, individual characters (made up of bits) are transmitted at irregular intervals, for example, when a user enters data.

第三是**传输模式**。传输模式包括**异步**和**同步**。在**异步传输模式**中，单独的字符（由比特组成）以不规则的时间间隔传输，例如当用户输入数据时。

3．非谓语动词多

有的段落中，多处使用非谓语动词。例如：

Packer-switched networks, the type usually **used** to connect computers, take an entirely different approach. In a packet-switched network, data **to be transferred** across a network is divided into small pieces **called** packets that **multiplexed** onto high capacity intermachine connections.

分组交换类型的网络通常用于连接计算机，它采取完全不同的方法。在一个分组交换网络中，网络上传输的数据被分成一个个小块，称为分组，分组被多路复用在大容量的机器间的连接上。

4．长句多

在专业英语文体中经常使用长句。这主要是因为在阐述科学事实、科学现象等事物的内在联系和解释一些科技术语或名词时，需要大量使用各种从句（尤其是定语从句）以及介词短语、形容词短语、分词短语或副词等作后置定语，分词短语表示伴随情况等。例如：

For example, a large file to be transmitted between two machines must be broken into many packets that are sent across the network one at a time, the network hardware delivers the packets to the specified destination, where software reassembles them into a single file again.

例如，一个要在两台机器间传送的大文件，必须被分成许多分组，在网络上一个一个地传送。网络硬件把分组传送到指定目的地，在那里，软件把它们重新组装成一个文件。

0.1.3 翻译的一般方法

1．直译与意译

1）直译

“直译”即基本保持原文表达形式及内容，改动较少或基本不改动，同时能做到语言通顺易懂，表述清楚明白。“直译”强调的是“形似”，主张将原文内容按照原文的形式（包括词序、语序、语气、结构、修辞方法等）直接表述出来。例如：

Multimedia systems combine a variety of information sources, such as voice, graphics, animation, images, audio and full-motion video, into a wide range of three industries: computing, communication and broadcasting.

多媒体系统综合了种类繁多的信息源，如语音、图形、动画、图像、声音和全动视频，涵盖了三大产业：计算机产业、通信产业和广播产业。

2）意译

“意译”，是将原文所表达的内容以一种释义性的方式把意义表达出来。“意译”强调“神似”，也就是不拘泥于原文在词序、语序、语法结构等方面的形式，用人们习惯的表达方式将原文的本意（真实含义）翻译出来。例如：

The primary colors consist of three hues from which we can theoretically mix all other hues.

从理论上说，原色包含了能够混合任何其他色调的三种色调。

总之，在翻译实践中，直译与意译不是两种完全孤立的翻译方法，我们不应该完全拘泥于某一种，要学会将直译与意译有机地结合起来，最终完整、准确、通顺地表达出原文意义。

2. 合译与分译

1）合译

“合译”就是把原文两个或两个以上的简单句或复合句在译文中用一个单句来表达。例如：

The advantage of a bitmap over a vector is that it can represent a much more complex range of colors and shades.

相对于向量图形，像素图形能够呈现更多复杂的颜色与阴影的变化。

2）分译

“分译”就是把原文的一个简单句中的一个词、词组或短语译成汉语的一个句子，这样，原文的一个简单句就被译成了汉语的两个或两个以上的句子。例如：

The sender is guaranteed that the samples can be delivered and reproduced because the circuit provides a guaranteed data path of 64 kb/s (thousand bits per second), the rate needed to send digitized voice.

发送方确信采样一定会被传输和重新生成，因为线路提供了一条被保证的 64kb/s（千比特每秒）数据路径，这个速率是发送数字化的语音所必需的。

3. 增译与省译

英语与汉语在表达上有着很大的差异。在汉译过程中，如果按原文英语句子一对一地翻译，译文很难符合汉语的表达习惯，会显得生搬硬套、牵强附会。在翻译过程中，译者应遵循汉语的习惯表达方式，在忠实原文的基础上，适当地进行增译或省译。

1）增译

所谓增译，就是在译文中增加英语原文省略或原文中无其词而有其义的词语，使译文既能准确地表达原文的含义，又符合汉语的表达习惯和修辞需要。例如：

Of course, there is still a long way to go before the ATM techniques is in general use, but a revolution is underway which will deeply affect the world of telecommunications, data processing and video. The impact of this upheaval will without any doubt be greater than the advent of digital techniques in analogue networks.

当然，在 ATM 技术普遍使用之前会有很长的路要走，但是这场正在进行的**技术**革命将会深刻地**影响**数据处理、视频处理和电信世界。这一革命**所产生的**影响无疑会比在模拟网络中出现数字技术**的影响**要大得多。（增译了“技术”“影响”“所产生的”等）

2）省译

通常情况下，翻译时需忠实原文，不允许对原文的内容有任何删略，但由于英、汉两种语言表达方式的不同，英语句子中有些词语如果硬要译成汉语，反而会使得译文晦涩难懂。为使译文通顺、准确地表达出原文的思想内容，有时需将一些词语省略不译。例如：

In an automatic speaker recognition system, the function of the system is to extract information from an incoming speech signal for the purpose of recognizing or identifying the speaker.

在自动语音识别系统中，系统的功能是从输入的语音信号中提取相关信息加以识别并确认讲话者的身份。（省译了名词 purpose）

0.2 词的翻译

0.2.1 词义的选择

1．根据上下文选择词义

英语中的同一个词、同一词类，在不同场合往往有不同的含义，常常要求我们根据上下文的联系和句型来确定某个词在特定场合下的词义。

以动词 develop 为例：

In 1995 the first generation of digital mobile phones came into being designed for voice calls only; and during 1996 and 1997 the second generation had been **developed** and was added to some other functions, such as replying e-mails or visiting web pages

1995 年，第一代数字手机出现，只用于通话，而 1996—1997 年**开发**的第二代数字移动通信在其基础上增加了数据接收功能，比如接收电子邮件或浏览网页。

In television, if the picture scan rate is made too low, moving scenes will **develop** a stop-and-go jerky movement in the same manner that slow-motion moving pictures do.

在电视中，如果图像扫描速率太低，移动景物会**出现**停停走走的抖动，好像慢动作的移动画面。

Other isolation methods are being **developed**.

目前正在**研究**其他隔离方法。

In **developing** the design, we must consider the feasibility of processing.

在**进行**设计时，必须考虑加工的可行性。

2．根据词的搭配来选择词义

英语的一词多义往往也体现在词与词的搭配上。不同搭配方式可以产生不同的词义。

以 large 为例：

large current　强电流

large pressure　高电压

a **large** amount of electric power　大量的电力

large loads　重载

large-screen receiver　宽屏电视接收机

large capacity　大容量

large growing　生长快的

3．根据学科和专业选择词义

同一个词在不同的学科领域或不同的专业中具有不同的词义，因此，在选择词义时，应考虑到阐述内容所涉及的概念属于哪个学科、何种专业。

以英语名词 carrier 为例：

邮政业：邮递员

军事：航空母舰

运输业：搬运工

生物化学：载体

医药学：带菌者

车辆制造：底盘

无线电：载波

0.2.2　名词的译法

名词可以分为普通名词、专有名词、集体名词、物质名词和抽象名词等。一般来说，物质名词、抽象名词和专有名词为不可数名词，普通名词和集体名词为可数名词。下面归纳几种名词翻译的方法。

1．直译

所谓"直译"，就是把原文中名词或名词短语本身的含义翻译出来。例如：

Packet-switched networks, the type usually used to connect computers, take an entirely different approach.

分组交换类型的网络通常用于连接计算机，它采取完全不同的方法。

2．重复

在英语句子中，常见的并列连词有 and, or, as well as, not only…but also, but, not…but, either…or, both…and, neither…nor 等。这些并列连词通常用于连接两个并列成分，其相应的核心名词在汉译时为了译文的通顺及语意完整，可以重复翻译该名词。例如：

There are two kinds of video capture devices, analog and digital.

视频采集设备有两种：模拟采集**设备**和数字采集**设备**。

3．增译

名词"增译"的目的是使译文更符合汉语的表达方式与习惯。例如：

ATM, much more than any other telecommunications technique, is able to meet the current and the future requirements of both operators and users. Compared with other techniques that may compete in certain applications, ATM is special mainly due to its universal nature, both in terms of bit rate and type of information transferred.

ATM 远比任何其他电信技术更能满足运营公司和用户对当前和未来业务的需求。与其他有可能在某些应用领域与 ATM 竞争的技术相比，ATM 主要由于其通用性，无论是比特率还是传输的信息类型都具有特殊的**优点**。（增译了"优点"）

4．转译

1）名词转译为动词

英语中有大量动词派生的名词和具有动作意义的名词，均可以转译为汉语的动词。例如：

According to consumer reports, prices for HDTV's are dropping sharply as the **selection** of HD sets, big screens especially, is growing, and display technologies are improving.

根据消费者的报告，高清晰度电视的价格大幅度下降而且显示技术不断改进，**选择**高清设备特别是大屏幕设备的人越来越多。（动词派生的名词 selection 转译为动词）

2）名词转译为形容词

例如：

In view of these needs, these satellite data are of great human **value**, provided they can be reduced to useful information both quickly and economically.

由于这些需要，如果这些卫星数据能被快速经济地精简为有用信息，那么它们会是很**有价值的**。（名词 value 转译为形容词）

3）名词转译为副词

例如：

Digital audio has emerged because of its **usefulness** in the recording, manipulation, mass-production, and distribution of sound.

数字音频的出现是因为它能**有效地**记录、处理、大批量生产和传播声音。（名词 usefulness 转译为副词）

4）其他类型的转译现象

例如：

Of course, there is still a long way to go before the ATM techniques is in general **use**, but a revolution is underway which will deeply affect the world of telecommunications, data processing and video.

当然，在 ATM 技术普遍**使用**之前会有很长的路要走，但是这场正在进行的技术革命将会深刻地影响数据处理、视频处理和电信世界。（名词 use 转译为动词）

5）省译

注意：名词多为实义词，一般情况下是不能省译的，但是在有的句子中，如果把名词译出来反而会使译文显得冗赘、啰唆，在能完整地表达原文意思的前提下，省译名词能使译文通顺流畅、简洁明晰。例如：

The major advantages of these reactors are excellent uniformity, large load size, and ability to accommodate large diameter wafers.

这些反应器的主要优点是均匀性好、负载量大，适用于大直径晶片。（省译了名词 ability）

0.2.3 冠词的译法

英语中的冠词有 3 种：一是定冠词（the Definite Article）；二是不定冠词（the Indefinite Article）；三是零冠词（Zero Article）。

英语中的定冠词就是 the。

不定冠词 a（an）与数词 one 同源，是“一个”的意思。不定冠词在句子中最大的语法功能是：用在可数名词的单数形式前表示泛指——表明一类人或事物区别于其他类。

零冠词是指名词前面没有不定冠词（a、an）、定冠词（the），也没有其他限定词的现象。

1．不定冠词

1）不定冠词的一般译法

英语的不定冠词 a 和 an 均用在单数名词之前，表示某一类人或事物中的“一个”。相当于汉语的“一”，但不强调数目概念。不定冠词表示数量时可以译为汉语的“一”，并相应加上汉语表示该事物的量词。请注意，英语的不定冠词不仅表示数量，还可表示分类。例如：

POF is **a** newer plastic-based cable which promises performance similar to glass cable on very short runs, but at a lower cost.

塑料光缆是**一种**新型的以塑料为基础的光缆，在短距离通信时其性能与石英玻璃光缆相似，但费用要低得多。

有时不定冠词可以放在某些名词（如 day, week, year, month 等）之前，和名词一起作状语，此时不定冠词译为“每一”或“一”。例如：

In half-duplex transmission, data can flow in both directions but in only one direction at **a** time.

在单工传输中，数据**每次**仅沿着一个方向流动。

2）不定冠词的省译

当不定冠词表示事物的类别时，常常可以省略不译。例如：

An ATM network can be considered, in a first approximation, as being three overlaid functional levels: a services and applications level, an ATM network level and a transmission level.

ATM 网络可以近似地看作是由三个覆盖功能层组成：业务和应用层、ATM 网络层、传输层。

2．定冠词

1）定冠词的一般译法

英语定冠词 the 表示某一类特定的人或事物中的“某一个”，可译为“该”“这个/种”“那个/种”等。如果定冠词后面是复数名词，the 就可以译为“这些”“那些”。例如：

The process is repeated until the entire screen is refreshed.

这个过程不断重复，直至整个屏幕被刷新。

2）定冠词的省译

通常情况下，定冠词与一个可数名词连用，代表一类人或事物，不含有特指的意义时，可省译。例如：

In **the** early days of television, a technique called “interlacing” was used to reduce **the** amount of information sent for each image.

在早期的电视技术中，有一种名为“隔行”的技术，用于减少发送给各个图像的信息。

0.2.4 代词的翻译

代词（pronoun，简称 pron.），是代替名词、名词词组甚至整个句子的一种词类。大多数代词具有名词和形容词的功能。英语中的代词，按其意义、特征以及在句中的作用分为人称代词、物主代词、指示代词、反身代词、相互代词、疑问代词、关系代词、连接代词和不定代词九种。为了清楚地表达代词本身的意义，以及所代替的词、词组或句子的意义，翻译时，首先要辨明代词所代替的先行词，以便正确地理解原文；其次要在译文中采用适当的翻译方法给以明确的表达，避免引起歧义。

1. 直译

“直译”就是直接翻译出代词本身的意思，此法常用于翻译人称代词。例如：

Key frame refers to such a frame whose contents are quite different from those of the precedent frame, so that **it** presents great changes in critical actions and contents.

所谓关键帧也就是该帧中的内容与先前帧中的内容有很大的区别，因而**它**呈现出关键性的动作及内容的变化。

2. 还原

所谓还原，就是把代词所代替的名词、名词词组的意思翻译出来。例如：

In addition, their potential speed for data communications is up to 10 000 times faster than **that** of microwave and satellite systems.

另外，它们实际的数据通信速度比微波和卫星系统的数据通信速度高10 000倍。（代词that还原为名词“数据通信速度”）

3. 互换

所谓互换，就是将英语原文句子中的名词（即先行词）与代替该名词的代词，在翻译时相互换位。具体的译法就是将代词译为该名词，将名词译为代词，使译文更符合汉语的表达习惯。这种译法多用于翻译英语中的偏正复合结构的句子。例如：

When **they** require the communicate resources, **the users and devices** require a relatively rapid response.

如果**用户和设备**用到这些通信资源，**他们**需要得到快速的响应。

4. 增译

增译代词通常是在译文的句首（有时在句中）增译“人们”“有人”“它们”“我们”等泛指代词，这完全是为了遵循汉语的表达习惯。代词的增译在被动句中更为常见，但是在翻译主动句时，为了行文的流畅，也可以增译代词。例如：

On the other hand, ATM retains all the flexibility of the packet mode, enabling only required information to be conveyed, offering a simple, unique multiplexing method irrespective of the bit rates of the different information flows, and allowing these bit rates to be varied.

另一方面，ATM保留了分组方式所有的灵活性，只传送所需要的信息，**为人们**提供简单、独特的复用方法而不管不同信息流的比特率，并且允许比特率变化。

5. 转译

转译就是将英语原文句子中的代词在汉译时翻译成其他词性的词，如转译为名词或副词，以使译文流畅，更符合汉语的表达习惯。例如：

All the other codecs discussed here are lossy which means a small part of the quality is lost.

这里提及的其他编解码方法**都**是有损的，即损失一小部分质量。（不定代词all转译成副词）

6．省译

省译代词的原则就是既要使译文符合汉语的表达习惯，又不能影响译文的准确性。英语中代词的使用较为频繁，翻译成汉语时，不需要全部译出。例如：

Layers provide a powerful way for you to organize and manage the various components of your image.

层为组织和处理图像的各种各样的组成部分提供了一种强有力的方式。（省译了人称代词 you 和 your）

0.2.5 数词、形容词和副词的翻译

0.2.5.1 数词的译法

“数词”用于表示数目的多少与顺序先后。数词与不定代词相似，其用法相当于形容词和名词，在句子中可作主语、表语、宾语和定语等成分。科技英语中用到数词的地方较多。理解和翻译数词本身并不难，但当数字在句子中表示各种概念时，要理解和翻译一些表示数量概念的句型结构时可能就会遇到困难，甚至有时候出现误解、错译等现象。

1．不定数量的译法

所谓不定数量，是指表示若干、许多、大量、不少、成千上万等概念的词组。英语中能够表示这些概念的词组主要有以下几类：

（1）在 number, lot, score, decade, dozen, ten, hundred, thousand, million 等词后加复数后缀-s。例如：

dozens of　几十，几打
hundreds of　几百，成百上千
hundreds of thousands of　数十万，几十万
lots of　许多，大量
millions of　千千万万，数以百万计
millions upon millions of　数亿，无数
numbers of　许多，若干
scores of　几十，许多
teens of　十几（13～19）
tens of thousands of　数万，成千上万
tens/decades of　数十，几十
thousands of　几千，成千上万

A typical bitmap is divided into a grid of thousands of tiny rectangles called “pixels” and each pixel can be assigned a different color or shade of gray.

一张典型的像素图像可以划分为由几千个称为“像素”的小矩形组成的网格，每个小矩形都具有不同的色彩或灰度。

（2）数字前加 above, more than, over, up to 等，可以译为“超过”“多达”等。例如：

A satellite is a solar-powered electronic device that has up to 100 transponders (a transponder is a small, specialized radio) that receive, and retransmit signal.

卫星利用太阳能，里面有多达100个转发机（转发机是一个小的、特制的无线电设备），用于接收、转发信号。

（3）“as…as”结构可以表示增减，也可译为“多达”。例如：

For example, the net result of the using the MPEG standard on the two audio channel of a stereo program is that each digitized audio signal, coming in at the rate of 768 kilobits per second (kb/s), is compressed to a rate as low as 16 kb/s.

例如，用MPEG标准对双声道立体声程序进行处理的结果是每个数字音频信号的数据率从768 kb/s压缩至16 kb/s。

（4）数字前加below, less than, under等，可译为“以下”“不足”等。例如：

As a result of the very high frequencies and the speed at which light travels (300 000 km per second), the wavelength is extremely short, **less than** one thousandth of a millimeter.

由于光的频率很高，速度快（光的传播速度为300 000 km/s），所以光的波长极短，**不足**千分之一毫米。

（5）数字前加about, around, close to, nearly, or so, some, toward(s)等可以译为“大约”“左右”“将近”等。例如：

Commonly used for storing uncompressed (PCM), CD-quality sound files, which means that they can be large in size—**around** 10MB per minute.

（wav格式）通常用于存储未压缩、CD音质的声音文件，这就意味着大容量存储——**大约**每分钟10MB。

（6）from…to, between…and等结构中可译为“从……到”“到”等。例如：

Each element of the table is an integer **from** 1 **to** 255 that represents the step size of the DCT coefficients, which, in turn, permits the representation of each quantized DCT coefficient by an 8-bit code word.

列表的每个元素是**从** 1～255 的整数，表示离散余弦变换系数的量化阶大小。这样，就能依次以8比特的码字形式来表示每个量化了的离散余弦变换系数。

2．倍数增减的译法

英汉两种语言表达倍数的方式截然不同，所以在汉译时，常常会发生理解和翻译上的错误。汉语里说“增加了几倍”是指纯增加量，但英语里表达倍数的方式很多，有的表示净增减，有的则表示包含基数的增减。汉译时，必须注意其中的差别。

1）倍数增加的译法

常用倍数增加表达法及译法有下列4种：

increase by *n* times “增加了*n*–1倍”或“增加到*n*倍”

increase *n* times “增加了*n*–1倍”或“增加到*n*倍”

increase by a factor of *n* “增加了*n*–1倍”或“增加到*n*倍”

increase to *n* times “增加到几倍”或“增加了*n*–1倍”

例如：

The input stage **amplifies** the input signal **by a factor of** 50.

在输入阶段，输入的信号被放大到50倍。

2）倍数比较的译法

常用倍数比较表达法及译法有下列 4 种：

A is n times larger than *B*. “*A* 是 *B* 的 *n* 倍”或“*A* 比 *B* 大 *n*–1 倍（净增 *n*–1 倍）”

A is n times as large as *B*. “*A* 相当于 *B* 的 *n* 倍”或“*A* 比 *B* 大 *n*–1 倍（净增 *n*–1 倍）”

A is larger than *B* by n times. “*A* 是 *B* 的 *n* 倍（净增 *n*–1 倍）”

A is n times *B*. “*A* 是 *B* 的 *n* 倍（净增 *n*–1 倍）”

例如：

Today, many planes can fly several times faster than sound.

现在，许多飞机能以**数倍于**声速的速度飞行。

3）倍数减少的译法

在英语里，人们可以说“减少了多少倍”和“成多少倍地减少”，但在汉语里却不能这样说。汉语里只能说“减少了几分之几”或“减少到几分之几”。因此，在翻译倍数的减少时，译文应更符合汉语的习惯。

常用的倍数减少表达法及译法有以下 5 种：

decrease by 3 times “减少了 2/3（减少到 1/3）”

decrease to 3 times “减至 2/3（减少了 1/3）”

decrease by a factor of 3 “减少了 2/3（减至 1/3）”

decrease 3 times “减至 1/3（减少了 2/3）”

3 times less than “减少了 3/4（减至 1/4）”

例如：

Switching time of the new-type transistor is **shortened by three times.**

新型晶体管的开关时间缩短了 2/3。

The speed of the machine was **decreased by a factor of five.**

该机器的运转速度降低了 4/5。

The wire is **two times thinner than** that.

这根导线比那根导线细 2/3。

3．百分数的译法

英语中用百分数表示数量增减的常用表达法主要有以下 5 种：

增减意义的动词+by+*n*%

表示减少的动词+to+*n*%

n%+比较级+than

a *n*% + increase

n% + (of)名词或代词

例如：

New engines can **increase** the pay load **by 50 percent.**

新型发动机能使有效负载**增加 50%。**

By using the new process the reject rate was **reduced to 3 percent.**

采用新的工艺方法，废品率**下降到 3%。**

At the same time, the infant mortality was **40 percent lower than** that of the 1960.

同时，婴儿死亡率比 1960 年**降低了 40%**。

The new-type pump wasted **10 percent less** energy supplied.
这台新型水泵少损耗所提供能量的 **10%**。

There is **a 30% increase** of our installed capacity with this year.
今年，我们的装机容量**增加了 30%**。

The cost of our power production is about **90%** that of theirs.
我们的发电成本大约是他们的 **90%**。

0.2.5.2 形容词的译法

英语的形容词不仅数量大，而且使用频率很高。英语的形容词可以单独使用，但更多的是用来修饰名词和某些代词，在句子中主要作定语、表语、宾语补足语，偶尔也作状语。

1．形容词的一般译法

1）直译（略）

2）转译

英语的形容词翻译时，可视情况转译成名词、动词和副词。

（1）转译成名词。例如：

The metal may be fluid, **plastic, elastic, ductile** or **malleable**.
金属具有流动性、塑性、弹性、延展性或韧性。

Insulators are by no means less **important** than conductors.
绝缘体的**重要性**绝不亚于导体。

（2）转译为动词。例如：

Designing a character can be difficult; not only have you got to be fresh and original, but you must also be **able** to communicate your ideas to others through your artwork.

设计出一个好的角色是很困难的，不仅要有新意、是原创，而且要**能够**通过你的作品向其他人传递你的思想。

（3）转译为副词。例如：

3G has witnessed the starting period of mobile communication business of the first generation and the rapid development of the digital mobile communication market of the second generation.

3G 通信见证了第一代移动通信业务的初始阶段和第二代数字移动通信市场的飞速发展。

3）形容词前增译名词

例如：

Computers have become smaller, faster, and cheaper.
计算机变得体积更小，速度更快，价格也更便宜。（形容词 smaller, faster, cheaper 前分别增译名词“体积”“速度”“价格”）

2．形容词作前置定语的译法

常译为“的”字结构，下面介绍其他译法。

1）译为短语

形容词作前置定语，除了直译为汉语的“的”字结构之外，许多情况下不能译出“的”字，而常常与被修饰的名词一起译成一个约定俗成的短语，尤其在科技英语中，常常应译为特定的专业技术术语。例如：

ATM (**Asynchronous Transfer Mode**) is both a multiplexing and switching technique.

ATM（**异步传输模式**）既是复用技术又是交换技术。

2）译为主谓结构

例如：

Further, the phosphors used in the receiver picture tubes have a relatively **low** persistence, allowing the picture to fade out between scans, and the scanning will produce a "flicker" at the picture rate.

而且接收机显像管所用的磷留存的时间相对较**短**，而使画面在两次扫描间消失，扫描将产生图像速率的抖动。

3．形容词作后置定语的译法

（1）"形容词+不定式短语"构成形容词短语作后置定语。例如：

STM provides fixed bandwidth channels, and therefore is not **flexible enough to handle** the different types of traffic typical in multimedia applications.

STM 提供固定带宽的通道，因此，对于处理多媒体应用中的不同类型的通信流显得不够灵活。

（2）当形容词修饰 some, something, any, anything, nothing, everything 等词时，形容词作后置定语。例如：

As she approached the mirror, she caught a glimpse of something dark on her left arm.

当她走近镜子时，她瞥见左臂上有一些黑色的东西。

（3）以 every, all, only 开头的句子，形容词作定语常常后置。可以顺译，也可将后置定语译在被修饰名词之前。例如：

Every object, **large or small**, has a tendency to move toward every other object.

每一个物体，**不论大小**，都有向其他物体移动的倾向。

All neutrons have no charge, either **positive or negative**.

中子不带**正**电，也不带**负**电。

4．作表语的形容词的译法

例如：

New networks and protocols are necessary to provide the high bandwidth, low latency, and low jitter required for multimedia.

新型网络及其协议必须支持多媒体所需要的高带宽、低延时和低抖动的要求。

Waveform diagrams representing even a few seconds of sound are, consequently, very big.

因此，即使是代表一小段时间的声音的波形图也是相当大的。

5．作状语的形容词的译法

英语形容词或形容词短语在科技英语中经常用来作原因、结果、让步、时间等状语。翻译时可将其置于句首，也可以置于句末。例如：

Small in size, the new generator is large in capacity with low coal consumption.

这台新的发电机虽然体积小，但是发电量大，耗煤量低。

0.2.5.3 副词的译法

1．"地"字法（略）

2．"上"和"下"字法

例如：

Typically, music has two structures: a melodic structure consisting of a time sequence of sounds, and a harmonic structure consisting of a set of simultaneous sounds.

在一般情况下，音乐有两种结构：由一串时序的声音组成的旋律和由一组同时发出的声音组成的和音。

3．转译法

1）副词转译成形容词

例如：

A signal is **formally** defined as a function of one or more variables that conveys information on the nature of a physical phenomenon.

“信号”的正式定义是：传达某种物理现象特性的信息的一个函数。

2）副词转译成名词

例如：

The process is shown **schematically** in Figure 7-1.

图7-1是表明这个过程的**简图**。

0.2.6 介词的翻译

0.2.6.1 介词的一般译法

1．直译

例如：

In the early computer, the only sound that we heard from a computer was a beep—often accompanied by an error message.

在早期的计算机上，我们能够听到计算机发出的声音就是伴随着出错信息的嘟嘟声。

2．转译

介词转译为动词是最常见的介词转译现象。例如：

Many challenging problems remain to be researched and resolved **for** the further growth of multimedia systems.

多媒体要**取得**进一步的发展，许多棘手的问题还有待研究和解决。

3．省译

根据汉语的表达需要，英语中许多介词在汉译时可以省略不译。例如：

The MPEG-1 audio layer 3 compression format is the most popular format **for** downloading and storing music.

MPEG-1音频标准第3层压缩格式是下载和存储音乐时最常用的格式。

0.2.6.2 介词短语的译法

1．介词短语作状语时的译法

1）译为状语从句

例如：

By placing an element on a separate layer, you can easily edit and arrange that element **without interfering with the other parts of the image.**

通过在一个单独的图层上放置一个单独的元素，能在不干涉图像其他部分的前提下，容易地编辑和安排那个元素。

2）译为句子的主语

例如：

When ice changes into water, it becomes smaller **in volume.**

冰变成水时**体积**变小了。

3）译为汉语的并列分句

例如：

In a generator mechanical power is put in and electrical power is taken out, instead of electrical power going in and mechanical power coming out as in a motor.

在发电机中是输入机械能、输出电能，**而在电动机中则是输入电能、输出机械能。**

2．介词短语作定语时的译法

1）译为汉语的“的”字结构

例如：

Documents **for the World Wide Web** are written in HTML.

万维网的文件是用 HTML（超文本标记语言）编写而成的。

2）译为动宾结构的句式

例如：

The big picture shows multimedia as the merging **of** computing, communications, and broadcasting.

多媒体的巨大的发展前景表明多媒体将是计算机产业、通信产业和广播产业三者结合而形成的一个新的产业。（动宾结构）

0.2.7 连词的翻译

0.2.7.1 连词的一般译法

1．直译

例如：

It should be object-oriented and capable of synchronizing data to be instantly available, **so** the user interface must be highly sophisticated and intuitive.

它是面向对象的，具有同步迅速获得数据的能力，**因此**用户界面高度复杂，但要求直观。

2．转译

并列连词有时可以转译成从属连词，而有时从属连词也可以转译成并列连词，或从属连词改变原有的功能转译成其他意义的从属连词。例如：

Every object, large **or** small, has a tendency to move toward every other object.

每一个物体，**不论**大小，都有向其他物体移动的倾向。（并列连词 or 转译成“不论”，表示让步）。

Save your file in a .psd format, and you can make changes to it at a later time.

将你的文件保存为.psd 格式，从而以后还可以对它进行修改。（并列连词 and 转译成“从而”，表示结果）

3．省译

为了符合汉语的表达和修辞习惯，有时连词也可以省略不译，例如：

Electric charges, positive **and** negative, which are responsible for electrical force, can wipe one another out and disappear.

产生电场力的正负电荷会相互抵消掉。（省译了 and）

0.2.7.2　常用主从连接词的译法

主从连接词（也称关联词）主要用来连接主句和从句。在科技英语中，使用主从连接词的频率非常高。

1．as 的译法

1）引导定语从句

（1）引导限制性定语从句。例如：

The TV system used in North American is the same as that used in Europe.

北美使用的电视系统**与欧洲使用的相同**。

（2）引导非限制性定语从句。例如：

As we know, multimedia systems combine a variety of information sources.

正如我们所知，多媒体系统综合了多种信息源。

2）引导时间状语从句

例如：

In essence, a waveform is a graph that charts minute changes in air pressure **as sound waves propagate.**

本质上，一个波形是**在声波传播过程中**空气压力随时间变化的曲线图。

3）引导原因状语从句

例如：

As they are based on the same concept, we don't tell the difference between them.

因为两者的概念是相似的，这里也就不做区分了。

4）引导方式状语从句

例如：

Though most audio file formats support only one audio codec, a file format may support multiple codecs, **as AVI does.**

尽管多数音频文件格式只能支持一个音频编码解码器，但是一个文件格式能够**像 AVI 一样**支持多种编码解码器。

5）引导比较状语从句

例如：

Waveforms aren't **as intimidating as they look.**

波形并不是它看上去的那样紧凑。

6）引导让步状语从句

例如：

Simple as it is, this method does not make very strong magnets.

这一方法虽然简单，但不能产生很强的磁铁。

2．as if 和 as though 的译法

例如：

Also, a .swf itself is a type of Sprite or movie clip, which is why you can load a .swf into another .swf and largely treat it as if it were just another nested Sprite or movie clip.

此外，.swf 文件本身就是一种影片 Sprite 或影片剪辑，这也就是为什么你可以把一个.swf 文件载入另一个.swf 文件中，而且多半能将其视为只是另一个内层影片 Sprite 或影片剪辑的原因所在。

3．as (so) long as 和 as (so) far as 的译法

as (so) long as, as (so) far as 都可引导条件状语从句，这时前者可译为“只要……就”，后者可译为“就……而言”等。例如：

A table lamp emits less light than a halogen lamp, but even a halogen source cannot be compared with bright sunlight, **as far as** luminosity is concerned.

就亮度而言，一个台灯可以发出不到 1 卤素的灯光，但即使 1 卤素的灯光也无法和明亮的太阳光相比较。

4．as soon as 的译法

as soon as 用来引导时间状语从句，可译为“一……就”“一旦……就”等。例如：

A class’s constructor is automatically run as soon as an instance of the class is created.

一旦类的实例创建，类的构造方法就会自动执行。

5．before 的译法

（1）before 作为从属连词，可以用来引导时间状语从句，译为“在……之前”“以前”等，或灵活处理。例如：

The fuse will go **before a short circuit can do much harm.**

在短路造成很大危害之前，保险丝就会熔断。

（2）在主句与从句中都含有 must, can 等情态动词的情况下，通常是强调主句动作与从句动作一先一后发生，翻译时需先译主句再译 before 引导的从句，常可译为“只有……才能”。例如：

Before sunshine can be used as a source of energy the storage problem must be solved.

只有解决了储存问题，阳光才能用作能源。

（3）before 从句有时可译为，“后……才”“才能”“未……就”等。例如：

As a result of the rapid progress in science and technology, great changes have taken place in the world **before we can realize them.**

由于科学技术的迅猛发展，有时**我们还未能意识到**，世界**就**已经发生了许多变化。

6．even if 和 even though 的译法

作为从属连词，even if 和 even though 用来引导让步状语从句，译为“即使”“尽管”等。例如：

Even if you are familiar with classes in ActionScript 2.0, there are some new things here.

即使你熟悉 ActionScript 2.0 的类，它还是有些新鲜的东西。

7．if 的译法

1）引导条件状语从句

例如：

If there were no electrical pressure in a conductor, the electron flow would not take place.

若导体上没有电压，就不会形成电流。

2）引导宾语从句、主语从句和表语从句

if引导宾语从句、主语从句和表语从句时，可译为“是否”“能否”等。例如：

They wonder **if users could interact with data the way they interact with other entities on a day-to-day basis.**

他们想知道**用户是否能用日常与其他实体交互的方式来进行数据交互**。（if 引导宾语从句）

8．in case 的译法

in case 可以引导条件状语从句，译为“如果”“假如”“万一”等，in case 引导的从句在句子中位置比较灵活，但译成汉语时多放在句首。例如：

In case a large leakage of steam occurs, the engine efficiency will fall.

如果蒸汽大量泄漏，引擎效率将会降低。

9．once 的译法

once 作为连接词，可以引导时间状语从句或条件状语从句，均可译为“一旦”“如果”等。例如：

Once the sketches are approved, you can spend more time creating better renderings of the concepts to flesh out your basic ideas.

一旦草案获得认可，你就可以用更多的时间更好地描绘你的想法，充实你的理念。

10．provided（that）的译法

provided（that）引导条件状语从句。可译为“如果”“假如”“若”“只要”等。同样的连接短语还有 providing（that）、supposing/suppose（that）、in the event（that）、on condition that 等。例如：

Variables are convenient placeholders for data in your code, and you can name them anything you'd like, provided the name isn't already reserved by ActionScript and the name starts with a letter, underscore, or dollar sign (but not a number).

变量是使代码中的数据便于使用的占位符（placeholder），只要名称不是 ActionScript 保留字，而且以字母、下画线或美元符号开头（不能是数字），就可以用任何喜欢的名称予以命名。

11．since 的译法

1）引导时间状语从句

Since 引导时间状语从句时，主句多用完成时态。但主句的内容涉及某个具体的事件发生（即表示时间点）时，一般从句可译为“自（从）……以来”，当主句中有表示时间段的词汇时，从句多译为“已……（年/天）”。例如：

Since 1989, when the first multimedia systems were developed, it has been possible to differentiate the three generations of multimedia systems.

自 1989 年第一代多媒体系统被开发成功**以来**，多媒体系统可以分为三代。

2）引导原因状语从句

since 引导原因状语从句时，从句在句中的位置比较灵活，通常可译为“由于……（所以)”“因为”“既然……那么”等。例如：

Since video is a series of still images, it makes sense to simply display each full image consecutively, one after another.

由于视频是一连串静态图像，也可以说成是一幅接一幅连续地呈现每一幅完整的图像。

12．that 的译法

1）引导名词性从句

that 作为从属连词，可以引导名词性从句，即主语从句、表语从句、宾语从句以及同位语从句，而 that 在这些从句中没有具体含义，只是起连接作用，故在翻译这些从句时，that 通常省略不译。例如：

The meaning of this high correlation is **that**, in an average sense, a video signal does not change rapidly from one frame to the next.

高度相关的意思是说，通常相邻两帧的视频信号没有很大的差别。

2）引导定语从句

如果定语从句不长，一般都将定语从句译为“的”字结构，作先行词的前置定语；如果定语从句较长，常常把定语从句译为一个并列分句或转译为其他状语从句。例如：

We need standard compression algorithms **that** enable the interoperability of equipment produced by different manufactures.

我们需要标准的压缩法则才能使由不同厂家生产的设备协同工作。

There are several technical terms used to describe waveforms **that** you should know.

有一些描述波形的技术术语是你应当了解的。

The frequency of a sound is the number of cycles **that** happen every second.

声音的频率是每秒的循环次数。

3）引导状语从句

（1）that 常用来引导表示目的和结果的状语从句，汉译时可适当增译表示这些概念的词语。例如：

Key frame refers to such a frame whose contents are quite different from those of the precedent frame, so that it presents great changes in critical actions and contents.

所谓关键帧，也就是该帧中的内容与先前帧中的内容有很大的区别，从而呈现出关键性的动作及内容的变化。(that 引导结果状语从句)

（2）that 还可以与 now 和 in 连用，即 now that 与 in that，引导原因状语从句，译为“由于”“既然”。例如：

For beginners, **now that** you know where to enter code, here is quick primer on terminology.

就初学者而言，**既然**你知道该把代码输入何处，以下就进行相关术语的快速入门。

13．unless 的译法

在英语中 unless 引导的条件状语从句通常位于主句之后，相当于“if…not”。汉译时，可译在句首也可译在句尾，常译为“除非”“如果不”等。例如：

Your design should fit the brief; for example, a conservative character would not sport a bright-pink Mohawk well, not **unless** the game guidelines ask for it.

你设计的角色需要与背景故事所描述的角色性格尽量相符，例如，一个比较保守的角色不会以鲜亮的粉红色的莫西干发型示人——**除非**游戏需要。

14．until 和 till 的译法

（1）主句为肯定句时，until 或 till 引导的从句一般译为“直到……（为止）”或“在……以前”等。例如：

This process is repeated **until** the entire screen is refreshed.

这个过程会一直重复，**直到**整个屏幕刷新完成。

（2）如果主句为否定句，一般应译为“（直）到……后……才”或“在……以前……不”等。例如：

People have **no** idea of what radioactivity was **until** Madame Curie discovered radium.

在居里夫人发现镭**以前**，人们并不了解放射性。

15．what 的译法

（1）仍然保留疑问含义，译为“什么”“那”“多少”等，这时 what 引导的名词从句作定语、表语、主语或宾语等。例如：

Finally, look at what the level/world builders in your game are doing.

最后，看一下设计游戏场景（也称游戏世界）的工作人员正在做些什么。（what 引导宾语从句，what 在该从句中作主语）

（2）what 从句译为“（所）……的（东西等）”。这时 what 相当 the thing(s) that。例如：

Now we have a rough idea of what's happening when we hear a sound, we can begin to make sense of **what** audio experts cryptically refer to as “waveform” diagrams.

我们对听到的声音有了一个大体认识，这样就可以了解音频专家所指的“波形”图的含义。

（3）在 what 引导的名词性从句（多数为主语从句）中，what 可译为相应的名词。例如：

Now what we're delivering to the home is exactly what we created.

而现在我们正在做的是向千家万户播送我们当时创造的东西。（what 引导主语从句，what 在该从句中作宾语）

16．when 的译法

1）引导名词性从句

可引导主语从句、表语从句和宾语从句。其译法可以保留疑问含义，译为“什么时候”“在什么时候”“何时”等。例如：

Special timing information, called *vertical sync*, is used to indicate **when** a new image is starting.

称为“垂直同步”的定时信息是用于指示一个新图像从何时开始。（when 引导宾语从句）

2）引导定语从句

when 作为关系副词引导定语从句，修饰表示时间的名词。这时，定语从句多为限制性定语从句，可译为“的”或“当……（时候）”等。例如：

The day will come **when coal and oil will be used as raw materials rather than as fuels.**

使用煤和石油作为原料而不是燃料的日子一定会到来。

3）引导时间状语从句

（1）表示时间，一般可译为“当……时候”“一……后就”“这时”等。例如：

When implementing a compression/decompression algorithm, the key question is how to partition between hardware and software in order to maximize performance and minimize cost.

实现压缩或者解压算法时，关键问题是如何在硬件与软件之间取得平衡，以实现最好的性能价格比。

（2）英语中 when 引导的状语从句除了表示时间外，在特定的上下文中可以有不同的逻辑含义。翻译时，完全可以按所理解的不同逻辑含义来分别处理。例如：

When electrons can move easily from atom to atom in a material, it is a conductor.

如果一种材料其电子能在原子间自由运动，这种材料便是导体。（表示条件）

17. where 的译法

1）引导名词性从句

例如：

This is **where** the new design has the advantage.

这就是新设计的优越之处。（where 引导表语从句）

2）引导状语从句

例如：

Where water is lacking, no plants can grow.

缺水的地方，植物不能生长。

3）引导定语从句

（1）Where 作为关系副词，可以引导定语从句，修饰表示地点的名词。如果是限制性定语从句，通常可译为汉语的“的”字结构的短语。例如：

Air moves from place **where the pressure is high** to places where the pressure is low.

空气从压力大的地方流向压力小的地方。

（2）如果 where 引导非限制性定语从句，那么定语从句可分译，而 where 可译成“其中”“在那里”等。例如：

For example, a large file to be transmitted between two machines must be broken into many packets that are sent across the network one at a time, the network hardware delivers the packets to the specified destination, **where** software reassembles them into a single file again.

例如，一个要在两台机器间传送的大文件，必须被分成许多分组，在网络上一个一个地传送。网络硬件把分组传送到指定目的地，**在那里，**软件把它们重新组装成一个文件。

（3）where 引导的定语从句无论是限制性定语从句还是非限制性定语从句，常常可以根据其在上下文中的逻辑意义翻译为原因状语从句、条件从句、结果从句、目的从句等。翻译时要注意从英语原句的字里行间去发现这种逻辑关系，并在翻译过程中加以体现。

18. whether 的译法

1）引导名词性从句

例如：

While there is some debate as to **whether** or not digital audio actually sounds better than analog audio, digital audio is certainly easier to reproduce and to manipulate without loss of quality.

虽然，仍然有一些关于数字音频声音质量是否比模拟音频高的争论，但是数字音频确实比较容易重现，并且以较少的质量损失处理音频文件。

2）引导让步状语从句

例如：

WAV, like any other uncompressed format, encodes all sounds, **whether** they are complex sounds or absolute silence, with the same number of bits per unit of time.

WAV 文件如同其他未压缩格式一样，对所有的声音编码，**不管**它们是复杂的声音或者只是静音，每个时间单位都有相同的字节数。

19．which 的译法

1）引导名词性从句

例如：

You'll have to do a little detective work to discover which vector and bitmap formats can be saved in one program and opened, placed, or imported in another.

哪些矢量和位图格式可以在某个程序中保存以及在另一个程序中打开、置入或导入，你得摸摸底。

2）引导定语从句

which 作为关系代词引导定语从句是科技英语中最为常见的。例如：

All the other codecs discussed here are lossy which means a small part of the quality is lost.

此处讨论的其他编解码器都是有损的，即质量有些降低。

20．while 的译法

1）while 作并列连词的译法

例如：

Waves are all of the same length, while they are in step.

所有波的波长都相同，而且同步。

2）while 作从属连接词的译法

(1）引导时间状语从句。

例如：

MP3 files are compressed to roughly one-tenth the size of an equivalent PCM file **while** maintaining good audio quality.

MP3 文件保持良好音质的同时，将文件大小压缩为原来 PCM 文件的 1/10。

(2）引导让步状语从句。

例如：

While there is some debate as to whether or not digital audio actually sounds better than analog audio, digital audio is certainly easier to reproduce and to manipulate without loss of quality.

虽然对于究竟数字声音听起来是否优于模拟音频还存在争议，但是数字音频的确更容易在无损音质的前提下进行复制和处理。

While researchers agree it has a long way to go before it reaches its potential, VR is in practical use today in various occupations in the military, entertainment, education, and business.

虽然研究人员认为在完全发掘它的潜力之前，虚拟现实还有很长的路要走，但是在实际使用中，虚拟现实被应用在诸多行业，如军事、娱乐、教育和商业。

（3）引导原因状语从句，相当于 since，译为“既然”。

21．why 的译法

1）引导名词性从句

why 可以引导主语从句、表语从句和宾语从句，可译为“为什么……（的原因）”“为何”等。

2）引导定语从句

why 引导定语从句，修饰表示原因的名词，可译为“为什么”“……的原因”“……的缘故”等。例如：

This is the reason **why** waveforms of audio recordings often look complicated and squiggly when you view them on your computer.

这就是你在计算机上看到的声音记录的波形经常复杂弯曲**的原因**。

0.3 长句的翻译

在书面语中，英语为了表达严谨，常使用长句。特别是在专业英语中，为了说理严谨，逻辑紧密，描述准确，会大量使用长句。因此，我们必须掌握长句的翻译技巧，才能准确地进行通篇翻译。

专业英语中的长句汉译，一般要拆成汉语短句，按照汉语的表达习惯和逻辑层次，重新排列顺序，组织成内容准确、逻辑分明、重点突出、通顺正确的译文。

在译文表达阶段，先将英语长句中每个短句译成汉语，然后按照汉语的语序，重新排列组合，以正确表达原意。在此阶段，语序不必与原句一致。最后加工润色。

1．分译法

英语长句中有些句子成分（主句、从句、短语、修饰语）的关系较复杂，与汉语多用短句的表达习惯不一致，按原句照译会使句子过于生硬累赘，故而可以把英语的某些句子成分单独译为一个句子。例如：

From what is stated above, it is learned that the sun's heat can pass through the empty space between the sun and the atmosphere that surrounds the earth, and that most of the heat is dispersed through the atmosphere and lost, which is really what happens in the practical case, but to what extent it is lost has not been found out.

由上述可知，太阳的热量可以穿过太阳与地球大气层之间的真空，而大多数热量在通过大气层时都扩散和消耗了。实际发生的情况正是如此，但是热量的损失究竟达到什么程度，目前尚未弄清。

2．顺译法

当英语长句的表达顺序与汉语的表达顺序一致时，可不改变原文语序和语法结构，译成汉语。但值得一提的是，并不是每一个词都按照原句顺序。事实上，一一对应是极少见的。例如：

Vibration in machines can be thought of as a combination of non-stationary periodic functions generated by a variety of imbalanced forces or disturbances, each of which has a characteristic repetition frequency.

可以把机器的振动视作一些非稳态周期性作用的综合反应，这些作用是由各种不平衡力或扰动引起的，每一种不平衡力或扰动都有一个特殊的重复频率。

3．倒译法

有些英语长句的表达顺序与汉语不同或相反。如主句后的一些状语从句（表示原因、时间、地点、方式等），或后置的定语从句、主语从句等。翻译时须交换顺序。例如：

The technical possibility could well exist, therefore, of nationwide integrated transmission network of high capacity, controlled by computers, inter-connected globally by satellite and submarine cable, providing speedy and reliable communications throughout the world.

本句主句为 The technical possibility could well exist，其余是很长的 of 介词短语用作 possibility 的定语。本句语义重心为“这种可能性从技术上来讲是完全存在的”，翻译成中文时应当放在句尾。

翻译：建立全国统一的大容量的通信网络，由计算机进行控制，通过卫星和海底电缆在全球从事快速可靠的通信服务，这种可能性从技术上来讲是完全存在的。

4．变序法

科技英语中，长句语法成分繁多，层次复杂，顺着原文的顺序或逆着原文顺序翻译成汉语都会显得牵强，以致难以理解。在这种情况下，须弄清语义层次及分句间的关系，做出合理的安排。例如：

One of the significant fringe benefits of the remarkably small size of integrated circuit systems is that effective external nuclear radiation shields are now feasible, where weight or cost ruled them out before.

集成电路装置的尺寸特别小，这就又增加了一个很大的优点——可以在外界给它提供有效的核辐射屏蔽，这在以前由于重量或价格的限制是不可能做到的。

0.4　文章的翻译

翻译不是单个句子翻译的简单相加，而是以篇章为基础的。

进行科技文章翻译时，首先要注意文章的逻辑关系，明确翻译的对象属于哪种文章，才能准确把握该文体结构特征和文章的语言逻辑。其次要注意文章的语篇衔接连贯。语篇衔接通过词汇或语法手段使文脉相通，形成语篇的有形网络。语篇连贯以信息发出者和接受者双方共同了解的情景为基础，通过推理来达到语义的连贯，这是语篇的无形网络。充分利用语篇的叙事次序（时间顺序、空间顺序等）和逻辑连接词有助于对文章的整体把握，做到传意达旨。

0.4.1　专业英语文章的文体特点及其翻译

科普文章内容上着重常识性、知识性和趣味性，语言上通俗易懂，深入浅出，语句简短，多用普通词汇。在翻译这类文章时应忠实于原文风格，用生动灵活、浅显易懂的汉语普及科学知识。

0.4.2　科技论文文体特点及其翻译

科技论文是科技研究人员研究成果的直接记录，因此论文内容专业性强，文字规范、严谨。论文侧重叙事和推理，具有很强的逻辑性：结构严谨，层次分明，前提完备，概念确切，推理严密，分析透辟，判断准确。

摘要是全篇论文的缩影，常常被专业期刊文献杂志编入索引资料或文献刊物，以便于学术之间的交流。摘要的英文术语原来有两个词汇：摘要（abstract）和概要（summary）。摘要不是正文的组成部分，但隶属于论文；而概要则是正文的一个组成部分。摘要放在正文前，概要一般安排在正文的最后。

科技论文摘要要求围绕着正文的论题，并就研究的目的、方法、结果、结论等主要环节进行概括性介绍，概要是正文各部分内容的综合性复述，因此翻译时要力求准确、简洁和清晰。准确是指内容上要忠实于原文。简洁要求使用标准术语，不用第一人称和第二人称，不混用时态。清晰指用有限的字数将文章的论题、论点、实验方法、实验结果表达出来，不使用带有感情色彩和意义不确定的词。

例如：

This chapter describes the recent advances made in broadband access network architectures employing Passive Optical Networks (PONs). The potential of PONs to deliver high bandwidths to users in access networks and their advantages over current access technologies have been widely recognized. PONs have made strong progress in terms of standardization and deployment over the past few years. In this chapter, we first review the Ethernet PON (EPON), which is currently being standardized by the IEEE 802.3ah task force. Next, we discuss the ATM PON (APON) and the Gigabit PON (GPON). We then review the technologies available for introducing wavelength-division multiplexing (WDM) in PONs, and the progress of research in this area. Finally, we examine the issues related to deploying PONs in access networks.

本章阐述了使用无源光纤网络（PONS）的宽带接入网结构最近的发展。利用无源光纤网络进行高带宽用户互联网传输的潜力以及其接入技术上的优点已被人们所广泛认识。在过去的几年中，PONS 在标准化过程和应用方面已经有了长足的进步。在本章中，我们将首先了解被 IEEE 802.3ah 工作团队标准化的以太网无源光纤（EPON）。其次我们将讨论 ATM PON（APON）和 Gigabit PON（GPON）。接下来我们将回顾引进波分复用技术（WDM）在无源光纤网络（PONS）中使用的可行的技术以及在该领域中的研究进程。最后我们将研究无源光纤网络技术在接入网中的应用。

0.4.3 产品说明书翻译

说明书文体包括各种产品的说明书、操作指南或使用手册、故障排除和维修保养方法等，从工业装置说明书到一般机械或家用电器的使用说明，以及旅游指南、食品说明、医药说明和服务须知等，均属此范围。

由于产品种类不同，说明书的内容及说明的方法也有所不同。机械说明书的内容一般包括产品特点、用途、规格、结构、性能、操作程序及注意事项等。家电产品说明书包括产品的安装方法、使用方法、常见问题的处理，以及日常的维护与保养。

尽管产品说明书的构成不同，各有特点，但是，产品说明书的性质决定了各种说明书的共性。一般来说，产品说明书包括产品的特征、功能和成分，安装和使用、服用、饮用、食用的方式方法，注意事项，主要性能指标及规格。

Part 1 Digital Media

Text 1: Multimedia—An Overview

Advances in distributed multimedia systems have begun to significantly affect the development of on-demand multimedia services. Researchers are working within established computer areas to transform existing technologies and develop new ones. The big picture shows multimedia as the merging of computing, communications, and broadcasting.

Multimedia systems combine a variety of information sources, such as voice, graphics, animation, images, audio, and full-motion video, into a wide range of three industries: computing, communication, and broadcasting.

Research and development efforts in multimedia computing fall into two groups. One group centers its efforts on the stand-alone multimedia workstation and associated software systems and tools, such as music composition, computer-aided learning, and interactive video. The other combines multimedia computing with distributed systems. This offers even greater promise. Potential new applications based on distributed multimedia systems include multimedia information systems, collaboration and conferencing systems, on-demand multimedia services, and distance learning.

The defining characteristic of multimedia systems is the incorporation of continuous media such as voice, video, and animation. Distributed multimedia systems require continuous data transfer over relatively long periods of time (for example, play out of a video stream from a remote camera), media synchronization, very large storage, and special indexing and retrieval techniques adapted to multimedia data types.

1. Technical Demands

A multimedia system can either store audio and video information and use it later in an application such as training, or transmit it live in real time. Live audio and video can be interactive, such as multimedia conferencing, or noninteractive, as in TV broadcasting. Similarly, stored still images can be used in an interactive mode (browsing and retrieval) or in a noninteractive mode (slide show).

The complexity of multimedia applications stresses all the components of a computer system. Multimedia requires great processing power to implement software codecs, multimedia file systems, and corresponding file formats. The architecture must provide high bus bandwidth and efficient I/O.

A multimedia operating system should support new data types, real-time scheduling, and fast-interrupt processing. Storage and memory requirements include very high capacity, fast access times, and high transfer rates. New networks and protocols are necessary to provide the high bandwidth, low latency, and low jitter required for multimedia. We also need new object-oriented, user-friendly software development tools, as well as tools for retrieval and data management—important for large, heterogeneous, networked and distributed multimedia systems.

Researchers are working within established computer areas to transform existing technologies, or develop new technologies, for multimedia. This research involves fast processors, high-speed networks, large-capacity storage devices, new algorithms and data structures, video and audio compression algorithms, graphics systems, human-computer interfaces, real-time operating systems, object-oriented programming, information storage and retrieval, hypertext and hypermedia, languages for scripting, parallel processing methods, and complex architectures for distributed systems.

2. Multimedia Compression

Audio, image, and video signals produce a vast amount of data. Compression techniques clearly play a crucial role in digital multimedia applications. Present multimedia systems require data compression for three reasons: the large storage requirements of multimedia data, relatively slow storage devices that cannot play multimedia data (specifically video) in real time, and network bandwidth that does not allow real-time video data transmission.

Digital data compression relies on various computational algorithms, implemented either in software or hardware. We can classify compression techniques into lossless and lossy approaches. Lossless techniques can recover the original representation perfectly. Lossy techniques recover the presentation with some loss of accuracy. The lossy techniques provide higher compression ratios, though, and therefore are applied more often in image and video compression than lossless techniques.

We can further divide the lossy techniques into prediction-, frequency-, and importance-based techniques. Predictive techniques (such as ADPCM) predict subsequent values by observing previous values. Frequency-oriented techniques apply the discrete cosine transform(DCT), related to fast Fourier transform. Importance-oriented techniques use other characteristics of images as the basis for compression; for example, the DVI technique employs color lookup tables and data filtering.

Hybrid compression techniques combine several approaches, such as DCT and vector quantization or differential pulse code modulation. Various groups have established standards for digital multimedia compression based on the existing JPEG, MPEG, and px64 standards, as shown in Table 1.

When implementing a compression/decompression algorithm, the key question is how to partition between hardware and software in order to maximize performance and minimize cost. Most implementations use specialized video processors and programmable digital signal processors (DSPs). However, powerful RISC processors are making software-only solutions feasible. We can classify implementations of compression algorithms into three categories: (1)a hardwired approach that maximizes performance (for example, C cube), (2) a software solution that emphasizes flexibility with a general-purpose processor, and (3) a hybrid approach that uses specialized video processors.

Table 1 Multimedia compression standards

Short name	Official name	Standards group	Compression
JPEG	Digital compression and coding of continuous-tone still images	Joint Photographic Experts Group	15 : 1 (full color still-frame applications)
H. 261 px64	Video coder/decoder for audio-visual services at px64 kb/s	Special Group on Coding for Visual Telephony	100 : 1 to 200 : 1 (video-based tele-communications)
MPEG	Coding of moving pictures and associated audio	Moving Picture Experts Group	200 : 1 motion-intensive application

3. Multimedia Networking

Many applications, such as video mail, video conferencing, and collaborative work systems, require networked multimedia. In these applications, the multimedia objects are stored at a server and played back at the clients' sites. Such applications might require broadcasting multimedia data to various remote locations or accessing large depositories of multimedia sources.

Traditional LAN environments, in which data sources are locally available, cannot support access to remote multimedia data sources for a number of reasons. Table 2 contrasts traditional data transfer and multimedia transfer.

Table 2 Traditional communications versus multimedia communications

Characteristics	Data Transfer	Multimedia transfer
Data rate	Low	High
Traffic pattern	Bursty	Stream-oriented highly bursty
Reliability requirements	No loss	Some loss
Latency requirements	None	Low, for example, 20ms
Mode of communication	Point-to-point	Multipoint
Temporal relationship	None	Synchronized transmission

Traditional networks do not suit multimedia. Ethernet provides only 10 Mb/s, its access time is not bounded, and its latency and jitter are unpredicatable. Token-ring networks provide 16 Mb/s and are deterministic; from this point of view, they can handle multimedia. However, the predictable worst-case access latency can be very high.

An FDDI network provides 100 Mb/s bandwidth, sufficient for multimedia. In the synchronized mode, FDDI has low access latency and low jitter. FDDI also guarantees a bounded access delay and a predictable average bandwidth for synchronous traffic. However, due to the high cost, FDDI networks are used primarily for backbone networks, rather than networks of workstation.

Less expensive alternatives include enhanced traditional networks. Fast Ethernet, for example, provides up to 100 Mb/s bandwidth. Priority token ring is another system.

Present optical network technology can support the Broadband Integrated Services Digital Network (B-ISDN) standard, expected to become the key network for multimedia applications. B-ISDN access can be basic or primary. Basic ISDN access supports 2B+D channels, where the transfer rate of a B channel is 64 kb/s, and that of a D channel is 16 kb/s. Primary ISDN access supports 23B+D in the US and 30B+D in Europe.

Proposed B-ISDN networks are in either synchronous transfer mode (STM) or asynchronous transfer mode (ATM), to handle both constant and variable bit-rate traffic applications. STM provides fixed bandwidth channels, and therefore is not flexible enough to handle the different types of traffic typical in multimedia applications. On the other hand, ATM is suitable for multimedia traffic; it provides great flexibility in bandwidth allocation by assigning fixed length packets called cells, to virtual connection. ATM can also increase the bandwidth efficiency by buffering and statistically multiplexing bursty traffic at the expense of cell delay and loss.

4. Multimedia Systems

Advances in several technologies are making multimedia systems technically and economically feasible. These advances include powerful workstations, high-capacity storage devices, high-speed networks, advances in image and video processing (such as animation and graphics), advances in audio processing (such as music synthesis and sound effects) , speech processing (speaker recognition and text-to-speech conversion), and advanced still, video, audio, and speech compression algorithms.

A multimedia system consists of three key elements: multimedia hardware, operating system and graphical user interface, and multimedia software development and delivery tools (referred to as authoring tools). Since 1989, when the first multimedia systems were developed, it has been possible to differentiate the three generations of multimedia systems (see Table 3).

Table 3 The three generations of multimedia systems

	First generation 1989—1991	Second generation 1992—1994	Third generation 1995—1996
Media	Text Black/white graphics Bit-mapped images Animation	Color bit-mapped images 16-bit audio Moving still images Full-motion video (15 frames/s)	Full-motion video (30 frames/s) (NTSC/PAL and HDTV quality)
Authoring capability	Hypertext Hypermedia	Object-oriented multimedia with text, graphics, sound, animation, still images, and full-motion video	Integration of object-oriented multimedia with operating system

续表

	First generation 1989—1991	Second generation 1992—1994	Third generation 1995—1996
Video compression technology	DCT JPEG	Motion JPEG MPEG-1	MPEG-2,3,4 Wavelets
Base platform	25MHz 386 (68030) 2MB DRAM 40MB Hard disk VGA Color (680×740) 500MB CD-ROM (100KB/s)	50MHz 486 (68040) 8～16MB DRAM 240MB Hard disk VGA with 256 colors(1024×768) 500MB CD-ROM (150KB/s) 1～2 1.5MB Floppies	50～100MHz Pentium (Power PC) 16～32MB DRAM 1～2 600MB Hard disk 20～30MB floppies SVGA (1280×960) 50MB Writable CD-ROM (300kb/s)
Operating system	DOS	DOS 5 Windows 3.x OS/2 Presentation Manager	Windows NT Pink (IBM/Apple)
Delivery mode	720KB Diskette 1.5MB Laser disk (R/O) 128MB CD-ROM (R/O)	500MB CD-ROM (R/O)	500MB WORM 128～500MB Magnet optic (R/W)
Local area network	Ethernet (10Mb/s) Token ring (16Mb/s)	FDDI (100Mb/s)	Ethernet,Token ring(100Mb/s) FDDI (500MB/s) Isochronous networks ATM

The first generation, based on Intel 80386 and Motorola 68030 processors, is characterized by bitmapped images and animation, JPEG video compression techniques, local area networks based on Ethernet and token ring, and hypermedia authoring tools. The second generation uses i80486 and MC68040 processors, moving and still images, 16-bit audio, JPEG and MPEG-1 video compression, FDDI networks, and object-oriented, multimedia authoring tools that incorporate text, graphics, animation, and sound.

We are presently at the transition stage from the second- to the third-generation systems, based on more powerful processors such as Pentium and PowerPC. The third generation will use full-motion, VCR-quality video, eventually moving to NTSC/PAL and HDTV. Compression algorithms will include MPEG-2, MPEG-3, and MPEG-4, and perhaps the wavelets method now in the research stage. The system will use enhanced Ethernet, token ring, and FDDI network, as well as new isochronous and ATM networks. The authoring tools will integrate object-oriented multimedia into the operating system.

5. Applications

Multimedia systems suggest a wide variety of potential applications. Three important applications already in use are multimedia mailing systems, collaborative work systems, and multimedia conferencing systems.

Multimedia mailing systems are more sophisticated than standard electronic mailing systems. They implement multiple applications, such as multimedia editing and voice mail, and require higher transmission rates than text-only system.

Collaborative work systems allow group members to discuss a problem and actually create something together. During a meeting, users can view, discuss, and modify multimedia documents.

Multimedia conferencing systems enable a number of participants to exchange various multimedia information via voice and data networks. Each participant has a multimedia workstation, linked to the other workstations over high-speed networks. Each participant can send and receive video, audio, and data, and can perform certain collaborative activities. The multimedia conference uses the concept of the shared virtual workspace, which describes the part of the display replicated, at every workstation.

Multimedia conferencing systems must provide a number of functions, such as multiple-call setup, conference status transmission, real-time control of audio and video, dynamic allocation of network resources, multiport data transfer, synchronization of shared workspace, and graceful degradation under fault conditions.

6. Research Directions

Research and development in high-speed networks will soon provide the bandwidth needed for distributed multimedia applications. Therefore, I envision tremendous growth in distributed multimedia systems and their applications.

Advances in distributed multimedia systems have begun to significantly affect the development of on-demand multimedia services, such as interactive entertainment, video news distribution, video rental services, and digital multimedia libraries. Various companies realized that fiber optic networks, coupled with improved computing and compression techniques, would soon be capable of delivering digital movies. Over the past year, a number of alliances have formed between entertainment, cable, phone, and computer companies, with the main focus on video-on-demand applications.

Many challenging problems remain to be researched and resolved for the further growth of multimedia systems. Multimedia applications make enormous demands on computer hardware and software resources. Therefore, one of the ongoing demands is to develop more powerful multimedia workstations. Multimedia workstations will also need multimedia operating systems (MMOSs) and advanced multimedia user interfaces. An MMOS should handle continuous media by providing preemptive multitasking, easy expandability, format-independent data access, and support for real-time scheduling. It should be object-oriented and capable of synchronizing data to be instantly available, so the user interface must be highly sophisticated and intuitive.

Integrating the user interface at the operating system level could eliminate many problems for application software developers. Other research challenges include developing new real-time compression algorithms (perhaps based on wavelets) , large storage devices, and multimedia data management systems. The constant challenge is further refinement of high-speed, deterministic networks with low latency and low jitter, as well as research in new multimedia synchronization algorithms.

New Words and Expressions

access time 存取时间
algorithm *n.* 算法
asynchronous transfer mode (ATM) 异步传输模式
bandwidth *n.* 带宽
browsing *n.* 浏览
channel *n.* 信道，通道
codec *n.* 编码译码器
compression *n.* 压缩
differential pulse code modulation 差分脉码调制
digital signal processor 数字信号处理
Fourier transform 傅里叶变换
heterogeneous *adj.* 不同种类的，杂散的
human-computer interface 人机接口
hybrid *n.* 混合
hypermedia *n.* 超媒体
hypertext *n.* 超文本
interactive video 交互视频
isochronous *adj.* 等时的
jitter *n.* 抖动，跳动
latency *n.* 等待时间
multimedia file system 多媒体文件系统
non-interactive *adj.* 非交互式的
partition *n.* 划分
real-time scheduling 实时调度
retrieval *n.* 检索
speech processing 语音处理
stand-alone multimedia workstation 独立多媒体工作站
still image 静止图像
synchronized mode 同步模式
synchronous transfer mode (STM) 同步传输模式
text-to-speech conversion 文本语音变换
token ring 令牌环网
transfer rate 传输速率
wavelet *n.* 子波，小波

Exercises to the Text

1. Translate the following words and phrases into English.

（1）独立多媒体工作站　（2）差分脉码调制　（3）人机接口　（4）数字信号处理　（5）实时调度

2. Read the following abbreviations and give the Chinese meaning of each one.

ADPCM　Adaptive Differential Pulse Code Modulation
ATM　Asynchronous Transfer Mode
BER　Bit Error Rate
B-ISDN　Broadband Integrated Service Digital Network
ITTCC　International Telegraph and Telephone Consultative Committee
Codec　Coder/Decoder
DCT　Discrete Cosine Transform
DPCM　Differential Pulse Code Modulation
DSP　Digital Signal Processor
DVI　Digital Visual Interface
FDCT　Forward Discrete Cosine Transform
FDDI　Fibre Distributed Data Interface
IDCT　Inverse Discrete Cosine Transform
JPEG　Joint Photographic Expert Group
MOS　Multimedia Operating System
MPEG　Moving Pictures Expert Group
NTSC　National Television Standards Committee
PAL　Phase Alternating Line
PER　Packet Error Rate
PTR　Priority Token Ring
QOS　Quality Of Service
RISC　Reduced Instruction Set Computer
STM　Synchronous Transfer Mode

3. Translate the following paragraphs into Chinese.

(1) Advances in distributed multimedia systems have begun to significantly affect the development of on-demand multimedia services. Researchers are working within established computer areas to transform existing technologies and develop new ones. The big picture shows multimedia as the merging of computing, communications, and broadcasting.

(2) The complexity of multimedia applications stresses all the components of a computer system. Multimedia requires great processing power to implement software codecs, multimedia file systems, and corresponding file formats. The architecture must provide high bus bandwidth and efficient I/O.

(3) Digital data compression relies on various computational algorithms, implemented either in software or hardware. We can classify compression techniques into lossless and lossy

approaches. Lossless techniques can recover the original representation perfectly. Lossy techniques recover the presentation with some loss of accuracy. The lossy techniques provide higher compression ratios, though, and therefore are applied more often in image and video compression than lossless techniques.

(4) Multimedia conferencing systems enable a number of participants to exchange various multimedia information via voice and data networks. Each participant has a multimedia workstation, linked to the other workstations over high-speed networks.

(5) Many challenging problems remain to be researched and resolved for the further growth of multimedia systems. Multimedia applications make enormous demands on computer hardware and software resources. Therefore, one of the ongoing demands is to develop more powerful multimedia workstations.

Text 2: An Introduction to Digital Image Processing

Interest in digital image processing methods stems from two principal application areas: improvement of pictorial information for human interpretation; and processing of image data for storage, transmission, and representation for autonomous machine perception.

1. What is Digital Image Processing?

An image may be defined as a two-dimensional function, f(x, y), where x and y are spatial (plane) coordinates, and the amplitude of f at any pair of coordinates (x, y) is called the intensity or gray level of the image at that point. When x, y, and the amplitude values of *f* are all finite, discrete quantities, we call the image a digital image. The field of digital image processing refers to processing digital images by means of a digital computer. Note that a digital image is composed of a finite number of elements, each of which has a particular location and value. These elements are referred to as picture elements, image elements, pels, and pixels. Pixel is the term most widely used to denote the elements of a digital image.

Vision is the most advanced of our senses, so it is not surprising that images play the single most important role in human perception. However, unlike humans, who are limited to the visual band of the electromagnetic (EM) spectrum, imaging machines cover almost the entire EM spectrum, ranging from gamma to radio waves. They can operate on images generated by sources that humans are not accustomed to associating with images. These include ultrasound, electron microscopy, and computer-generated images. Thus, digital image processing encompasses a wide and varied field of applications.

Sometimes a distinction is made by defining image processing as a discipline in which both the input and output of a process are images. We believe this to be a limiting and somewhat artificial boundary. For example, under this definition, even the trivial task of computing the average intensity of an image (which yields a single number) would not be considered an image processing operation. On the other hand, there are fields such as computer vision whose ultimate goal is to use computers to emulate human vision, including learning and being able to make inferences and take actions based on visual inputs. This area itself is a branch of artificial intelligence (AI) whose objective is to emulate human intelligence. The field of AI is in its earliest stages of infancy in terms of development, with progress having been much slower than originally anticipated. The area of image analysis (also called image understanding) is in between image processing and computer vision.

There are no clear-cut boundaries in the continuum from image processing at one end to computer vision at the other. However, one useful paradigm is to consider three types of computerized processes in this continuum: low-, mid-, and high-level processes. Low-level processes involve primitive operations such as image preprocessing to reduce noise, contrast enhancement, and image sharpening. A low-level process is characterized by the fact that both its inputs and outputs are images. Mid-level processing on images involves tasks such as segmentation (partitioning an image into regions or objects), description of those objects to reduce them to a form suitable for computer processing, and classification (recognition) of individual objects. A mid-level process is characterized by the fact that its inputs generally are images, but its outputs are attributes extracted from those images (e.g., edges, contours, and the identity of individual objects). Finally, higher-level processing involves "making sense" of an ensemble of recognized objects, as in image analysis, and, at the far end of the continuum, performing the cognitive functions normally associated with vision.

Based on the preceding comments, we see that a logical place of overlap between image processing and image analysis is the area of recognition of individual regions or objects in an image. Thus, what we call digital image processing encompasses processes whose inputs and outputs are images and, in addition, encompasses processes that extract attributes from images, up to and including the recognition of individual objects. As a simple illustration to clarify these concepts, consider the area of automated analysis of text. The processes of acquiring an image of the area containing the text, preprocessing that image, extracting (segmenting) the individual characters, describing the characters in a form suitable for computer processing, and recognizing those individual characters are in the scope of what we call digital image processing here.

2. Fundamental Steps in Digital Image Processing

It is helpful to divide the material covered in the following chapters into the two broad categories: methods whose input and output are images, and methods whose inputs may be images, but whose outputs are attributes extracted from those images. This organization is summarized in Figure 1. The diagram does not imply that every process is applied to an image. Rather, the intention is to convey an idea of all the methodologies that can be applied to images for different purposes and possibly with different objectives. The discussion in this section may be viewed as a brief overview of digital image processing.

Image acquisition is the first process shown in Figure 1. Note that acquisition could be as simple as being given an image that is already in digital form. Generally, the image acquisition stage involves preprocessing, such as scaling.

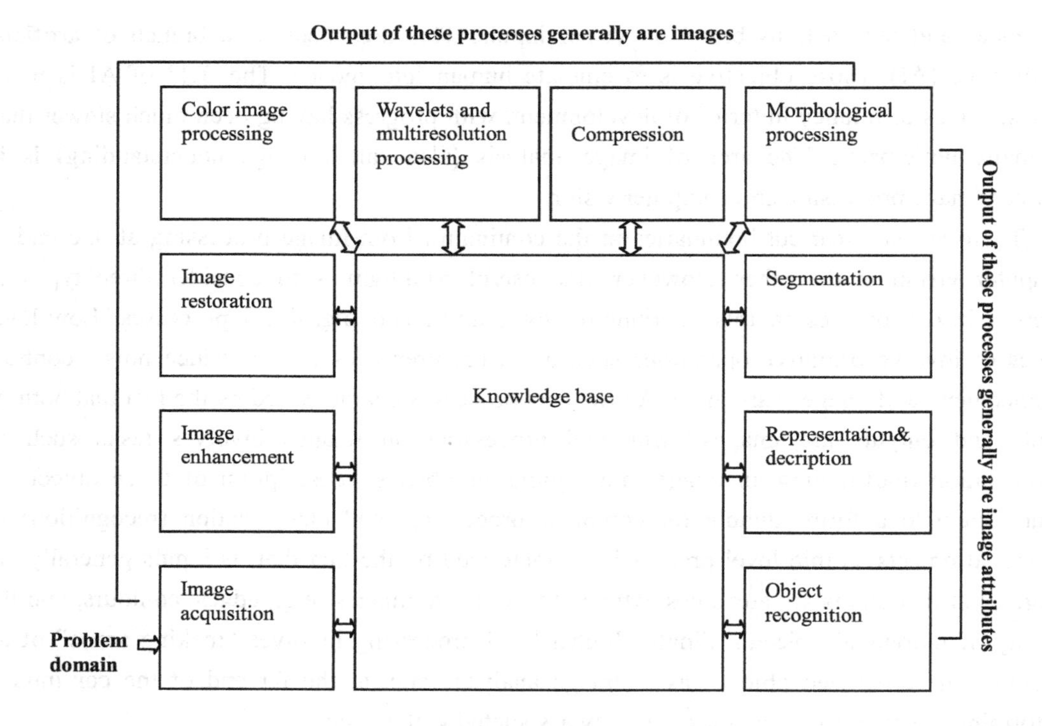

Figure 1　Fundamental steps in digital image processing

Image enhancement is among the simplest and most appealing areas of digital image processing. Basically, the idea behind enhancement techniques is to bring out detail that is obscured, or simply to highlight certain features of interest in an image. A familiar example of enhancement is when we increase the contrast of an image because "it looks better". It is important to keep in mind that enhancement is a very subjective area of image processing. We don't mean enhancement is more important than the other topics, rather, we use it as an avenue to introduce the other techniques. Thus, rather than having a chapter dedicated to mathematical preliminaries, we introduce a number of needed mathematical concepts by showing how they apply to enhancement. This approach allows the reader to gain familiarity with these concepts in the context of image processing. A good example of this is the Fourier transform.

Image restoration is an area that also deals with improving the appearance of an image. However, unlike enhancement, which is subjective, image restoration is objective, in the sense that restoration techniques tend to be based on mathematical or probabilistic models of image degradation. Enhancement, on the other hand, is based on human subjective preferences regarding what constitutes a "good" enhancement result.

Color image processing is an area that has been gaining in importance because of the significant increase in the use of digital images over the Internet.

Wavelets are the foundation for representing images in various degrees of resolution. This material is used for image data compression and for pyramidal representation, in which images are subdivided successively into smaller regions.

Compression, as the name implies, deals with techniques for reducing the storage required to save an image, or the bandwidth required to transmit it. Although storage technology has improved significantly over the past decade, the same cannot be said for transmission capacity. This is true particularly in uses of the Internet, which are characterized by significant pictorial content. Image compression is familiar (perhaps inadvertently) to most users of computers in the form of image file extensions, such as the jpg file extension used in the JPEG (Joint Photographic Experts Group) image compression standard.

Morphological processing deals with tools for extracting image components that are useful in the representation and description of shape.

Segmentation procedures partition an image into its constituent parts or objects. In general, autonomous segmentation is one of the most difficult tasks in digital image processing. A rugged segmentation procedure brings the process a long way toward successful solution of imaging problems that require objects to be identified individually. On the other hand, weak or erratic segmentation algorithms almost always guarantee eventual failure. In general, the more accurate the segmentation, the more likely recognition is to succeed.

Representation and description almost always follow the output of a segmentation stage, which usually is raw pixel data, constituting either the boundary of a region (i.e., the set of pixels separating one image region from another) or all the points in the region itself. In either case, converting the data to a form suitable for computer processing is necessary. The first decision that must be made is whether the data should be represented as a boundary or as a complete region. Boundary representation is appropriate when the focus is on external shape characteristics, such as corners and inflections. Regional representation is appropriate when the focus is on internal properties, such as texture or skeletal shape. In some applications, these representations complement each other. Choosing a representation is only part of the solution for transforming raw data into a form suitable for subsequent computer processing. A method must also be specified for describing the data so that features of interest are highlighted. Description, also called feature selection, deals with extracting attributes that result in some quantitative information of interest or are basic for differentiating one class of objects from another.

Recognition is the process that assigns a label (e.g., “vehicle”) to an object based on its descriptors.

So far we have said nothing about the need for prior knowledge or about the interaction between the knowledge base and the processing modules in Figure 1. Knowledge about a problem domain is coded into an image processing system in the form of a knowledge database. This knowledge may be as simple as detailing regions of an image where the information of interest is known to be located, thus limiting the search that has to be conducted in seeking that information. The knowledge base also can be quite complex, such as an interrelated list of all major possible defects in a materials inspection problem or an image database containing high-resolution satellite images of a region in connection with change-detection applications. In addition to guiding the operation of each processing module, the knowledge base also controls the interaction between modules. This distinction is made in Figure 1 by the use of double-headed arrows between the processing modules and the knowledge base, as opposed to single-headed arrows linking the processing modules.

Although we do not discuss image display explicitly at this point, it is important to keep in mind that viewing the results of image processing can take place at the output of any stage in Figure 1. We also note that not all image processing applications require the complexity of interactions implied by Figure 1. In fact, not even all those modules are needed in some cases. For example, image enhancement for human visual interpretation seldom requires use of any of the other stages in Figure 1. In general, however, as the complexity of an image processing task increases, so does the number of processes required to solve the problem.

3. Components of an Image Processing System

As recently as the mid-1980s, numerous models of image processing systems being sold throughout the world were rather substantial peripheral devices that attached to equally substantial host computers. Late in the 1980s and early in the 1990s, the market shifted to image processing hardware in the form of single boards designed to be compatible with industry standard buses and to fit into engineering workstation cabinets and personal computers. In addition to lowering costs, this market shift also served as a catalyst for a significant number of new companies whose specialty is the development of software written specifically for image processing.

Although large-scale image processing systems still are being sold for massive imaging applications, such as processing of satellite images, the trend continues toward miniaturizing and blending of general-purpose small computers with specialized image processing hardware. Figure 2 shows the basic components comprising a typical *general-purpose* system used for digital image processing. The function of each component is discussed in the following paragraphs, starting with image sensing.

With reference to sensing, two elements are required to acquire digital images. The first is a physical device that is sensitive to the energy radiated by the object we wish to image. The second, called a digitizer, is a device for converting the output of the physical sensing device into digital form. For instance, in a digital video camera, the sensors produce an electrical output proportional to light intensity. The digitizer converts these outputs to digital data.

Specialized image processing hardware usually consists of the digitizer just mentioned, plus hardware that performs other primitive operations, such as an arithmetic logic unit (ALU), which performs arithmetic and logical operations in parallel on entire images. One example of how an ALU is used is in averaging images as quickly as they are digitized, for the purpose of noise reduction. This type of hardware sometimes is called a front-end subsystem, and its most distinguishing characteristic is speed. In other words, this unit performs functions that require fast data throughputs (e.g., digitizing and averaging video images at 30 frames/s) that the typical main computer cannot handle.

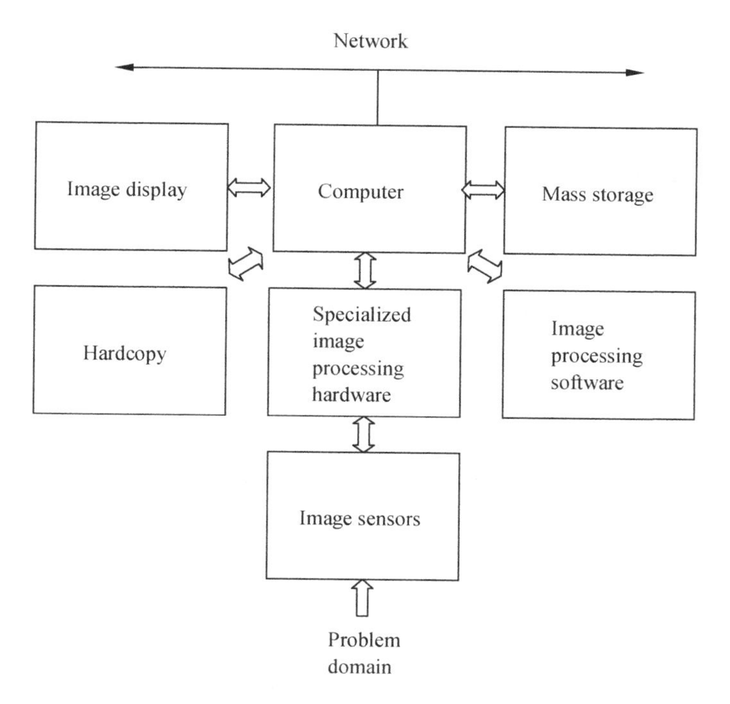

Figure 2 Components of a general-purpose image processing system

The computer in an image processing system is a general-purpose computer and can range from a PC to a supercomputer. In dedicated applications, sometimes specially designed computers are used to achieve a required level of performance, but our interest here is on general-purpose image processing systems. In these systems, almost any well-equipped PC-type machine is suitable for off-line image processing tasks.

Software for image processing consists of specialized modules that perform specific tasks. A well-designed package also includes the capability for the user to write code that, as a minimum, utilizes the specialized modules. More sophisticated software packages allow the integration of those modules and general-purpose software commands from at least one computer language.

Mass storage capability is a must in image processing applications. An image of size 1024 $\times$1024 pixels, in which the intensity of each pixel is an 8-bit quantity, requires one megabyte of storage space if the image is not compressed. When dealing with thousands, or even millions, of images, providing adequate storage in an image processing system can be a challenge. Digital storage for image processing applications falls into three principal categories: (1) short-term storage for use during processing, (2) on-line storage for relatively fast recall, and (3) archival storage, characterized by infrequent access. Storage is measured in bytes (eight bits), Kbytes (one thousand bytes), Mbytes (one million bytes), Gbytes (meaning giga, or one billion, bytes), and Tbytes (meaning tera, or one trillion, bytes).

One method of providing short-term storage is computer memory. Another is by Specialized boards, called frame buffers, that store one or more images and can be accessed rapidly, usually at video rates (e.g., at 30 complete images per second). The latter method allows virtually instantaneous image zoom, as well as scroll (vertical shifts) and pan (horizontal shifts). Frame buffers usually are housed in the specialized image processing hardware unit shown in Figure 2. On-line storage generally takes the form of magnetic disks or optical-media storage. The key factor characterizing on-line storage is frequent access to the stored data. Finally, archival storage is characterized by massive storage requirements but infrequent need for access. Magnetic tapes and optical disks housed in "jukeboxes" are the usual media for archival applications.

Image displays in use today are mainly color (preferably flat screen) TV monitors. Monitors are driven by the outputs of image and graphics display cards that are an integral part of the computer system. Seldom are there requirements for image display applications that cannot be met by display cards available commercially as part of the computer system. In some cases, it is necessary to have stereo displays, and these are implemented in the form of headgear containing two small displays embedded in goggles worn by the user.

Hardcopy devices for recording images include laser printers, film cameras, heat-sensitive devices, inkjet units, and digital units, such as optical and CD-ROM disks. Film provides the highest possible resolution, but paper is the obvious medium of choice for written material. For presentations, images are displayed on film transparencies or in a digital medium if image projection equipment is used. The latter approach is gaining acceptance as the standard for image presentations.

Networking is almost a default function in any computer system in use today. Because of the large amount of data inherent in image processing applications, the key consideration in image transmission is bandwidth. In dedicated networks, this typically is not a problem, but communications with remote sites via the Internet are not always as efficient. Fortunately, this situation is improving quickly as a result of optical fiber and other broadband technologies.

New Words and Expressions

interpretation *n.* 解释，说明，诠释
coordinate *n.* 坐标系，坐标
ultrasound *n.* 超声，超声波
perception *n.* 感知（能力），觉察（力），认识，观念，看法
paradigm *n.* 典范，范例，示例
infancy *n.* 婴儿期，幼年时代，未成年，初期，幼年期
continuum *n.* 连续统一体
overlap *vt. & vi.* 部分重叠 *n.* 重叠的部分

encompass　*vt.* 围绕，包围
methodology　*n.* 一套方法，方法学，方法论
preliminary　*adj.* 初步的，预备的，开端的 *n.* 准备工作，初步行动
ensemble　*n.* 整体，总效果
pyramidal　*adj.* 金字塔形的，锥体的
wavelet　*n.* 小浪，微波
erratic　*adj.* 不稳定的，无规律的，漂泊的
interrelated　*adj.* 相互关联的
explicitly　*adv.* 明白地，明确地
miniaturize　*vt.* 使小型化

Exercises to the Text

1. Translate the following words and phrases into English.

（1）人工智能　（2）增强对比　（3）图像锐化（清晰化）　（4）图像获取　（5）图像增强　（6）图像复原　（7）形态学处理　（8）特征选择　（9）专业图像处理硬件　（10）算术逻辑单元　（11）帧缓存　（12）前端子系统

2. Translate the following paragraphs into Chinese.

(1) There are no clear-cut boundaries in the continuum from image processing at one end to computer vision at the other. However, one useful paradigm is to consider three types of computerized processes in this continuum: low-level, mid-level, and high-level processes. Low-level processes involve primitive operations such as image preprocessing to reduce noise, contrast enhancement, and image sharpening.

(2) The processes of acquiring an image of the area containing the text, preprocessing that image, extracting (segmenting) the individual characters, describing the characters in a form suitable for computer processing, and recognizing those individual characters are in the scope of what we call digital image processing here.

(3) Image enhancement is among the simplest and most appealing areas of digital image processing. Basically, the idea behind enhancement techniques is to bring out detail that is obscured, or simply to highlight certain features of interest in an image.

(4) Mass storage capability is a must in image processing applications. An image of size 1024×1024 pixels, in which the intensity of each pixel is an 8-bit quantity, requires one megabyte of storage space if the image is not compressed.

Text 3: Digital Audio

In the early computer, the only sound that we heard from a computer was a beep—often accompanied by an error message. Now an entire range of sounds can be played through the computer, including music, narration, sound effects, and original recording of events such as a speech or a concert. The elements of sound used in computer often called digital audio, and it is fundamental to multimedia.

Now we have a rough idea of what's happening when we hear a sound, we can begin to make sense of what audio experts cryptically refer to as "waveform" diagrams. Here's what a waveform looks like (Figure 3).

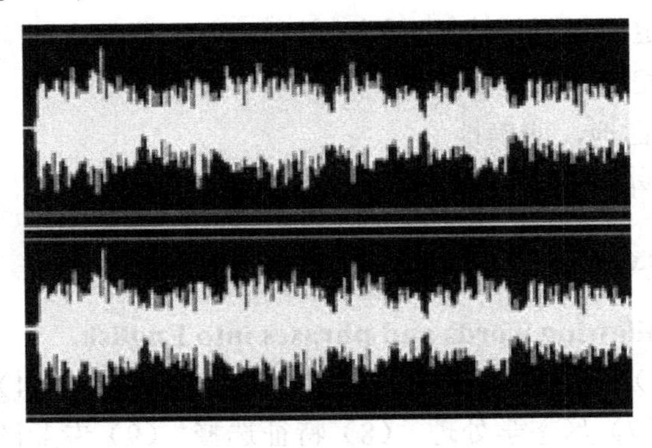

Figure 3 Waveform diagram

Waveforms aren't as intimidating as they look. In essence, a waveform is a graph that charts minute changes in air pressure as sound waves propagate. The y axis represents air pressure and the x axis represents time (Figure 4).

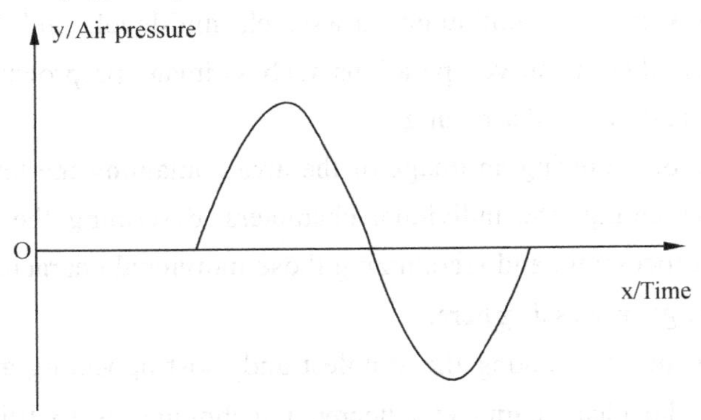

Figure 4 Waveform zoomed in

A sound wave arrives. The air pressure goes up and the line on the waveform rises. As the sound wave passes the air pressure falls again and the line falls as well. These changes in pressure happen very quickly—thousands of times as a second. Waveform diagrams representing even a few seconds of sound are, consequently, very big. This is the reason why waveforms of audio recordings often look complicated and squiggly when you view them on your computer.

There are several technical terms used to describe waveforms (Figure 5) that you should know. They'll come into use when we get to discussing digital audio.

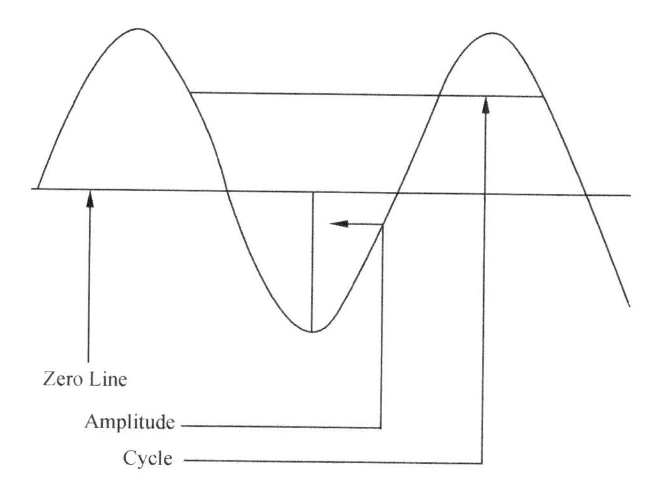

Figure 5 Waveform technical terms

Zero Line: The horizontal line running through the middle of the graph is called the zero line. It represents the rest state, when there is no compression or rarefaction.

Amplitude: Amplitude is the amount of compression or rarefaction at any point on the waveform. Graphically, it is the distance above or below the zero line. In general, the greater the amplitude is, the louder the volume is.

Cycle: A cycle is the amount of time it takes for the amplitude of the waveform to return to the same level.

Frequency: The frequency of a sound is the number of cycles that happen every second. The higher is the frequency, the higher is the perceived pitch of a sound. For humans, hearing is limited to frequencies between about 20 Hz and 20, 000Hz (20 kHz), with the upper limit generally decreasing with age. Other species have a different rage of hearing. The average dog can hear frequencies as high as 45,000 Hz. Cats can hear up to 63,000Hz, and the beluga whale can hear frequencies of up to 123, 000Hz.

Digital audio has emerged because of its usefulness in the recording, manipulation, mass—production, and distribution of sound. Modern distribution of music across Internet through on-line stores depends on digital recording and digital compression algorithms. Distribution of audio as data files rather than as physical objects has significantly reduced cost of distribution.

Digital audio uses digital signals for sound reproduction. This includes analog-to-digital conversion, digital-to-analog conversion, storage, and transmission.

A digital audio signal starts with an analog-to-digital converter (ADC) that converts an analog signal to a digital signal. The ADC runs at a sampling rate and converts at a known bit resolution. For example, CD audio has a sampling rate of 44.1 kHz (44,100 samples per second) and 16-bit resolution for each channel (stereo). If the analog signal is not already band limited then an anti-aliasing filter is necessary before conversion, to prevent aliasing in the digital signal.

Sound waves ripple past the microphone, causing a diaphragm inside to vibrate. The vibrating diaphragm creates a change in voltage in the wire that runs from the microphone to the computer. This fluctuating voltage is an analog representation of the sound, for it changes smoothly from one amplitude to the next, encompassing all values in-between. Inside of our computer the ADC (usually a part of the sound card), at regular intervals the ADC measures the microphone's analog signal and outputs a number representing the amplitude of the signal at that precise instant. This is called a "sample". Before long there are a huge number of samples all arranged in chronological order—a kind of "spot-map" of the original waveform(Figure 6).

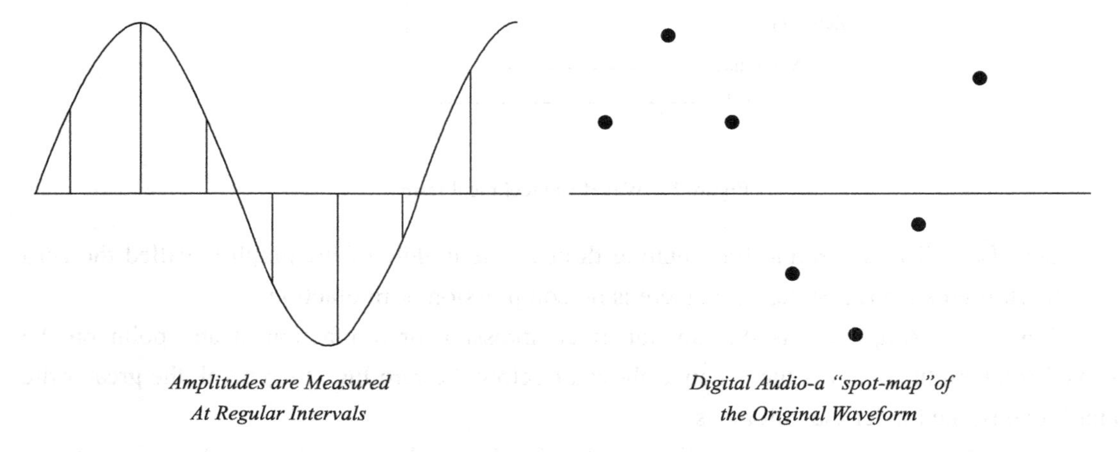

Figure 6 Spot-map of waveform

The sampling rate, sample rate, or sampling frequency defines the number of samples per second taken from a continuous signal to make a discrete signal. For time-domain signals, it can be measured in hertz (Hz). The three most sample rates are 11.025 kHz, 22.05 kHz and 44.1 kHz. The higher the sample rate is, the more samples that are taken and thus the better the quality of the digital audio.

A set of digital audio samples contains data that, when converted into an analog signal, provides the necessary information to reproduce the sound wave. The samples then code in the binary digit, it uses the bit depth. In digital audio, bit depth describes the number of bits of information recorded for each sample. Bit depth directly corresponds to the resolutions of each sample in a set of digital audio data. The two common bit depths are 8bits and 16bits, common examples of bit depth include CD audio, which is recorded at 16bits, and DVD-Audio, which can support up to 24 bits audio. The standard audio CD is said to have a data rate of 44.1kHz/16, implying the audio data is sampled 44,100 times per second, with a bit depth of 16. CD tracks are usually stereo, using a left and right track, so the amount of audio data per second is double that of mono, where only a single track is used. The bit rate is then 44 100 samples/second* 16bits/sample*2=1 411 200bit/s or 1.4Mb/s.

The digital audio signal may then be stored or transmitted. Digital audio storage can be on a CD, a MP3 player, a hard drive, an USB flash drive, a compact flash, or any other digital data storage device. Audio data compression techniques—such as MP3, advanced audio coding, or Flac—are commonly employed to reduce the file size. Digital audio can be streamed to other devices.

The last step for digital audio is to be converted back to an analog signal with a digital-to-analog converter (DAC). Like ADCs, DACs run at a specific sampling rate and bit resolution but through the processes of oversampling, upsampling, and downsampling, this sampling rate may not be the same as the initial sampling rate, Figure 7 shows the process of digital-to-analog conversion.

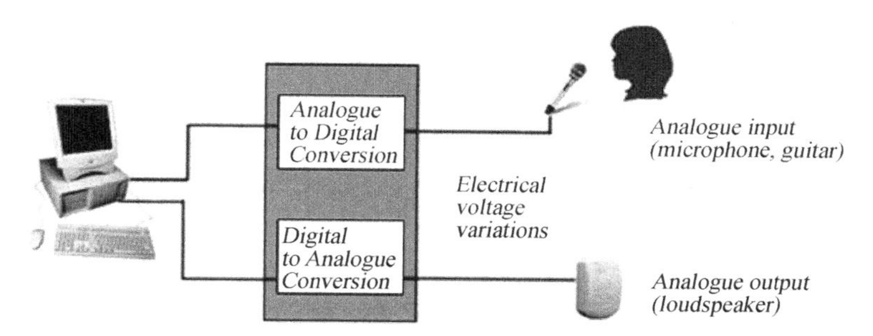

Figure 7 Process of digital-to-analog conversion

After sample and code the digital audio signal may then be stored or transmitted. This audio data can then be stored uncompressed or compressed to reduce the file size. An audio file format is a container format for storing audio data on a computer system.

It is important to distinguish between a file format and a codec. A codec performs the encoding and decoding of the raw audio data while the data itself is stored in a file with a specific audio file format. Though most audio file formats support only one audio codec, a file format may support multiple codecs, as AVI does.

There are three major groups of audio file formats.

- Uncompressed audio formats, such as WAV, AIFF, and AU.
- Lossless compression formats, such as FLAC, Monkey's Audio (filename extension APE), WavPack (filename extension WV), Shorten, Tom's lossless Audio Kompressor (TAK), TTA, Apple Lossless and lossless Windows Media Audio (WMA).
- Lossy compression formats, such as MP3, Vorbis, Musepack, lossy Windows Media Audio (WMA) and AAC.

There is one major uncompressed audio format, PCM, which is usually stored as a .wav on Windows or as .aiff on Mac OS. WAV is a flexible file format designed to store more or less any combination of sampling rates or bit rates. This makes it an adequate file format for storing and archiving an original recording. A lossless compressed format would require more processing for the same time recorded, but would be more efficient in terms of space used. WAV, like any other uncompressed format, encodes all sounds, whether they are complex sounds or absolute silence, with the same number of bits per unit of time. As an example, a file containing a minute of playing by a symphonic orchestra would be the same size as a minute of absolute silence if they were both stored in WAV. If the files were encoded with a lossless compressed audio format, the first file would be marginally smaller, and the second file taking up almost no space at all. However, to encode the files to a lossless format would take significantly more time than encoding the files to the WAV format, Recently some new lossless formats have been developed (for example TAK), whose aim is to achieve very fast coding with good compression ratio.

Some of the common sound file format types are shown in the following.

- WAV—Standard audio file container format used mainly in Windows PCs. Commonly used for storing uncompressed (PCM), CD-quality sound files, which means that they can be large in size—around 10MB per minute. Wave files can also contain data encoded with a variety of codecs to reduce the file size (for example the GSM or MP3 codecs). Wav files use a RIFF structure.
- FLAC—A lossless compression codec. This format is a lossless compression as like zip but for audio. If you compress a PCM file to FLAC and then restore it again it will be a prefect copy of the original. (All the other codecs discussed here are lossy which means a small part of the quality is lost). The cost of this losslessness is that the compression ratio is not good. Flac is recommended for archiving PCM files where quality is important (e.g. broadcast or music use).
- AIFF (Audio Interchange File Format—The standard file format used by Apple. It is like a WAV file for the Mac.
- RAW—A RAW file can contain audio in any codec but is usually used with PCM audio data. It is rarely used except for technical tests.
- AU—The standard audio file format used by Sun, Unix and Java.
- VOX—The VOX format most commonly uses the dialogic ADPCM (Adaptive Differential Pulse Code Modulation) codec. Similar to other ADPCM formats, it is compressed to 4-bit. Vox format files are similar to wave files except that the vox files contain no information about the file itself so the codec sample rate and number of channels must first to be specified in order to play a vox file.
- AAC—The Advanced Audio Coding format is based on the MPEG-2 and MPEG-4 standards. AAC files are usually ADTS or ADIF containers.
- MP3—The MPEG-1 audio layer 3 compression format is the most popular format for downloading and storing music. By eliminating portions of the audio file that are essentially inaudible, Mp3 files are compressed to roughly one-tenth the size of an equivalent PCM file while maintaining good audio quality.
- WMA—The popular Windows Media Audio format owned by Microsoft.
- RA—A Real Audio format designed for streaming audio over the Internet. The .ra format allows files to be stored in a self-contained fashion on a computer, with all of the audio data contained inside the file itself.

Today digital audio tends to be used more than analog audio as a method of storage. Records and cassette tapes continue to be used, but by a relatively small market. Why is digital audio more prevalent? While there is some debate as to whether or not digital audio actually sounds better than analog audio, digital audio is certainly easier to reproduce and to manipulate without loss of quality. Because of digital audio, it is much easier for both amateur and professional musicians today to produce studio-quality music.

New Words and Expressions

error *n.* 错误，误差
narration *n.* 叙述
waveform *n.* 波形
axis *n.* 轴，坐标轴
represent *vt.* 表现，描绘，声称，扮演 *vi.* 提出异议
complicated *adj.* 复杂的，难解的
squiggly *adv.* 弯弯曲曲地
horizontal *adj.* 地平线的，水平的
amplitude *n.* 振幅，丰富，广阔
volume *n.* 音量，卷，册，体积
cycle *n.* 周期，循环 *vi.* 循环，轮转 *vt.* 使循环
frequency *n.* 频率，周期，发生次数
pitch *n.* 斜度，程度，倾斜 *vt.* 投，掷，定位于 *vi.* 投掷，坠落，倾斜
converter *n.* 转换器
sample *n.* 采样，标本，样品 *vt.* 采样，取样，抽取……的样品，尝试
rate *n.* 比率，速度，等级，价格 *vt.* 估价，认定，鉴定等级 *vi.* 被评价
vibrate *v.*（使）振动，（使）摇摆
voltage *n.* 电压，伏特数
discrete *adj.* 离散的，不连续的
codec *n.* 编码解码器
uncompress *vt.* 未压缩
format *n.* 格式，形式，板式 *vt.* 格式化，安排……的格局
lossless *adj.* 无损的
lossy *adj.* 有损的
equivalent *adj.* 相等的，相当的，同意义的 *n.* 等价物，相等物

Exercises to the Text

1. Translate the following words and phrases into English.

（1）数模转换器 （2）无损压缩 （3）有损压缩 （4）采样频率

2. Translate the following paragraphs into Chinese.

(1) Digital audio has emerged because of its usefulness in the recording, manipulation, mass-production, and distribution of sound. Modern distribution of music across Internet through on-line stores depends on digital recording and digital compression algorithms. Distribution of audio as data files rather than as physical objects has significantly reduced cost of distribution.

(2) A set of digital audio samples contains data that, when converted into an analog signal, provides the necessary information to reproduce the sound wave. The samples then code in the binary digit, it uses the bit depth. In digital audio, bit depth describes the number of bits of information recorded for each sample.

(3) After sample and code the digital audio signal may then be stored or transmitted. This audio data can then be stored uncompressed or compressed to reduce the file size. An audio file format is a container format for storing audio data on a computer system.

(4) The MPEG-1 audio layer 3 compression format is the most popular format for downloading and storing music. By eliminating portions of the audio file that are essentially inaudible, MP3 files are compressed to roughly one-tenth the size of an equivalent PCM file while maintaining good audio quality.

(5) Why is digital audio more prevalent? While there is some debate as to whether or not digital audio actually sounds better than analog audio, digital audio is certainly easier to reproduce and to manipulate without loss of quality. Because of digital audio, it is much easier for both amateur and professional musicians today to produce studio-quality music.

Text 4: Asset Management

Television content creation is becoming more and more file-based. Broadcasters have to deliver more for less (money). Over-the-air broadcasting is now complemented by satellite, cable offers video on demand (VOD) and interactivity, and now competes with internet protocol television (IPTV) delivered over telco networks. Handheld devices can receive broadcast television and VOD.

Content has to be created for these different delivery channels in multiple formats. Today's over-the-air program is tomorrow's download to mobile. Content is now made available almost simultaneously over several media. The old paradigm of trickle releases to different media over months even years invites piracy. Publishers have learned to monetize program content while it is fresh, in any form the public will consume.

Content creators and publishers are looking to digital asset management (DAM) to improve productivity and to provide sensible management in a file-based production environment.

This article introduces the concepts of asset management. First it answers the question "What are assets?" The some of the related solutions are described—media management, document management and content management. Finally, it looks at why digital asset management is useful for any enterprise, not just media and publishing industries.

DAM has proved quite a technical challenge. It was not until the end of the twentieth century that affordable systems became available. The fall in the price of processing and disk storage has been the primary enabling factor, along with the ever-increasing bandwidths of computer networks, both within a local area and over long distances. Many of the products that are designed to manage the very large repositories can automate labor-intensive and repetitive tasks. These include cataloging and indexing the content, and the powerful search engines that we use to discover and find content.

To index and catalog digital assets, product developers have leveraged research into the understanding of speech, character recognition and conceptual analysis. The widespread application of these ideas requires considerable processing power to provide real-time ingesting of content, essential in many audio-video applications. The introduction of the low-cost and powerful computer workstations that we now take for granted makes such technology commercially viable.

The modern business creates media content in many guises. Traditionally each department would manage its own media content, usually with some formalized filing structure, as shown in Table 4. Global brand management demands the enterprise-wide sharing content if the corporation is not to descend into chaos, with the wrong logos, out-of-date corporate images, and link-rot on the website.

Table 4 Content in the enterprise

Department	Content	Formats	Distribution channel
Corporate Communications &Public Relations	Press releases MS Word	HTML, PDF	Web Mail
Investor Relations	Annual reports	InDesign, Quark	Print
	Quarterly earnings call	Windows Media, MPEG-4	Webcast
Marketing Communications	Brand management		
	Brochures	Adobe InDesign, Quark	Print
	Web pages	HTML	Web
	Exhibitions and shows	TIFF	Graphic panels
	Advertising	video, audio, TIFF	Television, radio, print
Sales	Presentations	MS PowerPoint, video	PC, projection
	Responses to RFP, RFQ	MS Word	Mail
	e-commerce		Web
Product Management	Guides & Handbooks	Adobe InDsign, QuarkXPress	shipped with product
	White pagers	MS Word, PDF	Mail, Web
Training	Audiovisual	MPEG, DV, QuickTime	CD-ROM
	E-learning	Windows Media, MPEG-4, MS PowerPoint	Web
	Manuals	Adobe InDsign, QuarkXPress	print

Content is now being delivered over a myriad of different channels. There are the electronic channels: television, ITV (interactive television), Internet and webcasting, cellular phone, and wireless personal digital assistants (PDAs). To this add the conventional print-based media: catalogs, brochures, direct mail, and display advertising. Repurposing the content to offer compelling viewing for these different channels presents an enormous challenge to content publishers and aggregators. The project-based methods familiar to the brochure designer or video producer are replaced with a cooperative web of talent. The potential for mistakes is ripe, with the large number of files and the disparate formats.

Content and media asset management (MAM) are already core back-office applications for the print industries. These systems are ideally suited to the control and management of the large content repositories that publish via the diversity of new technologies.

1. Digital Asset Management (DAM)

Document management has already proved its worth in the traditional print industries. Content management has enabled the efficient deployment of large web sites. As the number of distribution platforms increases, the needs of brand management and the desire to control costs lead naturally to a convergence of the creative and production processes. A large corporation can no longer run print, web, and multi-media production in isolation. Rich media is a term that has been applied to this converged content.

For some enterprises, this convergence may only apply to content creation. It can also extend to the management of marketing collateral and corporate communications. The corporation that communicates directly with the consumer may want an ITV commercial to link through to the company web site. The customer can then view web pages or request further information. All these different media must exhibit a seamless brand image.

The same principles used for content management can be used as the basis of systems for managing rich media asset. The complexity of rich media means that DAM is becoming vital to improve operational efficiency and to control costs.

Rich media production may start with conventional audio and video material, and be enhanced later with synchronized graphic and text elements. The production workflow will have more parallel processes than the traditional linear flow of television production using different creative talent. The pressure to drive down the cost of production is paramount, yet the market is demanding the publication of content in a plethora of format.

2. What are Assets?

A quick look at a dictionary will tell us that the word asset usually relates to property. The same association with property also applies to digital media content. If you have the intellectual property rights to content, then that content can represent an asset. So intellectual property management and protection (IPMP) and digital rights management (DRM) should form an integral part of an asset-management framework.

Content without the usage rights is not an asset (Figure 8).

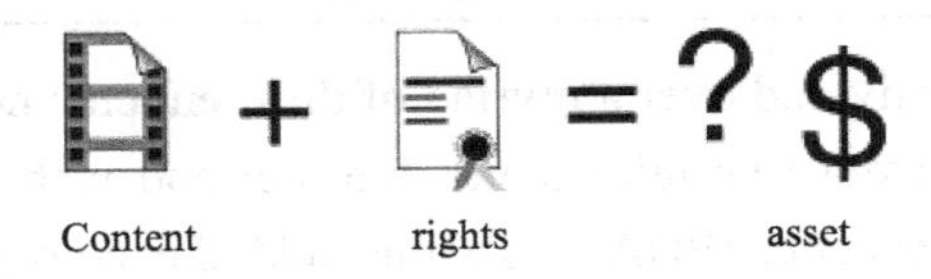

Figure 8 An asset is content with the right to use

Using metadata, DAM enables the linking of content with the rights information. The same systems that protect property rights can also be used to guard confidential information, so DAM can also be used for the internal distribution of commercially sensitive information.

Asset management is much more than rights management—it forms a small, but vital, part of the total solution. Perhaps the most important feature is that DAM provides a framework for the successful monetization of media assets.

There are potential drawbacks. If the system is to be dropped onto a mature corporation, there will be a vast legacy of existing content. What are the costs of indexing and cataloguing the archive? These costs may well outweigh any potential advantages of the online access to the older material. The deployment of DAM, like any other major business decision, should involve a cost-benefit analysis.

It may be that compromises can be made to reach a halfway house of partial digitization, rather than an across-the-board incorporation of every piece of content in the archive. As an example, one criterion could be the last-accessed date. Just because you have the rights to content, its value may well be negligible, so it could well remain in the vault. If the storage cost is high, you could even dispose of it.

What gives an asset value? If it can be resold, then the value is obvious. However, it can also represent a monetary asset, if it can be cost-effectively repurposed and then incorporated into new material. To cite just one example, a new advertising campaign can build on the lessons and experience of the past through access to the media archive. This can save a corporation both time and resources in such research projects.

3. What is Asset Management?

To be more specific, DAM provides for the sensible exploitation and administration of a large asset repository. An asset-management system provides a complete toolbox to the author, publisher, and the end users of the media to efficiently utilize the assets.

Media assets can be in a number of different formats: audio and video clips, graphic images, photographs, and text documents. There may be links and associations between the assets. The files may be shared across an enterprise-scale organization. The media may be traded; it can be syndicated, rented, or sold.

The system architecture must be flexible to provide for all these requirements. These are some of the features and facilities that may be found in a DAM system:

- Co-authoring
- Workflow
- Storage management
- Search tools
- Archiving
- Publishing tools
- Multiple formats
- Wide-area distribution
- Version control

Asset management can be extended through web-based access. This opens up the opportunity to preview and author at the desktop, remote from the content. This access can be within the enterprise, from any dial-up location, or over the Internet.

4. Why Use Asset Management?

Even the best-designed filing scheme has limitations. This is particularly evident when material is shared across a large enterprise in many formats. The optimum scheme for one user may not suit another. Support of multiple file formats on different platforms further complicates the scheme; files may be on Windows or UNIX servers, on dedicated video server, or videotape and CD-ROM.

The user may not know the file name. To help locate wanted material, the filing scheme will require a catalog. Imagine a request like "Find me a happy-looking child wearing a red dress on a crowded beach." Asset management provides the answer to such questions.

New Words and Expressions

asset *n.* 资产，有用的东西
VOD *abbr.* Video On Demand 视频点播
catalog *v.* 编目录
index *v.* 编入索引中，指出
analog *n.* 类似物，相似体，模拟
repository *n.* 储藏室，智囊团，知识库，仓库
PDA *abbr.* Personal Digital Assistant 个人数字助理
brochure *n.* 小册子
disparate *adj.* 全异的
booth *n.* 货摊，售货亭，棚
seamless *adj.* 无缝合线的，无伤痕的
monetary *adj.* 货币的，金钱的

Exercises to the Text

1. Translate the following words and phrases into English.

（1）数字资产管理 （2）个人数字助理 （3）知识产权管理与保护 （4）数字版权管理 （5）富媒体 （6）交互式网络电视 （7）索引和编目 （8）劳动密集型任务 （9）最后访问时间 （10）制作流程

2. Translate the following paragraphs into Chinese.

(1) Content is now being delivered over a myriad of different channels. There are the electronic channels: television, ITV (interactive television), Internet and webcasting, cellular phone, and wireless personal digital assistants (PDAs).

(2) DAM has proved quite a technical challenge. It was not until the end of the twentieth century that affordable systems became available. The fall in the price of processing and disk storage has been the primary enabling factor, along with the ever-increasing bandwidths of computer networks, both within a local area and over long distances. Many of the products that are designed to manage the very large repositories can automate labor-intensive and repetitive

tasks. These include cataloging and indexing the content, and the powerful search engines that we use to discover and find content.

(3) The user may not know the file name. To help locate wanted material, the filing scheme will require a catalog. Imagine a request like "Find me a happy-looking child wearing a red dress on a crowded beach." Asset management provides the answer to such questions.

Text 5: Virtual Reality and Applications

1. Virtual Reality

Virtual Reality (VR) is a computer technology that creates three dimensional real illusions in an artificial world. Virtual Reality is used in many real life applications, from business planning to manufacturing and entertainment.

1) History of Virtual Reality

Virtual Reality has a significant history that is both interesting and complex. A little over thirty years ago, a young cinematographer named Morton Heilig wanted to utilize the other seventy-two percent of the spectator's viewing field and create an ultimate full-view experience for the spectator.

Unable to obtain any financial support, Heilig was unable to create his dream; however, he created a unit called the "Sensorama Simulator", released in the early 1960's. A photo of the creation is shown in Figure 9. This virtual Workstation utilized 3D video, obtained with three 35 mm cameras mounted on the cameraman. The setup included stereo sound, integrated with the full 3D camera views. The viewer could ride a motorcycle while sensing the wind, simulated by a fan, and even potholes in the road. The machine was crude, but it opened the door for a multitude of ideas.

Figure 9 Sensorama Simulator

In 1966, Ivan Sutherland, a graduate student at the University of Utah, picked up where

Heilig had left of. He started the idea of the graphics accelerator, an integral part in modern virtual simulation. The military quickly recognized the potential of this idea in flight simulation, and spent most of the seventies designing helmets that could simulate a view of night. Also, National Aeronautics and Space Administration (NASA) began research on using the technology for space flight, and later, moon landings. Heilig's invention changed the world of computing, as well as the evolution of the computer itself.

The pivotal convergence of technologies that have made Virtual Reality possible has come about in the last ten years. The last ten years have seen advancements in areas that are absolutely crucial to the VR paradigm. These include the LCD and CRT display devices, high performance image generation systems, and tracking systems. As the world of the integrated circuits progressed into the MIPS era, high speed, high performance systems became affordable. As a result, non-military research was possible, and research migrated to other countries as well, Japan, France, and Germany in particular. Evolution of specific technologies included the display technology, as mentioned above, the human interface, and imaging. The evolution of display technology has played a vital role in the advancement of the Virtual Reality paradigm.

Virtual Reality can be divided into:

- The simulation of real environments such as the interior of a building or a spaceship often with the purpose of training or education.
- The development of an imagined environment, typically for a game or educational adventure.

Popular products for creating virtual reality effects on personal computers include Bryce, Extreme 3D, Ray Dream Studio, TrueSpace, 3D Studio MAX, and Visual Reality. The Virtual Reality Modeling Language (VRML) allows the creator to specify images and the rules for their display and interaction using textual language statements.

2) VRML

VRML (Virtual Reality Modeling Language) is an open, extensible, industry-standard scene description language for 3D scenes, or worlds, on the Internet. With VRML and Netscape's Live3D, users can author and view distributed interactive 3D worlds that are rich with text, images, animation, sound, music, and even video.

- Embed VRML into HTML document

Once VRML world is created, it can be embed within an HTML document by using the <EMBED>tag. Using the <EMBED>tag to place a VRML world in an HTML document is similar to using the<IMG>tag to place a 2D image in an HTML document. The following example embeds a VRML file called example.wrl into an HTML document:

```
<EMBED SRC="example.wrl" WIDTH=128 HEIGHT=128 BORDED=0 ALIGN=middle>
```

- Performance

Even the most enthusiastic user has limited patience for a slow Web page. This is a key concern for VRML authors, since VRML is based on computation-intensive 3D graphics and may incorporate other resource-intensive media. As with HTML documents, download time is an important factor in VRML world creation. A VRML world may require greater client system resources once downloaded. A fast browser will offset this to some extent, but it is important to construct VRML worlds efficiently. How to use the following elements will affect Web page's performance.

- Polygons

Shapes in a VRML world are made of polygons. The more complex a shape, the more polygons are required. A cube, for example, is typically comprised of just twelve polygons, since each side is made of two triangles. In contrast, a seemingly simple sphere requires more than 200 triangular polygons. As more objects are added to a world, the polygon count for that world increases. Each time a user's viewpoint changes in the VRML world, the browser has to redraw the scene. The more polygons the world contains, the longer the redraws take. Therefore, low polygon counts are one way to increase the user's navigation speed.

- Textures

VRML allows the textures to be mapped onto shapes. Textures used in a VRML world may increase its size considerably. This will affect both download and redraw times. Therefore, if textures are used, small textures are desirable as one way to keep download times low and navigation speed high. Also, textures used in VRML worlds will require fewer client resources if they use fewer colors.

- Instancing

If a user defines objects, the objects may be reused in a VRML world. This technique can help to keep a world file size small. This technique is called instancing. Though there are some limitations to instancing, its use can make VRML code easier to be written and maintained and the VRML worlds easier to be downloaded.

- Level of Detail

In the real world, as users get closer to an object, more details become visible. Level of Detail (LOD) makes this possible in VRML worlds. The LOD node determines which objects will be visible within defined ranges of coordinates within the VRML scene. This permits both special effects and realistic simulations.

- Inlines

Other world files may be "pulled into" a world to help create a VRML scene. When this technique is used, these files are called inlines. The WWWInline node is used to refer to a world file to be included and, optionally, to display a bounding box to show the user where the object, or objects, will be positioned before they are rendered.

- Compression

The larger the VRML world file, the longer it takes to be downloaded. World files may be compressed by using utilities such as GZIP. If a VRML browser recognizes the file type, it can automatically parse the compressed file to display the VRML world.

2. Virtual Reality Applications

Virtual reality may be described as a very special interaction between a human being and a computer. In VR, the person wears a "Head-Mounted Display (HMD)"—glasses, goggles, or a helmet with a tiny screen positioned in front of each eye (See Figure 10). Armed with a tracking system that links the HMD with the computer and some kind of navigational tool such as a 3D mouse, wand, or high-tech glove, the person is ready for a truly engaging computer adventure.

Once in the virtual world, the person sees the computer graphics in front, above, beside, below, and behind (See Figure 11). The traveler has a very convincing visual sense of being inside the computer's graphics.

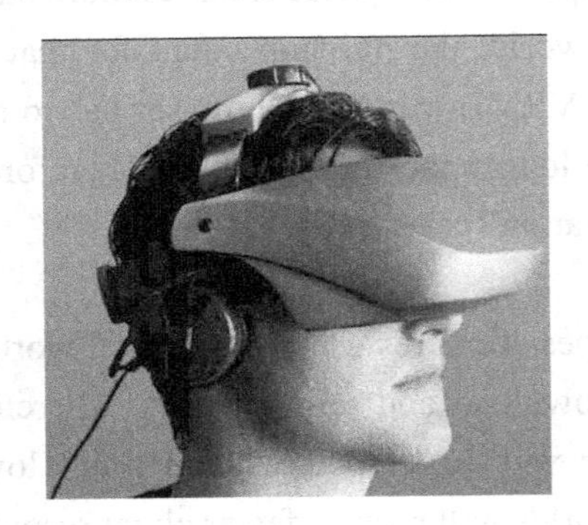

Figure 10 Person Wear a HMD

Figure 11 Visual Effect

The sort of systems common today had their beginnings in the 1960s. One of the first practical uses of virtual reality was the military's development of the heads-up display used by pilots.

While researchers agree it has a long way to go before it reaches its potential, VR is in practical use today in various occupations in the military, entertainment, education, and business. In education, gigantic growth in virtual reality is expected. Universities such as the University of Washington and the University of North Carolina devote resources to the study of virtual reality, its physical impact, and its application to the real world.

The next Step in education is to apply it to improve the learning process and to create new learning systems. The excitement and motivation that virtual reality creates ignites the imagination of far-sighted educators who perceive VR as the next logical step in computer-based instruction.

In business applications, using high-speed dedicated computer, multi-million-dollar flight simulators built by Singer, RediFusion, and others have led the way in commercial application of VR. Pilots of F-16s, Boeing 777s, and Rockwell space shuttles have made many dry runs before doing the real thing. At the California Maritime Academy and other merchant marine officer training schools, computer-controlled simulators teach the intricate loading and unloading of oil tankers and container ships. Virtual worlds and business intersect in the Virtual Worlds Consortium sponsored through the University of Washington. If current indicators pointing to virtual reality as a powerful learning tool are correct, future training in business and industry will take advantage of this ultimate visual learning method.

New Words and Expressions

manufacturing *n.* 制造业
cinematographer *n.* 电影摄影技师，放映技师
utilize *v.* 利用
spectator *n.* 观众
simulator *n.* 模拟器，假装者
workstation *n.* 工作站
accelerator *n.* 加速者，加速器
NASA *abbr.* National Aeronautics and Space Administration 美国国家航空和航天局
Pivotal *adj.* 枢轴的，关键的
convergence *n.* 集中，收敛
paradigm *n.* 范例
interior *n.* 内部
adventure *n.* 冒险，冒险的经历
polygon *n.* 多角形，多边形
texture *n.* 纹理
compression *n.* 浓缩，压缩
shuttle *n.* 往返汽车（列车、飞机），航天飞机，梭子，穿梭
marine *adj.* 海的，海产的，航海的，船舶的，海运的

Exercises to the Text

1. Translate the following words and phrases into English.

（1）虚拟现实 （2）三维图像 （3）虚拟模拟 （4）纹理 （5）压缩

2. Translate the following words and phrases into Chinese.

(1) virtual reality modeling language (2) polygon (3) inlines (4) heads-up display (5) simulator

3. Translate the following paragraphs into Chinese.

(1) The pivotal convergence of technologies that have made Virtual Reality possible has come about in the last ten years. The last ten years have seen advancements in areas that are absolutely crucial to the VR paradigm. These include the Liquid LCD and CRT display devices, high performance image generation systems, and tracking systems (to compute display areas into calculated machine coordinates).

(2) VRML (Virtual Reality Modeling Language) is an open, extensible, industry-standard scene description language for 3D scenes, or worlds, on the Internet. With VRML and Netscape's Live3D, users can author and view distributed interactive 3D worlds that are rich with text, images, animation, sound, music, and even video.

(3) The larger the VRML world file, the longer it takes to be downloaded. World files may be compressed by using utilities such as GZIP. If a VRML browser recognizes the file type, it can automatically parse the compressed file to display the VRML world.

Text 1: An Introduction to Photoshop

On opening Photoshop, you will see the following four items: Menu bar, Toolbox, Tool Options bar, Layers Palette. These will be used for shortcuts in many of the activities.

Menu bar—Click any of the words in Figure 12 to see a menu of options in that category—File and Edit are some of the commonly used for saving and opening files, and copying and pasting images.

Photoshop File Edit Image Layer Select Filter Analysis 3D View Window Help

Figure 12 Menu bar

Toolbox—The Toolbox contains tools that let you type, select, paint, draw, edit, move, and view images. Below are some of the tools that you may often use.

Rectangular Marquee tool—Used to select sections of a graphic.

Move tool—Used to nudge a selection.

Crop tool—Used to cut down the dimensions of a picture.

Paintbrush tool—Used to draw and touch up an image.

Paint Bucket tool—Used to fill in a selection.

Type tool—Used to add text.

Eyedropper tool—Used to select a specific color in a photo.

Zoom tool—Used to magnify an image.

Tool Options bar—Most tools have options that are displayed in the Tool Options bar. These options change as different tools are selected.

Layer Palette—Layers provide a powerful way for you to organize and manage the various components of your image. By placing an element on a separate layer, you can easily edit and arrange that element without interfering with the other parts of the image.

1. Creating a New Image

Choose File | New

Set the width and the height from inches to pixels which are standard dimensions for web images.

For this image, set the width and height to 600 and 100 respectively. Click OK.

2. Creating Text Graphics

Click the Create a New Layer in the Layer Palette (Figure 13).

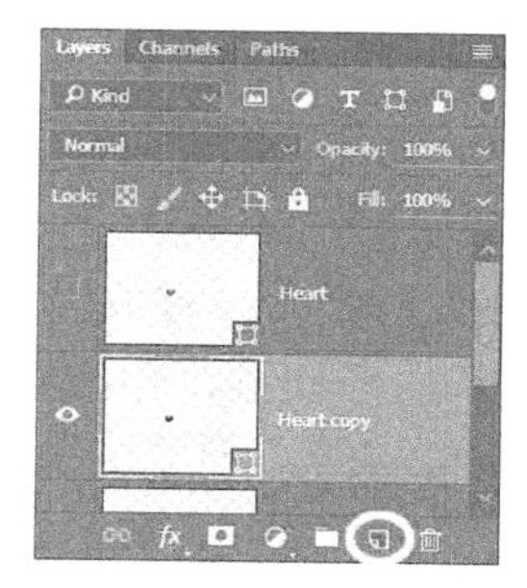

Figure 13　Click the Create a New Layer in the Layer Palette

Click on the Type tool.

Click on Set Foreground Color to bring up the Color Picker dialog box.

Check Only Web Colors.

Select your favorite color and click OK.

In the Tool Options bar, set the font to Arial Black and change the size to 48pt.

Type Photoshop.

Note: A quick way to set the foreground and background colors to black and white is by pressing the D key.

Note: When developing graphics for the web, it is usually a smart idea to stay with Only Web Colors. These colors usually look the same on various monitors.

3. Saving Images

Here are several options for saving an image in Photoshop.

File | Save—Save a file, repeated use of Save will overwrite the original file with changes.

File | Save As—Save a file with a new name and in a new location, and you then are working on the new file you just named.

File | Save for the Web—When using this option, it optimizes your file for web use.

For your purposes here, choose File | Save As. Now we need to select a file format. Here are the most commonly used formats to save images.

.psd—Photoshop Document Format, saves your image and all your layer effects so that you can edit it at a later time.

.gif—Graphical Interchange Format, use this format for saving images for the web. This will compress the file for faster downloads on the web.

.jpg—Joint Photographic Experts Group, great for photographs, however, Web browsers only use 216 colors.

Save your file in a .psd format so you can edit and make changes to it at a later time.

New Words and Expressions

palette　*n.* 调色板，控制面板

item　*n.* 项目

category　*n.* 种类，类型

contain *vt.* 包含，容纳；相当于
nudge *vt.* 轻推，推进 *n.* 轻推，推动
dimension *n.* 尺寸，维
magnify *vt.* 放大，扩大 *vi.* 放大，扩大；有放大能力
arrange *vt.* 整理，布置，安排 *vi.* 作安排，作准备
optimize *vt.* 使最优化，使尽可能有效，使完美 *vi.* 持乐观态度
compress *vt.* 压，压缩；归纳，精简
download *vt.* [计算机]下载

Exercises to the Text

1. Translate the following words and phrases into English.

（1）“矩形选择”工具 （2）“移动”工具 （3）“裁切”工具 （4）“画笔”工具 （5）“油漆桶”工具 （6）“文本”工具 （7）“吸管”工具 （8）“放大镜”工具 （9）前景色 （10）背景色 （11）“颜色拾取”对话框

2. Translate the following paragraphs into Chinese.

(1) On opening Photoshop, you will see the following four items: Menu bar, Toolbox, Tool Options bar, Layers Palette. These will be used for shortcuts in many of the activities.

(2) Layer Palette—Layers provide a powerful way for you to organize and manage the various components of your image. By placing an element on a separate layer, you can easily edit and arrange that element without interfering with the other parts of the image.

(3) When developing graphics for the web, it is usually a smart idea to stay with Only Web Colors. These colors usually look the same on various monitors.

Text 2: Making Movie Clips with Flash

This text is intended for introducing the basics of the making of Flash movie clip (animated cartoon), including movie clip theory, flash work environment and an example for design of simple movie clips.

1. Basic Theory of Making Movie Clips

Before we study design of movie first, we should talk about a topic, that is, what is movie clip? Actually, a sequence of static cartoons moving continuously will present animated cartoons; like notion pictures; movie clips make use of persistence of vision in human eyes to produce a sense of movement in our mind. In Flash, animated cartoons are called movies; as they are based on the same concept, we don't tell the difference between them.

Every static picture in movie clips is called “frame”. The rate movie clips are played back is represented by how many frames are played back in a second, that is fps (Frame Per Second), and in computers, the usual rate is 8-12 frames/second; especially in network, the frame rate cannot be set too high. If the frame rate is set too high, the network frequency cannot match up with the audio/video data in transmission, as a result, the movie clips are set in a sequence orderly. On this point, some friends who have not engaged in design of computerized movie clips may be skeptic—if we make movie clips played back at a rate of 100 frames/second, shall we first make 100 pieces of pictures? Is the job too tedious? Well, Flash sees the difficulty and makes use of the powerful computing function of computers, and adopts a technique called "key frame", tremendously reducing the work for making movie clips.

Now let's study the important concept—key frame. Key frame refers to such a frame whose contents are quite different from those of the precedent frame, so that it presents great changes in critical actions and contents.

2. Flash Work Environment

To facilitate study and to avoid misunderstanding of technical terms, here presented is a brief introduction of Flash work environment. Standard Flash work environment includes menu bar, tool bar, stage, time-axis window, and tool palette. Besides these several major parts, with window menu opened, some small windows like material window can be called on.

1) Work Area

Work area refers to a Flash work platform. It is a rather big area, actually covering all stages as mentioned below and work object for drawing pictures or editing movie clips.

2) Stage

Stage is a platform for demonstrating all elements of Flash movie clip, displaying the content of the currently selected frame. Different from work area, only the content of stage is visible after the movie clip is played back, while content in the work area beyond the stage is invisible, just like players and work staffs in the backstage which are invisible to the audience.

3) Scene

Just like drama having several scenes, the stage can have several scenes. Note, on the upper right part of the stage, there are two small buttons; of them, the first one is a button for switching scenes. With interactivity between different scenes, very complex work can be created.

4) Time-axis Window

The time-axis window is for performing operations on two Flash basic element layers and frames. In default configuration, the time-axis window appears above the stage, close to the upper border, in a form of editing bar. It is possible to drag the editing bar with the mouse to other positions of Flash interface, or even outside the border, to make it a freely floating window. The time-axis window is divided into left and right areas used as layer control area and as time-axis control area, respectively.

In concept about layer, Flash and Photoshop are similar. Both are transparent to the viewer, except that in Photoshop it refers to graph layer, while in Flash it refers to movie clip one upon another orderly. The layer control area is at the left section of the time-axis window; it is a major area for layer demonstration and operation, consisted of several illustrative bars and buttons for layer functional operations. Name, type and state of all layers of current work being edited in the stage are all queued in an illustrative bar, in a same order as layers are placed. In the layer control area, not only the layers and relevant information of current work are displayed, but also operations on one or more layers, including adding layers, deleting layers and rearranging their order, etc. Here we remind the reader that the use of layers dose not increase the size of work, instead, reasonable use of layers can help design a work with distinct layers and ease editing work; therefore, for the novice, It is better to get familiar with the use of layers.

The lower section of the time-axis control area, containing several lines corresponding to the sequence of frames at the left layer, illustrative bars, information prompt bar, as well as tool buttons for controlling playing back and operation of movie clip. It is used for effective control on playing back and placing frames.

Using time-axis window to make good arrangement of spatial position and time setting for material can produce better animation effect shown in the work.

5) Tool Palette

If you cannot find the tool palette, you can select menu Window->Toolbar. In Flash tool palette, all Flash selecting tools and drawing tools are included. The toolbar includes two sections: one is selection section on the upper part of the tool palette, used for selecting tools; another is property section, used for setting properties of the tools selected.

6) Material Library Window

In Flash, material window plays the role of organizing and managing all basic elements of movie clip. Every Flash file contains various elements and all the basic elements are stored in the material library, so as to enable the user to search, edit and set these elements easily.

The material library window can change its size, to adapt to the current need of the user. When the narrow state button is pressed, the material library window will appear in narrow state, displaying only the name of materials. When the wide state, displaying information of the name, type of use and times of use, etc. Above the display state button there is a queuing key which allows the bar to be placed in either increasing or decreasing orders.

In editing Flash works, it is also possible to call for material symbol library of other Flash works, under the prerequisite that the works have not been transformed into movie clip show file. To call for symbol library of other works, you should open those works first.

7) Menu Bar and Tool Bar

Most commands, except drawing command, can be realized in menu bar. This is basically similar to the use of menu bar in Windows, so long as you know its name. In the tool bar, some frequently used buttons are placed for facilitating operations.

There are several parts that are important to understand the Flash interface.

- Standard toolbar: icons for open, save, new, print and a few icons unique to Flash.
- Drawing toolbar: drawing tools similar to many other drawing software packages.
- Status bar: display a brief description explaining the use of a tool when you rest your mouse cursor over one of the drawing tools.
- Stage: actual workspace for drawing individual frames of the movie, you can draw directly onto the stage or arrange imported artwork.
- Layers: think of layers as a stack of acetate transparencies. You can put different

elements on different layers to make them easier to work with. The number of layers does not affect the file size of the movie.

- Frame: An animation is made up of a series of frames. Each image changes from frame to frame. To that when the frames are passed quickly before the eyes, viewers see the figure in motion. If the change in each image is slight and the frames move quickly, the motion-like that in a movie-appears smooth. A flash movie is series of frames.

3. The Drawing Tools

In order to draw, edit, select, and arrange objects on the stage you will use the drawing toolbar. The drawing tools are:

arrow tool—selecting and moving objects on the stage, also use to reshape objects.
line tool—drawing straight lines, you can adjust color, thickness, and style.
T text tool—add text to a frame.
oval tool—draws an oval with selected stroke and fill color.
rectangle tool—draws an rectangle with selected stroke and fill color.
pencil tool—draws lines like your cursor is the point of a pencil.
paintbrush tool—works like the pencil and allows broad strokes and gradient fills.
ink bottle tool—changes stroke color of selected line or shape.
paint bucket—changes the fill color of the a selected shape.
eyedropper—selects the fill color of one object and makes that color the current fill color.
eraser tool—eraser lines and fills.
hand tool—use to move entire stage.
magnifier tool—use to zoom in on stage.

4. Symbol and Instance

For most novices of Flash, symbol and instance are a pair of special concepts, but actually, they are not difficult for understanding. In simplified words, symbol is graph that can be repeatedly used as button, or movie clip; while instance is concrete incarnation on stage of symbols. When a movie clip symbol is created and put on the stage, then this movie clip on the stage is called instance.

Introduction of the concept of symbol is beneficial to designers of movie clip. First, when handling a great number of elements, they wouldn't be trapped in chaos. When there are a great number of repetitive elements in the movie and they should be revised, it doesn't need to do revision on every instance but revise the symbol; as a result, editing work can be simplified. Besides, the use of symbol can evidently reduce the size of files, and the space for storing all the data. For example, when many scenes use the same background symbol, it doesn't need to store all the background graphs for all the scenes, instead, only information for a few calls needs to be stored. Using symbols in playing back movie clip on-line, another merit is that the customer doesn't need to download repeatedly multiple same elements but only once for one symbol; as a result, the rate can be tremendously speed up.

5. Design Process of Multimedia Movie Clip

Before we study movie clip design, it is necessary to make a brief introduction of steps involved in multimedia movie clip design. It is nothing exaggerated to say that design of multimedia movie clip is quiet a complex system process, especially when the work is rather big and rather important. Designers must be competent for the overall process and should first make an overall plan; otherwise, the design must become tedious and time wasting. When we study design of movie clip, we must first make an overall plan before we embark on the design.

In design of movie clip, energy and time invested in the work will depend on requirement for its quality of content. As experience shows, the design process of multimedia movie clip, involves three phases:

The first phase is the most important one, in which the designer should first consider the purpose, investment, content and time involved in the movie clip, also he should consider the way of how to use the movie clip, its after-effect and interactivity, etc. With these preparations well done, the subsequent work will be very smooth.

The second phase involves collection of materials as planned in the first phase and processing on the materials, for example, compilation of file, audio recording, making of movie clip, etc. In collection of materials, stringent requirement is necessary for ensuring quality.

In the third phase, designer uses Flash to integrate and test the materials collected. In this phase, test should be performed constantly and problems discovered should be timely revised, to ensure an elaborately made movie clip is born.

If several persons participate in the making of multimedia movie clip, taking different jobs, every participant should coordinate with each other. By personal experience in the making process, everyone can accumulate experience and will have more profound understanding of the making of multimedia movie clip.

New Words and Expressions

cartoon *n.* 卡通画，漫画，动画片
persistence *n.* 坚持，持续
vision *n.* 视觉
skeptic *n.* 无神论者，怀疑者
backstage *adv.* 在后台
stage *n.* 舞台，活动场景
scene *n.* 情景，现场
drama *n.* 戏剧
default *n.* 默认（值），缺省（值）
transparent *adj.* 明显的
viewer *n.* 阅读器
demonstration *n.* 示范
illustrative *adj.* 说明性的

facilitate *vt.* 使便利；促进，推进
reshape *vt.* 改造，再成形
thickness *n.* 厚度，浓度，稠密
eyedropper *n.* 滴眼药器，滴管
magnifier *n.* 放大器，放大镜
zoom *n.&vi.* 缩放
instance *n.* 实例，情况
exaggerate *vt.&vi.* 夸大，夸张
investment *n.* 投资
elaborately *adv.* 苦心经营地，精巧地
accumulate *vt.&vi.* 累加，堆积
profound *adj.* 深刻的，意义深远的
match *n.&vt.&vi.* 匹配

Exercises to the Text

1. Translate the following words and phrases into English.

（1）时间轴窗口 （2）上边距 （3）工具面板 （4）元件和实例 （5）素材库窗口

2. Translate the following paragraphs into Chinese.

(1) Every static picture in movie clips is called "frame". The rate movie clips are played back is represented by how many frames are played back in a second, that is fps (Frame Per Second), and in computers, the usual rate is 8-12 frames/second; especially in network, the frame rate cannot be set too high.

(2) Now let's study the important concept—key frame. Key frame refers to such a frame whose contents are quite different from those of the precedent frame, so that it presents great changes in critical actions and contents.

(3) Stage is a platform for demonstrating all elements of Flash movie clip, displaying the content of the currently selected frame. Different from work area, only the content of stage is visible after the movie clip is played back, while content in the work area beyond the stage is invisible, just like players and work staffs in the backstage which are invisible to the audience.

(4) In flash, material window plays the role of organizing and managing all basic elements of movie clip. Every Flash file contains various elements and all the basic elements are stored in the material library, so as to enable the user to search, edit and set these elements easily.

Text 3: An Introduction to Dreamweaver

Adobe Dreamweaver is a professional Web site development program for creating both standard Web pages and dynamic applications. In its latest incarnation, Dreamweaver has rededicated itself to Web standards and sharpened its focus. In addition to creating standards-based HTML pages with enhanced Cascading Style Sheet (CSS) rendering, it is also suitable for coding a wide range of Web formats, including JavaScript, XML, and ActionScript—even those incorporating Web 2.0 methods, such as Ajax. Among its many other distinctions, it was the first Web authoring tool capable of addressing multiple server models. This feature makes it equally easy for developers of ASP, ColdFusion, or PHP to use it.

Dreamweaver is truly a tool designed by Web developers for Web developers. Designed from the ground up to work the way professional Web designers do, it speeds site construction and streamlines site maintenance. Because Web designers rarely work in a vacuum, Dreamweaver integrates smoothly with the leading media programs, Adobe Photoshop, Adobe Fireworks, and Adobe Flash. This article describes the philosophical underpinnings of the program and provides a sense of how Dreamweaver blends traditional HTML and other Web languages with cutting-edge server-side techniques and CSS design standards. You also learn some of the advanced features that it offers to help you manage a Web site.

Dreamweaver is a program very much rooted in the real world. Web applications are developed for a variety of different server models, and Dreamweaver writes code for the most widely used ones. Because the real world is also a changing world, its extensible architecture opens the door for custom or third-party server models as well. The latest version, for example, provides new tools for developing the rich, interactive Web 2.0 pages.

Moreover, Dreamweaver recognizes the real-world problem of incompatible browser commands and addresses that by producing code that is compatible across browsers. It includes browser-specific HTML validation so that you can see how your existing or new code works in a particular browser.

Dreamweaver extends this real-world approach to the workplace. Dreamweaver's CSS rendering is top-of-the-line and lets you design with Web standards like no other program. The advanced Design view makes it possible to quickly structure whole pages during the production stage, while maintaining backward compatibility with browsers when the pages are published. Features such as the Assets panel streamline the production and maintenance process on large Web sites. Dreamweaver's Commands capability enables Web designers to automate their most difficult Web creations, and its Server Behavior Builder enables them to easily insert frequently used custom code.

1. Connectivity

Connectivity is more than a buzzword in Dreamweaver; it's an underlying concept. Dreamweaver makes it possible to connect to any data source supported by the most widely used application servers: ASP, ASP.NET, ColdFusion, PHP, JSP, and even XML. Moreover, the actual connection type is quite flexible; developers can opt for a connection that is easier to implement but less robust or one that requires slightly more server-side savvy and offers greater scalability. A special set of features is available for transforming XML data into a browser-ready format using Extensible Stylesheet Transformation (XSLT) technology. Dreamweaver offers a choice of languages for a number of applications servers and a collection of ready-to-use CSS standard designs, as shown in Figure 14.

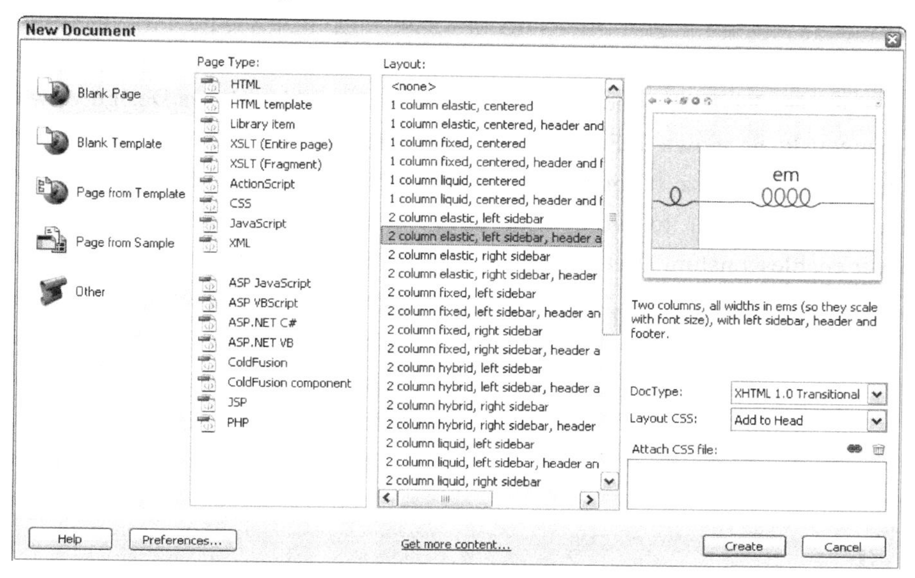

Figure 14 Get a jump-start on building your Web pages by choosing a page—as well as the corresponding Document Type and CSS file—from Dreamweaver's extensive collection.

Dreamweaver accesses standard recordsets—subsets of a database—as well as more sophisticated data sources, such as session or application variables and stored procedures. Through their implementation of cookies and server-side code, Web applications designed in Dreamweaver can track visitors or deny them entrance.

You also find support in Dreamweaver for high-end technologies such as Web services, JavaBeans, and ColdFusion components. Dreamweaver enables you to introspect elements of all technologies, enabling coders to quickly grasp the syntax, methods, and functions required.

2. True Data Representation

One of Dream weaver's truly innovative features integrates the actual data requested with the Web page—while still in the design phase. Live Data view sends the page-in-process to the application server to depict records from the data source within the page. All elements on the page remain editable; you can even alter the dynamic data's formatting and see those changes instantly applied. Live Data view shortens the work cycle by showing the designer exactly what the user will see. In addition, the page can be viewed under different conditions through the Live Data Settings feature.

3. Integrated Visual and Text Editors

In the early days of the World Wide Web, most developers hand-coded their Web pages using simple text editors such as Notepad and SimpleText. The second generation of Web authoring tools brought visual design or WYSIWYG (what you see is what you get) editors to market. What these products furnished in ease of layout, they lacked in completeness of code. Professional Web developers were required to hand-code their Web pages, even with the most sophisticated WYSIWYG editor.

Dreamweaver acknowledges this reality and has integrated a superb visual editor with its browser-like Document view. You can work graphically in Design view, or programmatically in Code view. You even have the option of a split-screen view, which shows Design view and Code view simultaneously, as shown in Figure 15. Any change made in the Design view is reflected in the Code view and vice versa. If you prefer to work with a code editor you're more familiar with, Dreamweaver enables you to work with any text editor. Whichever route you choose, Dreamweaver enables a natural, dynamic flow between the visual and code editors.

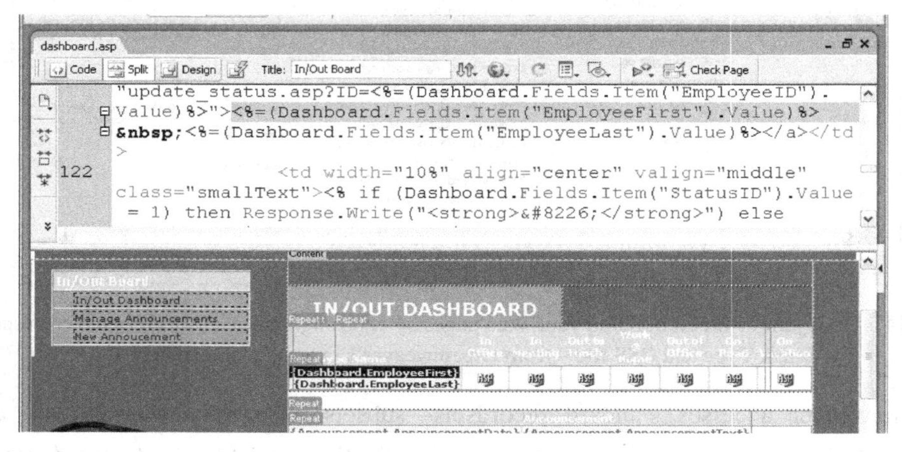

Figure 15 Dreamweaver enables you to work with a visual WYSIWYG editor and a code editor simultaneously.

Dreamweaver further tightens the integration between the visual design and the underlying code with the Quick Tag Editor. Web designers frequently adjust the HTML code minutely—changing an attribute here or adding a single tag there. The Quick Tag Editor, which appears as a small pop-up window in the Design view, makes these code tweaks quick and easy.

4. World-class Code Editing

Coding is integrally tied to Web page development, and Dreamweaver's coding environment is second-to-none. If you're hand-coding, you'll appreciate the Code Hints (see Figure 16), code collapse, and code completion features that Dreamweaver offers. Many of these elements have been encapsulated into a Coding toolbar displayed along the side of Code view. Not only do all these features speed development of HTML pages, but Dreamweaver's underlying Tag Libraries also extend their use to the full range of other code formats such as JavaScript, ActionScript, and XML.

Dreamweaver's Code view is easy on the eyes as well with syntax coloring that can be turned off and on at will. To get around the page quickly, use either the standard line-numbering facility or the advanced Code Navigation feature; Code Navigation lists all the functions found on a page and instantly jumps to that code when a function is selected.

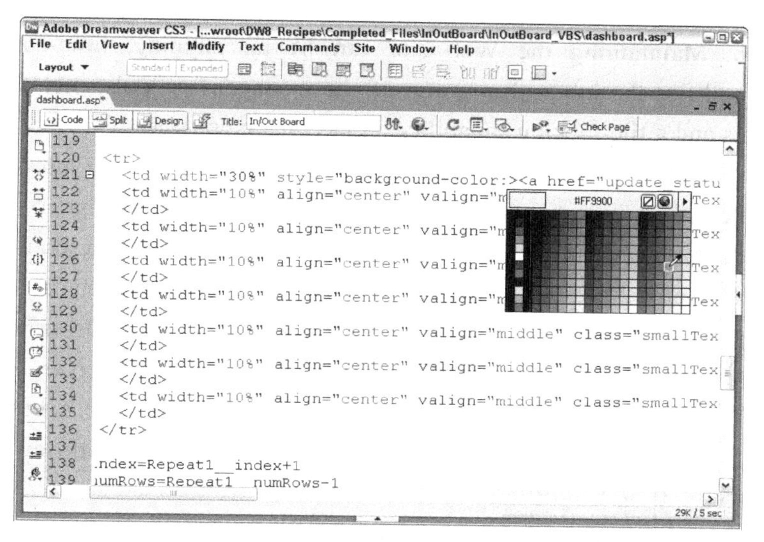

Figure 16 Code Hints speed hand-coding by displaying all the attributes available for a specific tag, including color.

Veterans and novices alike find Dreamweaver's Tag Chooser and Tag inspector indispensable. As the name implies, the Tag Chooser enables the coder to select a tag from a full list of tags in the various Web markup languages including HTML, CFML, PHP, ASP, ASP.NET, and more.

The Tag inspector gives a complete overview of all the aspects of a selected tag. Not only do you get to see a full array of all the associated properties—far more than could ever fit in the Property inspector—but you can also modify their values in place. Any applied JavaScript behaviors are also displayed in the Tag inspector. Perhaps the most innovative feature of this inspector is a CSS-related one, which displays any style impacting on a tag with completely modifiable properties and values. Select a CSS style and the Tag inspector becomes the Rule inspector for quick and easy CSS editing.

Code is far more than just a series of individual tags, of course. Dreamweaver's Snippets panel stores the most commonly used sections of code just a drag-and-drop away. Dreamweaver comes with hundreds of snippets ready to use—and gives you a way to add your own at any time.

5. Roundtrip HTML

Most Web authoring programs modify any code that passes through their system—inserting returns, removing indents, adding <meta> tags, uppercasing commands, and so forth. Dreamweaver's programmers understand and respect the fact that all Web developers have their own particular coding styles. An underlying concept, Roundtrip HTML, ensures that you can move back and forth between the visual editor and any HTML text editor without your code being rewritten.

6. Web Site Maintenance Tools

Dreamweaver's creators also understand that creating a site is only a part of the

Webmaster's job. Maintaining the Web site can be an ongoing, time-consuming chore. Dreamweaver simplifies the job with a group of site management tools, including a library of repeating elements and a file-locking capability for easy team updates.

Dreamweaver's built-in FTP transfer engine is quite robust and now better fits the designer's workflow with its capability to work in the background. Designers are now free to begin a large publishing operation and return to Dreamweaver to continue crafting pages while the FTP transfer is in process or bring up the log at any time to view the details.

Speed is another essential aspect in Web site maintenance. With Dreamweaver's siteless editing mode, you can make changes as quickly as you can connect to a server. You don't define an entire site if you only want to alter a couple of pages; just set up a server connection. Dreamweaver lets you access, edit, and publish the page in one smooth workflow.

7. Team-oriented Site Building

Until now, individual Web developers have been stymied when attempting to integrate Dreamweaver into a team-development environment. File-locking was all too easily subverted, enabling revisions to be inadvertently overwritten; site reports were limited in scope and only output to HTML; and, most notable of all, version control was nonexistent. Dreamweaver CS3 addresses all these concerns while laying a foundation for future connectivity.

Dreamweaver CS3 supports two industry-standard source control systems: Visual SourceSafe (VSS) and WebDAV. Connecting to a Visual SourceSafe server is well integrated into Dreamweaver; simply define the VSS server as your remote site and add the necessary connection information. WebDAV, although perhaps less well known than VSS, offers an equally powerful and more available content-management solution. More importantly, Adobe has developed the source-control solution as a system architecture, enabling other third-party content-management or version-control developers to use Dreamweaver as their front end.

ColdFusion developers have long enjoyed the benefits of Remote Development Services (RDS)—and now, RDS connectivity is available in Dreamweaver. Through RDS, teams of developers can work on the same site stored on a remote server. Moreover, you can connect directly to an RDS server without creating a site.

Extensible architecture also underlies Dreamweaver's site reporting facility. Dreamweaver ships with the capability to generate reports on usability issues (such as missing Alt text) or workflow concerns (such as who has what files checked out). Users can also develop custom reports on a project-by-project basis.

New Words and Expressions

incarnation *n.* 具体化，化身，体现

render *vt.* 报答；归还；给予；呈递；提供；开出

incorporate *vt.*（使）合并，并入，合编，组成公司，具体表现，使混合，使加入

distinction *n.* 区别；差别；不同之处；特征；特性；个性

streamline *vt.* 把……做成流线型；简化使效率更高

maintenance *n.* 维持；维护；保养；维修
vacuum *n.* 真空（度）；真空状态；空虚；空白；空处
underpinning *n.* 基础，支柱，支撑
validation *n.* 证明正确，批准，确认
buzzword *n.*（报刊等的）时髦术语，流行话
introspection *n.* 内省，反省，自省
syntax *n.* 句法
superb *adj.* 卓越的，杰出的，极好的
veteran *n.* 经验丰富的人；老兵
novice *n.* 新手，初学者
snippet *n.*（尤指讲话或文字的）小片，片段，零星的话
indent *vt.* 切割……使呈锯齿状，缩进排版
webmaster *n.* 网管
chore *n.* 零星工作（尤指家常杂务）
stymied *adj.* [美]被侵袭的
subvert *vt.* 颠覆，破坏（政治制度、宗教信仰等）
underlie *vt.* 位于或存在于（某物）之下；构成……的基础（或起因）；引起
inadvertently *adv.* 不注意地

Exercises to the Text

1. Translate the following words and phrases into English.

（1）层叠样式表 （2）第三方服务器模型 （3）可扩展样式表语言转换 （4）网页标记语言 （5）远程开发服务 （6）站点开发工具 （7）高级代码导航特性 （8）网站维护工具 （9）面向团队的站点构建

2. Translate the following paragraphs into Chinese.

(1) Dreamweaver is truly a tool designed by Web developers for Web developers. Designed from the ground up to work the way professional Web designers do, it speeds site construction and streamlines site maintenance. Because Web designers rarely work in a vacuum, Dreamweaver integrates smoothly with the leading media programs, Adobe Photoshop, Adobe Fireworks, and Adobe Flash.

(2) Dreamweaver recognizes the real-world problem of incompatible browser commands and addresses that by producing code that is compatible across browsers. It includes browser-specific HTML validation so that you can see how your existing or new code works in a particular browser.

(3) Dreamweaver acknowledges this reality and has integrated a superb visual editor with its browser-like Document view. You can work graphically in Design view, or programmatically in Code view. You even have the option of a split-screen view, which shows Design view and Code view simultaneously.

(4) Dreamweaver's creators also understand that creating a site is only a part of the

Webmaster's job. Maintaining the Web site can be an ongoing, time-consuming chore. Dreamweaver simplifies the job with a group of site management tools, including a library of repeating elements and a file-locking capability for easy team updates.

(5) Speed is another essential aspect in Web site maintenance. With Dreamweaver's siteless editing mode, you can make changes as quickly as you can connect to a server. You don't define an entire site if you only want to alter a couple of pages; just set up a server connection. Dreamweaver lets you access, edit, and publish the page in one smooth workflow.

Text 4: The Model Types of 3ds Max

Modeling is the process of pure creation. Whether it is sculpting, building with blocks, construction work, carving, architecture, or advanced injection molding, many different ways exist for creating objects. Max includes many different model types and even more ways to work with them.

You can climb a mountain in many ways, and you can model one in many ways. You can make a mountain model out of primitive objects like blocks, cubes, and spheres, or you can create one as a polygon mesh. As your experience grows, you'll discover that some objects are easier to model using one method and some are easier using another. Max offers several different modeling types to handle various modeling situations.

1. Parametric Objects Versus Editable Objects

All geometric objects in Max can be divided into two general categories—parametric objects and editable objects. Parametric means that the geometry of the object is controlled by variables called parameters. Modifying these parameters modifies the geometry of the object. This powerful concept gives parametric objects lots of flexibility. For example, the sphere object has a parameter called Radius. Changing this parameter changes the size of the sphere. Parametric objects in Max include all the objects found in the Create menu.

Editable objects do not have this flexibility of parameters, but they deal with subobjects and editing functions. The editable objects include Editable Spline, Mesh, Poly, Patch, and NURBS (Non-Uniform Rational B-Splines). Editable objects are listed in the Modifier Stack with the word Editable in front of their base object (except for NURBS objects, which are simply called NURBS Surfaces). For example, an editable mesh object is listed as Editable Mesh in the Modifier Stack.

Actually, NURBS objects are a different beast altogether. When created using the Create menu, they are parametric objects, but after you select the Modify panel, they are editable objects with a host of subobject modes and editing functions.

Editable objects aren't created; instead, they are converted or modified from another object. When a primitive object is converted to a different object type like an Editable Mesh or a NURBS object, it loses its parametric nature and can no longer be changed by altering its base parameters. Editable objects do have their advantages, though. You can edit subobjects such as vertices, edges, and faces of meshes—all things that you cannot edit for a parametric object. Each editable object type has a host of functions that are specific to its type.

Several modifiers enable you to edit subobjects while maintaining the parametric nature of an object. These include Edit Patch, Edit Mesh, Edit Poly, and Edit Spline.

Max includes the following model types:

- **Primitives:** Basic parametric objects such as cubes, spheres, and pyramids. The primitives are divided into two groups consisting of Standard and Extended Primitives. The AEC Objects are also considered primitive objects.
- **Shapes and splines:** Simple vector shapes such as circles, stars, arcs, and text, and splines such as the Helix. These objects are fully renderable. The Create menu includes many parametric shapes and splines. These parametric objects can be converted to Editable Spline objects for more editing.
- **Meshes:** Complex models created from many polygon faces that are smoothed together when the object is rendered. These objects are available only as Editable Mesh objects.
- **Polys:** Objects composed of polygon faces, similar to mesh objects, but with unique features. These objects are also available only as Editable Poly objects.
- **Patches:** Based on spline curves; patches can be modified using control points. The Create menu includes two parametric Patch objects, but most objects can also be converted to Editable Patch objects.
- **NURBS:** Stands for Non-Uniform Rational B-Splines. NURBS are similar to patches in that they also have control points. These control points define how a surface spreads over curves.
- **Compound objects:** A miscellaneous group of model types, including Booleans, loft objects, and scatter objects. Other compound objects are good at modeling one specialized type of object such as Terrain or BlobMesh objects.
- **Particle systems:** Systems of small objects that work together as a single group. They are useful for creating effects such as rain, snow, and sparks.
- **Hair and fur:** Modeling hundreds of thousands of cylinder objects to create believable hair would quickly bog down any system, so hair is modeled using a separate system that represents each hair as a spline.
- **Cloth systems:** Cloth—with its waving, free-flowing nature—behaves like water in some cases and like a solid in others. Max includes a specialized set of modifiers for handling cloth systems.

Hair, fur, and cloth are often considered effects or dynamic simulations instead of modeling constructs, so their inclusion on this list should be considered a stretch.

With all these options, modeling in Max can be intimidating, but you learn how to use each of these types the more you work with Max. For starters, begin with primitive or imported objects and then branch out by converting to editable objects. A single Max scene can include multiple object types.

2．Converting to Editable Objects

Of all the commands found in the Create menu and in the Create panel, you won't find any menus or subcategories for creating editable objects.

To create an editable object, you need to import it or convert it from another object type. You can convert objects by right-clicking on the object in the viewport and selecting the Convert To submenu from the popup quad-menu, or by right-clicking on the base object in the Modifier Stack and selecting the object type to convert to in the pop-up menu.

Once converted, all the editing features of the selected type are available in the Modify panel, but the object is no longer parametric and loses access to its common parameters such as Radius and Segments. However, Max also includes specialized modifiers such as the Edit Poly modifier that maintain the parametric nature of primitive objects while giving you access to the editing features of the Editable object.

If a modifier has been applied to an object, the Convert To menu option in the Modifier Stack pop-up menu is not available until you use the Collapse All command.

The pop-up menu includes options to convert to editable mesh, editable poly, editable patch, and NURBS. If a shape or spline object is selected, then the object can also be converted to an editable spline. Using any of the Convert To menu options collapses the Modifier Stack.

Objects can be converted between the different types several times, but each conversion may subdivide the object. Therefore, multiple conversions are not recommended.

Converting between object types is done automatically using Max's best guess, but if you apply one of the Conversion modifiers to an object, several parameters are displayed that let you define how the object is converted. For example, the Turn to Mesh modifier includes an option to Use Invisible Edges, which divides polygons using invisible edges. If this option is disabled, then the entire object is triangulated. The Turn to Patch modifier includes an option to make quads into quad patches. If this option is disabled, all quads are triangulated.

The Turn to Poly modifier includes options to Keep Polygons Convex, Limit Polygon Size, Require Planar Polygons, and Remove Mid-Edge Vertices. The Keep Polygons Convex option divides any polygon that is concave, if enabled. The Limit Polygon Size option lets you specify the maximum allowable polygon size. This can be used to eliminate any pentagons and hexagons from the mesh. The Require Planar Polygons option keeps adjacent polygons as triangles if the angle between them is greater than the specified Threshold value. The Remove Mid-Edge Vertices option removes any vertices caused by intersections with invisible edges.

All Conversion modifiers also include options to preserve the current subobject selection (including any soft selection) and to specify the Selection Level. After a Conversion modifier is applied to an object, you must collapse the Modifier Stack in order to complete the conversion.

New Words and Expressions

sculpt *v.* 雕刻，雕塑

primitive *adj.* 原始的，早期的
sphere *n.* 球（体）
polygon *n.* 多边形，多角形
parametric *adj.* 参（变）数的，参（变）量的
geometric *adj.* 几何的，几何学的
arc *n.* 弧，弧线
helix *n.* 螺旋结构
miscellaneous *adj.* 不同种类的，多种多样的；混杂的
intimidate *vt.* 恐吓，威胁
segment *n.* 部分，片段
collapse *vi.* 倒塌，塌下
triangulate *vt.* 分成三角形，对……作三角测量
convex *adj.* 凸的，凸面的
planar *adj.* 平面的，平坦的
vertices *n.* 制高点，头顶
pentagon *n.* 五边形；五角形
hexagon *n.* 六边形；六角形
intersection *n.* 横断；交叉
threshold *n.* 门槛，入口，开端[计算机]阈；阈值

Exercises to the Text

1. Translate the following words and phrases into English.

（1）几何对象 （2）高级注塑物 （3）基本体对象 （4）复合对象 （5）粒子系统 （6）织物系统 （7）修改器堆栈 （8）阈值 （9）参数化对象

2. Translate the following paragraphs into Chinese.

(1) Actually, NURBS objects are a different beast altogether. When created using the Create menu, they are parametric objects, but after you select the Modify panel, they are editable objects with a host of subobject modes and editing functions.

(2) With all these options, modeling in Max can be intimidating, but you learn how to use each of these types the more you work with Max. For starters, begin with primitive or imported objects and then branch out by converting to editable objects. A single Max scene can include multiple object types.

(3) Once converted, all the editing features of the selected type are available in the Modify panel, but the object is no longer parametric and loses access to its common parameters such as Radius and Segments. However, Max also includes specialized modifiers such as the Edit Poly modifier that maintain the parametric nature of primitive objects while giving you access to the editing features of the Editable object.

Text 5: Premiere Pro's Windows

The Adobe Premiere Pro user interface is a combination of a video-editing studio and an electronic image-editing studio. If you're familiar with film, video editing, or audio editing, you'll feel right at home working within Premiere Pro's Project, Monitor, and Audio windows. If you've worked with such programs as Adobe After Effects, Adobe Live Motion, macromedia Flash, or Macromedia Director, Premiere Pro's timeline, digital tools, and palettes will seem familiar. If you're completely new to video editing and computers, don't worry; premiere Pro palettes, windows, and menus are efficiently designed to get you up and running quickly.

To help get you started, this chapter provides an overview of Premiere Pro windows, menus and palettes. Consider it a thorough introduction to the program's workspace and a handy reference for planning and producing your own digital video productions.

After you first launch Premiere Pro, several windows automatically appear onscreen, each vying for your attention. Why do you need more than one window open at once? A video production is a multifaceted undertaking. In one production, you may need to capture video, edit video, and create titles, transitions, and special effects. Premiere Pro windows help keep these tasks separated and organized for you.

Although the Premiere Pro program's primary windows open automatically onscreen from time to time, you may want to close one of them. To close a window, simply click its close window X icon. If you try to close the Project window, Premiere Pro assumes that you want to close the entire project and prompts you to save your work before closing. If you want to open the Timeline, Monitor, Audio Mixer, History, Info, or Tools windows, choose the Window menu and then click the desired window name. If you have more than one timeline on the screen, you'll see it listed in the Window⇨Timelines submenu.

If you have your windows and palettes set up in specific positions at specific sizes, you can save this configuration by choosing Window⇨Workspace⇨Save Workspace. After you name your workspace and save it, the name of the workspace appears in the Window⇨Workspace submenu. Any time you want to use that workspace, simply click its name.

1. The Project Window

If you've ever worked on a project with many video and audio clips as well as other production elements, you'll soon appreciate the Premiere Pro program's Project window, which is shown in Figure 17. The Project window provides a quick bird's eye view of your production elements and enables you to preview a clip right from the Project window.

As you work, Premiere Pro automatically loads items into the Project window. When you import a file, the video and audio clips are automatically loaded into a Project window *bin* (a folder in the Project window). If you import a folder of clips, Premiere Pro creates a new bin for the clips, using the folder name as the bin name. When you capture sound or video, you can quickly add the captured media to a Project window bin before closing the clip. Later you can

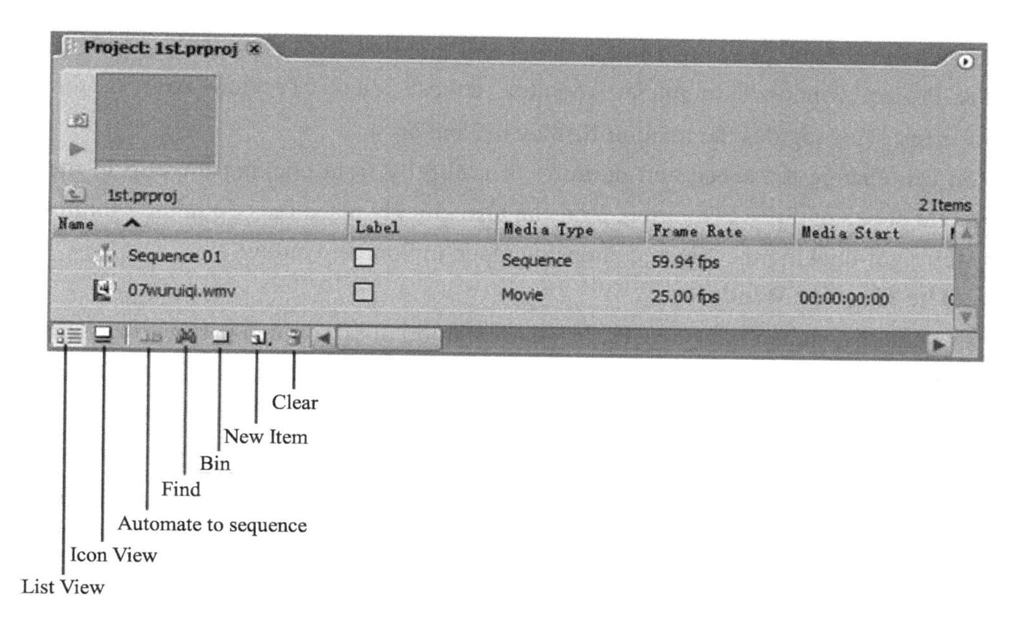

Figure 17 Premiere Pro's Project window stores production elements.

create your own bins by clicking the Bin button, at which point you can drag production elements from one bin to another. The New Item button enables you to quickly create a new title or other production element, such as a color matte or bars and tone (used to calibrate color and sound when editing). The Project window also includes a New Item button that allows you to add new sequences, offline files, titles, bars and tone, black video, color mattes, or universal counting leaders. If you click the Icon button, all production elements appear as icons onscreen rather than in list format. Clicking the List button returns the display of the Project Window to List view. If you want to quickly add Project window elements to the Timeline, you can simply select them and then click the Automate to Sequence button.

If you expand the Project window by clicking and dragging the Window border, you'll see that Premiere Pro lists the Start and Stop time as well as the in and out points and the duration of each clip. In the Project window, production elements are grouped according to the current sort order. You can change the order of production elements so that they are arranged by any of the column headings. To sort by one of the column categories, simply click it. The first time you click, production items are sorted in ascending order. To sort in descending arrow, click again on the column heading. The sort order is represented by a small triangle. When the arrow points up, the sort order is ascending. When it points down, the sort order is descending.

To keep your production materials well organized, you can create bins to store similar elements. For example, you may create a bin for all sound elements or a bin for all interview clips. If the bin gets stuffed, you can see more elements at one time by switching from the default thumbnail view to List view, which lists each item but doesn't show a thumbnail image.

If you want to play a clip in the thumbnail monitor in the Project window, click the clip, then click the tiny right arrow next to the thumbnail monitor. If you want to preserve space and hide the Project window's thumbnail monitor, choose View➪Preview Area in the Project window menu. This toggles the monitor display off and on.

You can change the speed and duration of a clip by right-clicking the clip in the Project window and choosing Speed/Duration. You can also quickly place the clip in the Monitor Source window by right-clicking the clip and choosing Open in Source window.

2. The Timeline Window

The Timeline, shown in Figure 18, is the foundation of your video production. It provides a graphic and temporal overview of your entire project. Fortunately, the Timeline is not for viewing only—it's interactive. Using your mouse, you can build your production by dragging video and audio clips, graphics, and titles from the Project window to the Timeline. Using Timeline tools, you can arrange, cut, and extend clips. By clicking and dragging the work area Markers at either end of the work area bar—edges of the light gray bar at the top of the Timeline—you specify the portion of the Timeline that Premiere Pro previews or exports. The thin, colored bar beneath the work area bar indicates whether a preview file for the project exists. A red bar indicates no preview, and a green bar indicates that a video preview has been created. If an audio preview exists, a thinner, light green bar appears. (To create the Preview file, choose Sequence➪Render Work Area, or press Enter to render the work area.)

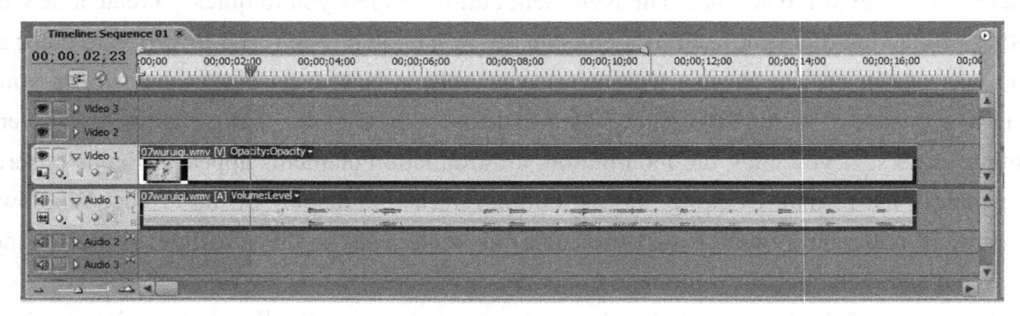

Figure 18 The Timeline window provides an overview of your project and enables you to edit clips.

Rendering the work area helps ensure that your project is played back at the project frame rate. Also, if you create video and audio effects, the Preview file stores the rendered effects. Thus, the next time you play back the effect, Premiere Pro does not have to process the effect again.

Undoubtedly, the most useful visual metaphor in the Timeline window is its representation of video and audio tracks as parallel bars. Premiere Pro provides multiple, parallel tracks so that you can both preview and conceptualize a production in real time. For instance, parallel video and audio tracks enable you to view video as audio plays. The parallel tracks also enable you to create transparency effects where a portion of one video track can be seen through another. The Timeline also includes icons for hiding or viewing tracks. Clicking the Toggle Track Output button (Eye icon) hides a track while you preview your production; clicking it again makes the track visible.

Clicking the audio Toggle Track Output button (Speaker icon) turns audio tracks on and off. Beneath the Eye icon is another icon that sets the display mode for clips in the track. Clicking the Set Display icon allows you to choose whether you want to see frames from the actual clip in the Timeline or only the name of the clip. At the bottom left of the window, the Time Zoom Level slider enables you to change the Timeline's time intervals. For example, zooming out shows your project over less Timeline space, and zooming in shows your work over a greater area of the Timeline. Thus, if you are viewing frames in the Timeline, zooming in reveals more frames. You can also zoom in and out by clicking the edges of gray bar at the top of the Timeline.

The other buttons—Track Options dialog box, Toggle Snap to Edges, Toggle Edge Viewing, Toggle Shift Tracks Options, and Toggle Sync Mode—at the bottom left of the window enable you to change options for syncing tracks and for making edges snap together.

3. The Monitor Window

The Monitor window, shown in Figure 19, is primarily used to preview your production as you create it. When previewing your work, click the Play button to play the clips in the Timeline and click the Loop button to start from frame 1. You can click and drag the shuttle slider to jump to a specific clip area. As you click, the time readout in the Monitor window indicates your position in the clip. The Monitor window can also be used to set the in and out points of clips. The in and out points determine which part of a clip appears in your project.

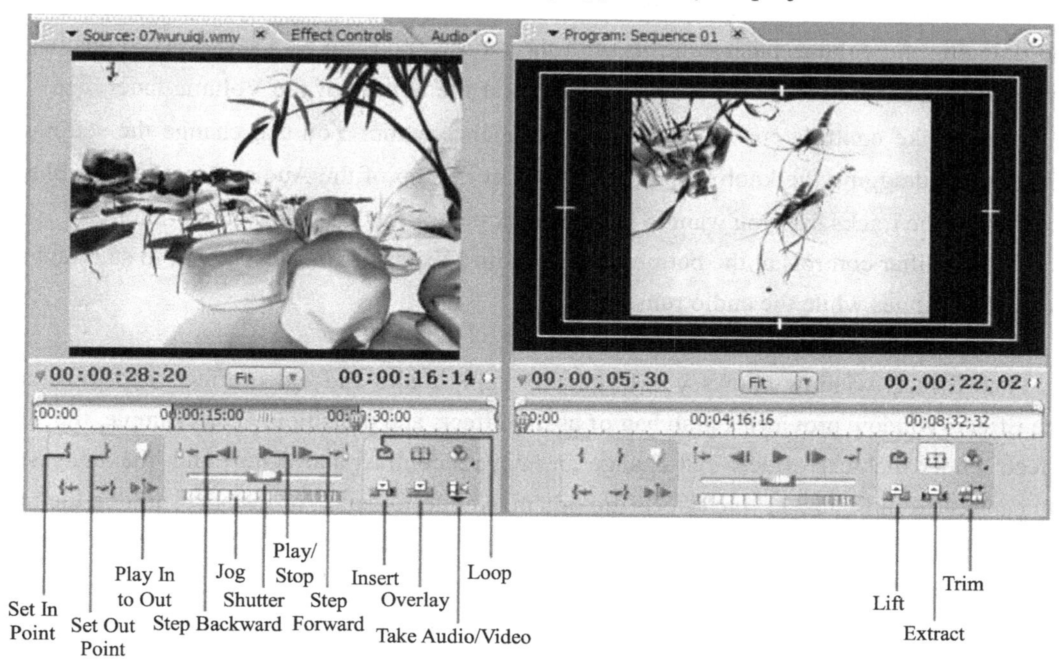

Figure 19 The Monitor window can be used to set in and out points of clips while you edit.

The Monitor window provides three viewing modes. To switch the mode, choose Dual View, Single View, or Trim View in the Monitor Window menu. (You can also open Trim View by clicking the Trim icon at the bottom of the Monitor window.)

- **Dual View.** The Monitor window is set up similarly to a traditional videotape-editing studio. The source clip (footage) appears on one side of the Monitor window, and the program (edited video) appears on the other side of the window. This mode is primarily used when creating three- and four-point edits.
- **Single View.** The window displays one monitor as you preview your production. Using this mode is similar to viewing your production on a television monitor.
- **Trim View.** This mode allows for precision editing. One clip appears in one Monitor window, and the other clip appears in the second Monitor window.

All the Monitor window modes provide icons (refer to Figure 19) that enable you to quickly set in and out edit points as well as step through the video frames.

4. The Audio Mixer Window

The Audio Mixer window, shown in Figure 20, enables you to mix different audio tracks, and to create cross fades and pans. (*Panning* enables you to balance stereo channels or shift sound from the left to the right stereo channels, and vice versa). Users of earlier versions of Premiere Pro can appreciate the Audio Mixer's capability to work in real time, which means that you can mix audio tracks while viewing video tracks.

Using the palette controls, you can raise and lower audio levels for three tracks by clicking and dragging the volume fader controls with the mouse. You can also set levels in decibels by typing a number into the dB level indicator field (at the bottom of the Volume fader area). The round knob-like controls enable you to pan or balance audio. You can change the settings by clicking and dragging the knob icon. The buttons at the top of the Audio Mixer let you play all tracks, pick the tracks that you want to hear, or pick the tracks that you want to mute.

The familiar controls at the bottom of the Audio Mixer window enable you to start and stop recording changes while the audio runs.

5. The Effects Window

The Effects window allows you to quickly apply audio and video effects and transitions. The Effects window provides a grab bag of useful effects and transitions. For example, the Video Effects folder includes effects that change an image's contrast and distort and blur images. As you can see from Figure 21, the effects are organized into folders. For instance, among the many folders in the Effects window is the Distort folder, which features effects that distort clips by bending or pinching them.

Applying an effect is simple—just click and drag the effect from its palette to a clip in the Timeline. Typically, doing this opens a dialog box in which you specify options for the effect.

The Effect window allows you to create your own folders and move effects into them so that you can quickly access the effects you want to use in each project.

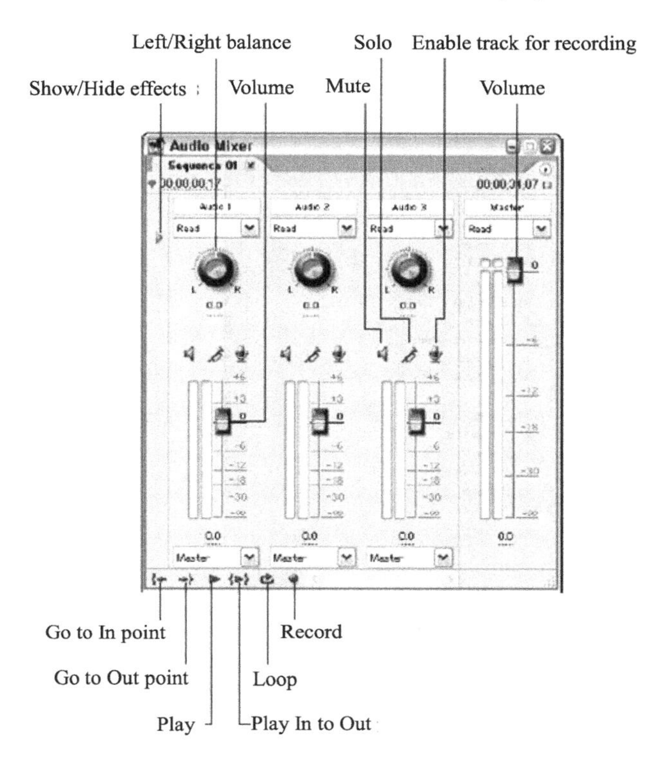

Figure 20 Use the Audio Mixer to mix audio and create audio effects.

Premiere Pro's Transitions folder, which also appears in the Effects window, features more than 70 transitional effects. Some effects, such as the Dissolve group, can provide a smooth transition from one video clip to another. Other transitions, such as page peel, can be used as a special effect to dramatically jump from one scene to another.

If you'll be using the same transitions throughout a production, you can create a folder, name it, and keep the transitions in the custom folder for quick access.

6. The Effect Controls Window

The Effect Controls window, shown in Figure 22, allows you to quickly create and control audio and video effects and transitions. For instance, you can add an effect to a clip by selecting it in the Effect window and then dragging the effect over the clip in the Timeline or directly into the Effect Controls palette. As you can see from Figure 22, the Effect Controls palette includes its own version of the Timeline as well as a slider control for zooming into the Timeline. By clicking and dragging the Timeline and changing effect settings, you can change effects over time. As you change settings, you create key frames along the Timeline.

If you create effects for a clip, you can see the different effects by selecting the clip and opening the Effect Controls window.

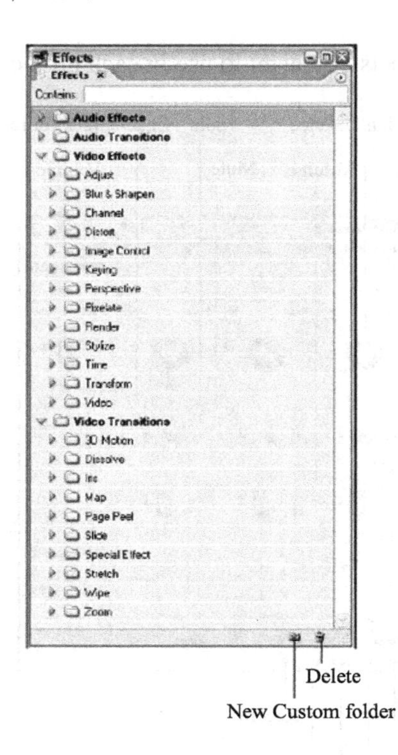

Figure 21　Use the Effect windows to apply transitions and special effects.

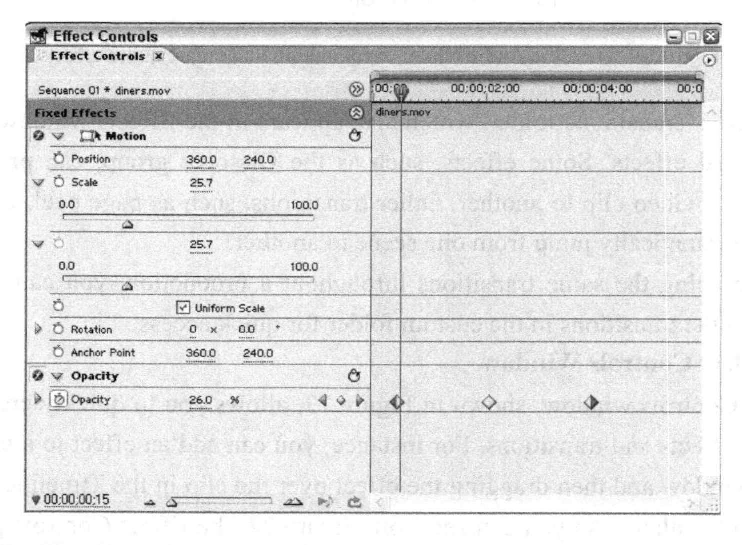

Figure 22　The Effect Controls palette enables you to quickly display and edit video and audio effects.

New Words and Expressions

Timeline　*n.* 时间线

palette　*n.* 调色板

onscreen　*adv, adj.* 在银幕上（的）

icon *n.* 图标，肖像，偶像
configuration *n.* 构造，结构，配置，外形
clip *n.* 剪辑，从电影胶片或录像带剪出的片段
matte *adj.* 不光滑的
bar *n.* 条，棒
ascend *v.* 攀登，上升
descend *v.* 下来，下降
default *n.* 默认（值），缺省（值）
thumbnail *adj.* 极小的，极短的
monitor *n.* 班长，监听器，监视器，监控器
duration *n.* 持续时间，期间
drag *v.* 拖，拖曳，缓慢而费力地行动
render *v.* 渲染
parallel *adj.* 平行的，相同的，类似的，并联的
metaphor *n.* 隐喻，暗喻
conceptualize *v.* 概念化
zoom *v.* 突然扩大，急速上升，摄像机移动
toggle *v.* 切换
snap *vt.* 猛咬，突然折断
slider *n.* 滑动器；滑子[块，板，座]
trim *v.* 整理，修整，装饰
mixer *n.* 混频器，混合器
pan *vt.* 摇动（镜头），使拍摄全景；*n.* 摇镜头，拍全景
distort *v.* 扭曲
transition *n.* 转变，转换，跃迁，过渡，变调

Exercises to the Text

1. Translate the following words and phrases into English.

（1）窗口菜单 （2）时间线 （3）监视器 （4）特效 （5）效果控制窗口 （6）音频混合器窗口 （7）折叠夹 （8）关键帧 （9）面板 （10）倒计时向导 （11）双杠（12）滚动条

2. Translate the following paragraph into Chinese.

(1) To keep your production materials well organized, you can create bins to store similar elements. For example, you may create a bin for all sound elements or a bin for all interview clips. If the bin gets stuffed, you can see more elements at one time by switching from the default thumbnail view to List view, which lists each item but doesn't show a thumbnail image.

(2) Rendering the work area helps ensure that your project is played back at the project frame rate. Also, if you create video and audio effects, the Preview file stores the rendered effects. Thus, the next time you play back the effect, Premiere Pro does not have to process the

effect again.

(3) The Effects window allows you to quickly apply audio and video effects and transitions. The Effects window provides a grab bag of useful effects and transitions. For example, the Video Effects folder includes effects that change an image's contrast and distort and blur images. As you can see from Figure 21, the effects are organized into folders. For instance, among the many folders in the Effects window is the Distort folder, which features effects that distort clips by bending or pinching them.

Text 6: Workspaces and Panels of After Effects

How much time and effort was wasted over the years by After Effects artists moving palettes around to see what they needed? We'll never know, but the days of juggling windows are banished to the sands of time for anyone who upgrades to 7.0.

Before we look at how to use this interface, it is helpful to define what is here. Figure 23 shows the Standard workspace that appears when you first open After Effects. The interface consists of one main application window; on the Mac, this window contains the name of the open project, and on Windows it also includes the menu bar. It is possible to create additional floating windows; more on this in a bit.

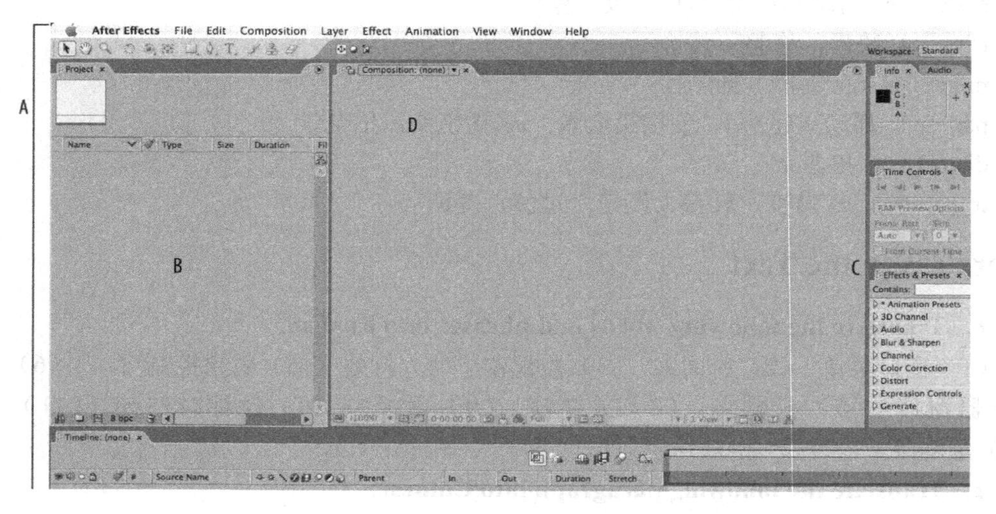

Figure 23 The Standard workspace layout is all contained in a single application window (A). The frame containing the Project panel (B) is currently active, as indicated by the yellow highlight around the panel's border. Dividers such as the long one (C) between the Composition panel and the smaller panels at the right separate the frames. The tab of the Composition viewer (D) includes a pull-down menu for choosing a particular composition, and a lock icon for keeping that composition forward regardless of what else is clicked.

Both types of windows contain frames, separated by dividers. Each frame contains one or more panels. If a frame contains multiple panels, the tab of each panel can be seen at the top, but only the contents of the forward tab can be seen. To bring a panel forward, you click on its tab.

Some panels are viewers; these include a pull-down menu in the tab that lets you choose the content displayed. These also include a lock option (specified via a small lock icon, also in the tab) that prevents that panel from switching to a different display automatically.

Now, to put these definitions in context, look more closely at Figure 23. The Standard workspace contains the bones of an entire project:

- Project panel: Contains all of the resources used in your compositions (source footage, stills, solids that you create, audio, even the compositions themselves).
- Composition panel: Is the viewer where you perform the predominant visual work of assembling a shot.
- Timeline panel: Organizes the elements that go into the individual composition (otherwise known as a shot).
- Info, Audio, Time Controls, and Effects & Presets panels: Help you work with your compositions.

All available panels in After Effects are listed under the Window menu; some even list preset keyboard shortcuts for rapid access. These panels used to be actual floating windows, but Adobe decided not to change the name of the menu, as it is standard across many applications.

The basic workflow of After Effects, then, is to create a new composition, typically containing items from the Project panel, and to work with it in the Composition and Timeline viewers.

To see the full story of the After Effects workflow, however, you can use the Workspace pull-down to switch to the All Panels layout. This reveals lots of little collapsed panels for various tools such as the paint tools (Paint and Brush Tips), the type tools (Character and Paragraph), the motion tracker (Tracker Controls), and so on.

Look closely at the workspace, and you will see the other key pieces of the After Effects workflow: the Effect Controls, for adding specific effects to layers; the Layer and Footage palettes, for working with an individual layer in a composition or viewing source footage; the Flowchart view, which provides a node-based overview of a project; and the Render Queue, where you output your work.

In this discussion, it is assumed that you have not yet customized any of the After Effects Workspaces. If what you see when you choose a Workspace doesn't match what is described here, you can use Window | Workspace | Reset to bring a given workspace back to its previous settings (although if you have saved over any of the defaults, you will see the saved version).

The All Panels layout is not particularly useful, it's crowded with panels you may never use but it provides a glimpse of what all is available in the After Effects user interface. Adobe set other default workspaces to anticipate common usages of After Effects: Animation, Effects, Motion Tracking, Paint, and Text. The Minimal workspace consists only of Timeline and Composition panels (where you'll do most of your work).

Workspaces used to be an underused feature in After Effects, but now they have taken center stage. Quite likely, however, you will end up customizing your workspaces to differ, slightly or radically, from what ships with version 7.0. The best way to do this is to mess around with the user interface yourself, but here are some pointers you might miss with such an approach.

In case you haven't already figured it out, each panel in After Effects is labeled via a tab at the upper left, which contains the title of the panel, an x to close the panel (to the right of the title), and a little grip area at the left of the title (Figure 24). At the upper right is a smaller tab with another grip area and a triangular icon for the panel's context menu.

Figure 24 The active tab, for the Effect Controls panel, includes grip areas at the left of each tab section for dragging the panel and an x to close it. At the far right is the panel menu icon, which reveals a context menu. Because this panel is used to contain effects for any layer, it includes a pull-down menu (to specify which one) and a lock (to keep the current one active). Also in this frame is the Project panel, which does not contain these extras.

Those grip areas are your target for clicking and dragging a panel to another location in the user interface (UI). In any workspace, try dragging a panel.

If you hold your pointer on the border between two panels, you will see it become a divider dragger (Figure 25). This allows you to resize the adjacent panels. But if you want the panel you are working with to take up a much larger amount of space for a moment, there is a better option.

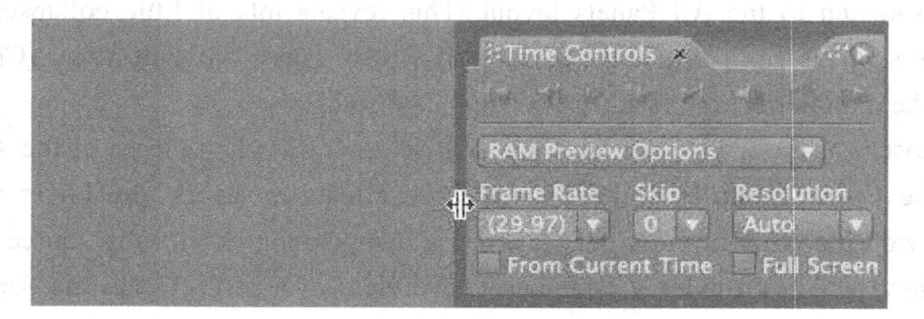

Figure 25 The border icon appears to indicate that you can offset the border between sets of frames, causing some panels to expand and others to contract.

The tilde key (~) toggles whichever panel is currently active (indicated by the yellow line around its border) to take over the entire After Effects window. You will use this shortcut a lot as you work with compositions, regardless of your monitor size; its most prominent use is to reveal more of the image in your Composition or Layer view, free of the rest of the interface.

When too many panels are grouped together in one area to see all of their titles, a slider bar appears above them, allowing you to scroll back and forth among them. It's very small and can easily escape notice (Figure 26).

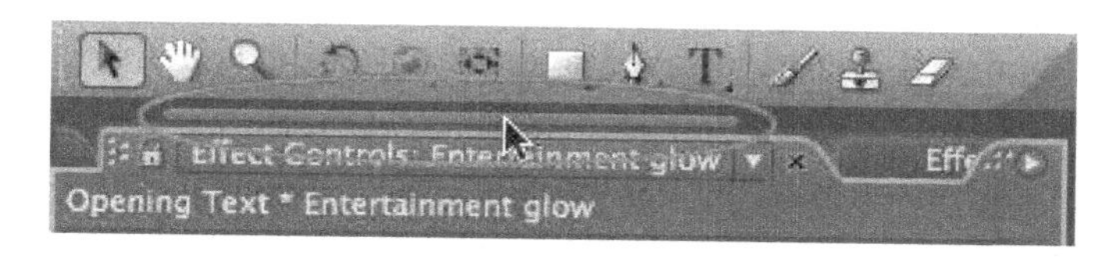

Figure 26 Highlighted is the scrollbar, which subtly appears atop a frame that contains too many panels to be displayed together horizontally.

There is no undo available when you rearrange panels. You can reset any given workspace, however, by choosing the Reset option at the bottom of the Workspace menu.

If you've switched back and forth between custom workspaces, you have probably noticed that all of the changes you make to a given workspace persist with it. If you make a complete mess of Standard and switch to Effects, then back to Standard, it's still the exact mess that you left it. If you're used to the old workspaces of version 6.5 and earlier, this is at first dismaying, until you notice that Reset option.

If you never choose Reset, your customizations will remain, at least until your preferences are reset or you inadvertently change them. For that reason, if you come up with a customization you like, save it (using the New Workspace option in the Workspace menu). You'll probably want to save at least one workspace whose proportions and orientation to your monitor setup suits you best, and if you change your mind, you can always save over it or delete it.

New Words and Expressions

Juggling *v.* 欺骗，杂耍；*adj.* 欺骗的，欺诈的，变戏法（似）的
Composition *n.* 成分；合成；作品，著作，结构；构图；布置，布局
collapse *vi.* 倒塌，崩溃，瓦解
predominant *adj.* 占主导地位的，显著的
context *n.* 背景，环境；上下文，语境
dismay *vt.* 使灰心，使沮丧；使惊愕
inadvertent *adj.* 疏忽的；漫不经心的；非有意的；因疏忽所致的

Exercises to the Text

1. Translate the following words and phrases into English.

（1）浮动窗口 （2）快捷键 （3）运动跟踪 （4）用户界面 （5）保存和复位工作空间

2. Translate the following paragraphs into Chinese.

(1) All available panels in After Effects are listed under the Window menu; some even list preset keyboard shortcuts for rapid access. These panels used to be actual floating windows, but Adobe decided not to change the name of the menu, as it is standard across many applications.

(2) In this discussion, it is assumed that you have not yet customized any of the After Effects Workspaces. If what you see when you choose a Workspace doesn't match what is described here, you can use Window | Workspace | Reset to bring a given workspace back to its

previous settings (although if you have saved over any of the defaults, you will see the saved version).

(3) If you hold your pointer on the border between two panels, you will see it become a divider dragger. This allows you to resize the adjacent panels. But if you want the panel you are working with to take up a much larger amount of space for a moment, there is a better option.

Part 3 Communication and Telecommunication Technology

Text 1: Signals and Systems

This textbook provides an introduction to the tools and mathematical techniques necessary for understanding and analyzing both continuous-time and discrete-time linear systems. We have attempted to give an insight into the application of these tools and techniques for solving practical engineering problems. Our philosophy has been to adopt a systems approach throughout the book for the introduction of continuous-time signal and system analysis, rather than use the framework of traditional circuit theory. We believe that the systems viewpoint provides a more natural approach to introducing this material in addition to broadening the horizons of the student. Furthermore, the topics of discrete time signal and system analysis are most naturally introduced from a systems viewpoint, which lends overall consistency to the development. We have, of course, relied heavily on the students' circuit theory background to provide illustrative examples.

The organization of the book is straightforward. The first six chapters deal with continuous-time linear systems in both the time domain and the frequency domain. The principal tool developed for time-domain analysis is the convolution integral. Frequency-domain techniques include the Fourier and the Laplace transforms. An introduction to state variable techniques is also included. The remainder of the book deals with discrete-time systems including z-transform analysis techniques, digital filter analysis and synthesis, and the discrete Fourier transform and fast Fourier transform (FFT) algorithms.

This organization allows the book to be covered in two three-semester-hour courses, with the first course being devoted to continuous-time signals and systems and the second course being devoted to discrete-time signals and systems. Alternatively, the material can be used as a basis for three quarter length courses. With this formats the first course would cover time and frequency-domain analysis of continuous-time systems. The second course would cover state variables, sampling, and an introduction to the z-transform and discrete-time systems. The third course would deal with the analysis and synthesis of digital filters and provide an introduction to the discrete Fourier transform and its applications.

The assumed background of the student is mathematics through differential equations and the usual introductory circuit theory course or courses. Knowledge of the basic concepts of matrix algebra would be helpful but is not essential. Appendix A is included to bring together the pertinent matrix relations that are used in Chapters 5 and 6. We feel that in most electrical engineering curricula the material presented in this book is best taught at the junior level.

We begin the book by introducing the basic concepts of signal and system models and system classifications. The idea of spectral representations of periodic signals is first introduced in Chapter 1 because we feel that it is important for the student to think in terms of both the time and the frequency domains from the outset.

The convolution integral and its use in fixed, linear system analysis by means of the principle of superposition are treated in Chapter 2. The evaluation of the convolution integral is treated in detailed examples to provide reinforcement of the concepts. Calculation of the impulse response and its relation to the step and ramp responses of a system are discussed. Chapter 2 also contains optional sections and examples regarding writing the governing equations for lumped, fixed, linear systems and the solution of linear, constant coefficient differential equations. These are intended as review and may be omitted without loss of continuity.

The Fourier series and Fourier transform are introduced in Chapter 3. We have emphasized the elementary approach of approximating a periodic function by means of a trigonometric series and obtaining the expansion coefficients by using the orthogonality of sines and cosines. We do this because this is the first time most of our students have been introduced to Fourier series. The alternative generalized orthogonal function approach is included as a nonrequired reading section at the end of this chapter for those who prefer it. The concept of the transfer function in terms of sinusoidal steady-state response of a system is discussed in relation to signal distortion. The Fourier transform is introduced next, with its applications to spectral analysis and systems analysis in the frequency domain. The concept of an ideal filter, as motivated by the idea of distortion-less transmission, is also introduced at this point. The Gibbs Phenomenon, window functions, and convergence properties of the Fourier coefficients are treated in optional closing sections.

The Laplace transform and its properties are introduced in Chapter 4. Again, we have tried to keep the treatment as simple as possible because this is assumed to be a first exposure to the material for a majority of students, although a summary of complex variable theory is provided in Appendix B so that additional rigor may be used at the instructor's option. The derivation of Laplace transforms from elementary pairs is illustrated by example, as is the technique of inverse Laplace transform using partial fraction expansion. Optional sections on the evaluation of inverse Laplace transforms by means of the complex inversion integral and an introduction to the two-sided Laplace transform are also provided.

The application of the Laplace transform to network analysis is treated in detail in Chapter 5. The technique of writing Laplace transformed network equations by inspection is covered and used to review the ideas of impedance and admittance matrices, which the student will have learned in earlier circuit courses for resistive networks. The transfer function is treated in detail, and the Routh test for determining stability is presented. The chapter closes with a treatment of Bode plots and block diagram algebra for fixed, linear systems.

In Chapter 6, the concepts of a state variable and the formulation of the state variable approach to system analysis are developed. The state equations are solved using both time-domain and Laplace transform techniques, and the important properties of the solution are examined. Finally, as an example, we show how the state-variable method can be applied to the analysis of circuits.

The final three chapters provide coverage of the topics of discrete-time signal and system analysis.

A complete solutions manual, which contains solutions to all problems, is available from the publisher as an aid to the instructor. Answers to selected problems are provided in Appendix E as an aid to the student.

The authors wish to express their thanks to the many people who have contributed, both knowingly and unknowingly, to the development of this textbook. First, thanks go to our long-suffering students, who have been forced to study from our notes, often while they were still in various stages of development. Their many comments and criticisms have been invaluable and are gratefully appreciated. Many of our colleagues in the Electrical Engineering Department at the University of Missouri-Rolla taught courses that used the book in note form and provided many suggestions for improvement. In this regard, we thank Professors Gordon E. Carlson, Kenneth H. Carpenter, and some others. Professor Carlson critically reviewed much of the manuscript and provided valuable suggestions for improvement. Additionally, we would like to thank the reviewers at other institutions who provided valuable criticism. However, any shortcomings of the final result are solely the responsibility of the authors. A most sincere thanks goes to our secretaries whose great care and expert typing skills allowed us to generate the final manuscript with a minimum of headaches. The National Engineering Consortium is also due thanks since it was through their series of seminars that much of the material in Chapters 7 and 8 was originally taught.

Last, but not least, we thank our wives and families for putting up with a project whose end at times seemed nonexistent.

New Words and Expressions

algorithm *n.* 算法
coefficient *n.* 系数
convergence *n.* 收敛，集中
convolution *n.* 卷积
derivation *n.* 推导
evaluation *n.* 求值，评估
impedance *n.* 阻抗
invariance *n.* 不变性
lumped *adj.* 集总的
matrix *n.* 矩阵
motivate *v.* 激发，促动
orthogonal *adj.* [数学]直角的，矩形的，直交的
pertinent *adj.* 恰当的

prototype *n.* 原型，样机
reinforcement *n.* 加强
synthesis *n.* 综合，合成

Exercises to the Text

1. Translate the following words and phrases into English.

（1）复变量 （2）围线积分 （3）差分方程 （4）傅里叶级数 （5）冲击响应 （6）拉氏逆变换 （7）斜坡响应 （8）频谱表示 （9）阶跃响应 （10）时域 （11）波特图

2. Translate the following paragraphs into Chinese.

(1) We begin the book by introducing the basic concepts of signal and system models and system classifications. The idea of spectral representations of periodic signals is first introduced in Chapter 1 because we feel that it is important for the student to think in terms of both the time and the frequency domains from the outset.

(2) The Laplace transform and its properties are introduced in Chapter 4. Again, we have tried to keep the treatment as simple as possible because this is assumed to be a first exposure to the material for a majority of students, although a summary of complex variable theory is provided in Appendix B so that additional rigor may be used at the instructor's option. The derivation of Laplace transforms from elementary pairs is illustrated by example, as is the technique of inverse Laplace transform using partial fraction expansion. Optional sections on the evaluation of inverse Laplace transforms by means of the complex inversion integral and an introduction to the two-sided Laplace transform are also provided.

(3) In Chapter 6, the concepts of a state variable and the formulation of the state variable approach to system analysis are developed. The state equations are solved using both time-domain and Laplace transform techniques, and the important properties of the solution are examined. Finally, as an example, we show how the state-variable method can be applied to the analysis of circuits.

Text 2: Data Communication

The need to communicate is part of man's inherent being. Since the beginning of time man has used different techniques and methods to communicate. Circumstances and available technology have dictated the method and means of communications. Data communications concerns itself with the transmission (sending and receiving) of information between two parties. Now let's learn the foundation knowledge of data communication.

1. Signals

1s and 0s can't be sent as such across network links. They must be further converted into a form that transmission media can accept. Transmission media works by conducting energy along a physical path. So, a data stream of 1s and 0s must be turned into energy in the form of electromagnetic signals.

2. Analog and Digital

Both data and the signals that represent them can take either analog or digital form. Analog refers to something that is continuous—a set of specific points of data and all possible points between. Digital refers to something that is discrete.

Information can be analog or digital. Analog information is continuous. Digital information is discrete.

Signals can be analog or digital. Analog signals can have any value in a range; digital signals can have only a limited number of values.

3. Characteristics of Communications Channels

First is transmission rate. The transmission rate of a communication channel is determined by its bandwidth and its speed. The bandwidth is the range of frequencies that a channel can carry. Since transmitted data can be assigned to different frequencies, the wider the bandwidth, the more frequencies, and the more data can be transmitted at the same time.

The speed at which data is transmitted is usually expressed as bits per second or as a baud rate. Bits-per second (b/s) is the number of bits that can be transmitted in one second. The baud rate is the number of times per second that signal being transmitted changes. Usually only one bit is transmitted per signal change and, thus, the bits per second and the baud rate are the same.

The second is direction of transmission. The direction of data transmission is classified as simplex, half duplex, or full duplex. In simplex transmission, data flows in one direction only. Simplex is used only when the sending device, such as radio, never requires a response from the computer. In half-duplex transmission, data can flow in both directions but in only one direction at a time. Half-duplex is often used between terminals and a central computer. For example, interphone. In full-duplex transmission, data can be sent in both directions at the same time. A normal telephone line is an example of full-duplex transmission. Both parties can talk at the same time. Full-duplex transmission is used for most interactive computer applications and for computer-to-computer data transmission.

Third is transmission mode. The transmission mode includes asynchronous and synchronous. In asynchronous transmission mode, individual characters (made up of bits) are transmitted at irregular intervals, for example, when a user enters data. To distinguish where one character stops and another starts, the asynchronous communication mode used a start and a stop bit. An additional bit called a parity bit is sometimes included at the end of each character. Parity bits are used for error checking, asynchronous transmission mode is used for lower speed data transmission and is used with most communications equipment designed for personal computers.

4. Serial and Parallel Transmission

Data travels in two ways: in serial and in parallel, in serial data transmission, bits flow in a series or continuous stream, like cars crossing a one-lane bridge. Serial transmission is the way most data is sent over telephone lines. For this reason, external modems typically connect to a microcomputer through a serial port. More technical names for the serial port are RS-232C connector and asynchronous communications port.

With parallel data transmission, bits flow through separate lines simultaneously. In other words, they resemble cars moving together at same speed on a multilane freeway. Parallel transmission is typically limited to communications over short distances and typically is not used over telephone lines. It is, however, a standard method of sending data from the system unit to a printer.

New Words and Expressions

technique *n.* 方法，技术
electromagnetic *adj.* 电磁的
analog *n.* 模拟 *adj.* [计算机]模拟的
continuous *adj.* 继续的，连续的，持续的，延伸的
discrete *adj.* 离散的，分立的，不连续的
characteristic *adj.* 表示特性的，典型的，特有的
frequency *n.* 屡次，频繁，频率
baud *n.* 波特
duplex *adj.* [电信、计算机]双工的，双向的
simplex *n.* 单工
interphone *n.* 对讲机
asynchronous *adj.* 不同时的，异步的
synchronous *adj.* 同时发生的，同步的
parity *n.* 同等，平等，[计算机]奇偶校验
serial *adj.* 连续的
parallel *adj.* 平行的，并行的
resemble *v.* 看起来像
simultaneously *adv.* 同时地

Exercises to the Text

1. Translate the following words and phases into English

（1）信号 （2）波特率 （3）单工 （4）同步 （5）信道 （6）模拟 （7）数字 （8）电磁的 （9）串行口 （10）带宽

2. Translate the following words and phrases into Chinese.

(1) baud (2) asynchronous (3) full duplex (4) parallel (5) electromagnetic (6) discrete (7) simultaneously

3. Translate the following paragraph into Chinese.

Third is transmission mode. The transmission mode includes asynchronous and synchronous. In asynchronous transmission mode, individual characters (made up of bits) are transmitted at irregular intervals, for example, when a user enters data. To distinguish where one character stops and another starts, the asynchronous communication mode used a start and a stop bit. An additional bit called a parity bit is sometimes included at the end of each character. Parity bits are used for error checking, asynchronous transmission mode is used for lower speed data transmission and is used with most communications equipment designed for personal computers.

Text 3: Data Transmission Media

To go from here to there, data must move through something. A telephone line, cable, or the atmosphere can be called transmission medium or channel. But before the data can be communicated, it must be converted into a form suitable for communication. The three basic forms into which data can be converted for communication are:

- Electrical pulses or charges (used to transmit voice and data over telephone lines);
- Electromagnetic waves (similar to radio waves);
- Pulses of light.

The form or method of communications affects the maximum rate at which data can be moved through the channel and the level of noise that will exist—for example, light pulses travel faster than electromagnetic waves, and some types of satellite transmission systems are less noisy than transmission over telephone wires, Obviously, some situations require that data be moved as fast as possible; others don't. Channels that move data relatively slowly, like telegraph lines, are narrow-band channels. Most telephone lines are voice band channels, and they have a wider bandwidth than narrow band channels. Broadband channels (like coaxial cable, fiber optic cable, microwave circuits, and satellite systems) transmit large volumes of data at high speeds.

The transmission media used to support data transmission are telephone lines, coaxial cables, microwave systems, satellites systems, and fiber optic cables. Understanding how these media function will help you sort out the various rates and charges for them and determine which is the most appropriate in a given situation.

1. Telephone Lines

The earliest type of telephone line was referred to as open wire-unsheathed copper wires strung on telephone poles and secured by glass insulators. Because it was uninsulated, this type of telephone line was highly susceptible to electromagnetic interference; the wires had to be spaced about 12 inches apart to minimize the problem. Although open wire can still be found in a few places, it has almost entirely been replaced with cable and other types of communications media. Cable is insulated wire. Insulated pairs of wires twisted around each other—called twisted-pair wire or cable—can be packed into bundles of a thousand or more pairs. These wide-diameter cables are commonly used as telephone lines today and are often found in large buildings and under city streets. Even though this type of line is a major improvement over open wire, it still has many limitations. Twisted-pair cable is susceptible to a variety of types of electrical interference (noise), which limits the practical distance that data can be transmitted without being garbled. (To be received intact, digital signals must be "refreshed" every one to two miles through the use of an amplifier and related circuits, which together are called repeaters.

Although repeaters do increase the signal strength, which tends to weaken over long distances, they can be very expensive.) Twisted-pair cable has been used for years for voice and data transmission; however, more advanced media are replacing it.

2. Coaxial Cable

More expensive than twisted-pair wire, coaxial cable (also called shielded cable) is a type of thickly insulated copper wire that can carry a larger volume of data—about 100 million bits per second, or about 1800 to 3600 voice calls at once. The insulation is composed of a nonconductive material covered by a layer of woven wire mesh and heavy-duty rubber or plastic. Coaxial cable, which is laid underground and underwater, is similar to the cable used to connect your TV set to a cable TV services. Coaxial cables can also be bundled together into a much larger cable; this type of communications line has become very popular because of its capacity and reduced need for signals to be "refreshed", or strengthened, every two to four miles. Coaxial cables are most often used as the primary communications medium for locally connected networks in which all computer communication is within a limited geographic area, such as in the same building. Coaxial cable is also used for undersea telephone lines.

3. Microwave Systems

Instead of using wire or cable, microwave systems can use the atmosphere as the medium through which to transmit signals. These systems are extensively used for high-volume as well as long-distance communication of both data and voice in the form of electromagnetic waves similar to radio waves but in a higher frequency range. Microwave signals are often referred to as "line of sight" signals because they cannot bend around the curvature of the earth; instead, they must be relayed from point to point by microwave towers, or relay stations, placed 20 to 30 miles apart. The distance between the towers depends on the curvature of the surface terrain in the vicinity. The surface of the earth typically curves about 8 inches every mile. The towers have either a dish or a horn-shaped antenna. The size of the antenna varies according to the distance signals must cover. A long-distance antenna could easily be 10 feet or larger in size; a dish of 2 to 4 feet in diameter, which you often see on city buildings, is large enough for small distances. Each tower facility receives incoming traffic, boots the signal strength, and sends the signal to the next station.

The primary advantage of using microwave systems for voice and data communication is that direct cabling is not required. (Obviously, telephone lines and coaxial cable must physically connect all points in a communication system.) More than one half of the telephone systems now use microwave transmission. However, the saturation of the airwaves with microwave transmissions has reached the point where future needs will have to be satisfied by other communications methods, such as fiber optic cables or satellite systems.

4. Satellite Systems

Satellite communications systems transmit signals in the gigahertz range—billions of cycles per second. The satellite usually must be placed in a geosynchronous orbit, 22 330 miles above the earth's surface, so it revolves once a day with the earth. To an observer, it appears to be fixed over one place at all times. A satellite is a solar-powered electronic device that has up to 100 transponders (a transponder is a small, specialized radio) that receive, and retransmit signal; the satellite acts as a relay station between satellite transmission stations on the ground (called earth station).

Although establishing satellite systems is costly (owing to the cost of a satellite and the problems associated with getting it into orbit above the earth's surface and compensating for failures), satellite communications systems have become the most popular and cost-effective method for moving large quantities of data over long distances. The primary advantage of satellite communications is the amount of area that can be covered by a single satellite. Three satellites placed in particular orbits can cover the entire surface of the earth, with some overlap.

However, satellite transmission does have some problems:

- The signals can weaken over the long distances, and weather conditions and solar activity can cause noise interference.
- A satellite is useful for only 7 to 10 years, after which it loses its orbit.
- Anyone can listen in on satellite signals, so sensitive data must be sent in a secret or encrypted form.
- Depending on the satellite's transmission frequency, microwave stations on earth can "jam", or prevent transmission by operating at the same frequency.

Of course there is a very important data transmission media—Fiber optics, we will learn in more detail next.

New Words and Expressions

coaxial *adj.* 同轴的，共轴的
unsheathed *adj.* 未覆盖的
insulator *n.* 绝缘体，绝热器
interference *n.* 冲突，干涉
susceptible *adj.* 易受影响的，易感动的；容许……的
intact *adj.* 完整无缺的
conductive *adj.* 传导的
curvature *n.* 弯曲，曲率
vicinity *n.* 邻近，附近，接近
antenna *n.* 天线
saturation *n.* 饱和（状态），浸润，浸透
gigahertz *n.* 千兆赫
geosynchronous *adj.* 与地球的相对位置不变的，相对地球是静止的
transponders *n.* 异频雷达收发机
encrypt *v.* 加密，将……译成密码

Exercises to the Text

1. Translate the following words and phrases into English.

（1）声波信道 （2）双绞线 （3）微波塔 （4）电脉冲和电荷 （5）同步轨道

2. Translate the following words and phrases into Chinese.

(1) pulses of light (2) coaxial cable (3) electromagnetic interference (4) twisted-pair cable (5) the saturation of the airwaves

3. Translate the following paragraphs into Chinese.

(1) Obviously, some situations require that data be moved as fast as possible; others don't. Channels that move data relatively slowly, like telegraph lines, are narrow-band channels. Most telephone lines are voice band channels, and they have a wider bandwidth than narrow band channels. Broadband channels (like coaxial cable, fiber optic cable, microwave circuits, and satellite systems) transmit large volumes of data at high speeds.

(2) More than one half of the telephone systems now use microwave transmission. However, the saturation of the airwaves with microwave transmissions has reached the point where future needs will have to be satisfied by other communications methods, such as fiber optic cables or satellite systems.

Text 4: Switching Technologies

Whether they provide connection between one computer and another or between terminals and computers, communication networks can be divided into two basic types: circuit-switched (sometimes called connection oriented) and packet-switched (A variation of message switching is packet switching, sometimes called connectionless). Circuit-switched networks operate by forming a dedicated connection (circuit) between two points. The U.S. telephone system uses circuit switching technology, a telephone call establishes a circuit from the originating phone through the local switching office, across trunk lines, to a remote switching office, and finally to the destination telephone. While a circuit is in place, the phone equipment samples the microphone repeatedly, encodes the samples digitally, and transmits them across the circuit to the receiver. The sender is guaranteed that the samples can be delivered and reproduced because the circuit provides a guaranteed data path of 64 kb/s (thousand bits per second), the rate needed to send digitized voice. The advantage of circuit switching lies in its guaranteed capacity: once a circuit is established, no other network activity will decrease the capacity of the circuit. One disadvantage of circuit switching is cost: circuit costs are fixed, independent of traffic. For example, one pays a fixed rate for a phone call, even when the two parties do not have a talk.

In message switching, the transmission unit is a well-defined block of data called a message. In addition to the text to be transmitted, a message comprises a header and a checksum. The header contains information regarding the source and destination addresses as well as other control information; the checksum is used for error control purpose. The switching element is a computer referred to as a message processor, with processing and storage capabilities. Messages travel independently and asynchronously, finding their own way from source to destination. First the message is transmitted from the host to the message processor to which it is attached. Once the message is entirely received, the message processor to which it is attached. Once the message is entirely received, the message processor examines its header, and accordingly decides on the next outgoing channel on which to transmit it. If this selected channel is busy, the message waits in queue until the channel becomes free, at which time transmission begins. At the next message processor, the message is again received, stored, examined, and transmitted on some outgoing channel and the same process continues until the message is delivered to its destination. This transmission technique is also referred to as the store-and-forward transmission technique.

A variation of message switching is packet switching. Here the message is broken up into several pieces of a given maximum length, called packets. As with message switching, each packet contains a header and a checksum. Packets are transmitted independently in a store-and-forward manner.

Packet-switched networks, the type usually used to connect computers, take an entirely different approach. In a packet-switched network, data to be transferred across a network is divided into small pieces called packets that multiplexed onto high capacity inter-machine connections. A packet, which usually contains only a few hundred bytes of data, carries identification that enables the network hardware to know how to send it to the specified destination. For example, a large file to be transmitted between two machines must be broken into many packets that are sent across the network one at a time, the network hardware delivers the packets to the specified destination, where software reassembles them into a single file again. The chief advantage of packet-switching is that multiple communications among computers can proceed concurrently, with inter-machine connections shared by all pairs of machines that are communicating.

With circuit switching, there is always an initial connection cost incurred in setting up the circuit, it is cost-effective only in those situations where once the circuit is set up there is a guaranteed steady flow of information transfer to amortize the initial cost. This is certainly the case with voice communication in the traditional way, and indeed circuit switching is the technique used in the telephone system. Communication among computers, however, is characterized as bursty. Burstiness is a result of high degree of randomness encountered in the message-generation process and message size, and of the low delay constraint required by the user. The users and devices require the communicate resources relatively infrequently; but when they do, they require a relatively rapid response. If a fixed dedicated end-to-end circuit were to be set up connecting the end users, then one must assign enough transmission bandwidth to the circuit in order to meet the delay constraint with the consequence that the resulting channel utilization is low. If the circuit of high bandwidth were set up and released at each message transmission request, then the set-up time would be large compared to the transmission time of the message, resulting again in low channel utilization. Therefore, for bursty users (which can also be characterized by high peak-to-average data rate requirements), store-and-forward transmission techniques offer a more cost-effective solution, since a message occupies a particular communications link only for the duration of its transmission on that link; the rest of the time it is stored at some intermediate message switch mid the link is available for other transmissions. Thus the main advantage of store-and-forward transmission over circuit switching is that the communication bandwidth is dynamically allocated, and the allocation is done on the fine basis of a particular link in the network and a particular message (for a particular source - destination pair).

Packet switching achieves the benefits discussed so far and offers added disadvantage. The disadvantage, of course, is that as activity increases, a given pair of communicating computers receives less of the network capacity. That is, whenever a packet switched network becomes overloaded, computers using the network must wait before they can send additional packets.

Despite the potential drawback of not being able to guarantee network capacity, packet-switched networks have become extremely popular. The motivations for adopting packet switching are cost and performance. Because multiple machines can share the network hardware, fewer connections are required and cost is kept now. Because engineers have been able to build high speed network hardware, capacity is not usually a problem. So many computer interconnections use packet-switching that, throughout the remainder of this text, the term network will refer only to packets-switched networks.

New Words and Expressions

dedicated *adj.* [计算机]专用的
destination *n.* 目的地，终点
microphone *n.* 麦克风，话筒，扩音器（也作 mike）
guaranteed *n.* 有保证的，被担保的
decrease *v.* 减少，变少，降低
comprise *v.* 包括，包含，构成
checksum *n.* 检验[校验]和，核对和
outgoing *adj.* 往外去的，即将离任的，好交往的
maximum *n.& adj.* 最大量（的），最大值（的）
multiplex *adj.* 复合的，多重的
identification *n.* 认明，识别，鉴定
concurrent *adj.* 同时发生的，同时存在的
incurred *v.* 招致，遭受
amortize *v.* 摊销，摊还，分期偿付
burst *v.* 爆炸，胀裂
constraint *n.* 约束，限制
randomness *n.* 随意，无安排
relatively *adv.* 相对地，比较地

utilization *n.* 利用
intermediate *adj.* 中间的，居中的
whenever *adv.* 随便什么时候
motivation *n.* 动机
remainder *n.* 剩余物，其余（的人）

Exercises to the Text

1. Translate the following words and phrases into English.

（1）交换技术 （2）存储转发技术 （3）数据块 （4）分组交换 （5）电路交换 （6）带宽 （7）多路通信 （8）面向连接

2. Translate the following paragraph into Chinese.

Packet-switched networks, the type usually used to connect computers, take an entirely different approach. In a packet-switched network, data to be transferred across a network is divided into small pieces called packets that multiplexed onto high capacity intermachine connections. A packet, which usually contains only a few hundred bytes of data, carries identification that enables the network hardware to know how to send it to the specified destination. For example, a large file to be transmitted between two machines must be broken into many packets that are sent across the network one at a time, the network hardware delivers the packets to the specified destination, where software reassembles them into a single file again. The chief advantage of packet-switching is that multiple communications among computers can proceed concurrently, with inter-machine connections shared by all pairs of machines that are communicating.

Text 5: ATM

ATM is based on the efforts of the ITU-T Broadband Integrated Services Digital Network (B-ISDN) standard. It was originally conceived as a high-speed transfer technology for voice, video and data over public networks.

ATM is a cell-switching and multiplexing technology that combines the benefits of circuit switching (guaranteed capacity and constant transmission delay) with those of packet switching (flexibility and efficiency for intermittent traffic). It provides scalable bandwidth from a few megabits per second (Mbps) to many gigabits per second (Gbps). Because of its asynchronous nature, ATM is more efficient than synchronous technologies, such as time-division multiplexing (TDM).

ATM (Asynchronous Transfer Mode) is both a multiplexing and switching technique. It was initially intended to handle high bit rates, but it has in fact proved to be a universal technique for transporting and switching any type of digitized information at a wide variety of bit rates.

ATM transfers information in short packets called "cells" with a fixed length of 48 bytes plus five header bytes, irrespective of the underlying type of transmission. Cell routing is based on the principle of logical channels with dual identification: the cell header contains the identifier of the basic connection to which the cell belongs—called a virtual circuit (VC) and the identifier of the group of VCs to which the connection belongs—called a virtual path (VP).

ATM is related to both circuit and packet modes. Because of the simplicity of the protocol used, the transfer of cells to the network nodes can be handled entirely by hardware, which leads to very short transit time and high usage of transmission paths, even at bit rates of several hundred megabits a second. On the other hand, ATM retains all the flexibility of the packet mode, enabling only required information to be conveyed, offering a simple, unique multiplexing method irrespective of the bit rates of the different information flows, and allowing these bit rates to be varied.

An ATM network can be considered, in a first approximation, as being three overlaid functional levels: a services and applications level, an ATM network level and a transmission level. The applications provide an end-to-end service. They use the logical connections of ATM network level which in turn multiplexes and logically mutes the information flow as ATM cells go through the transmission links shared by logical connections called virtual connections. The transmission level provides these physical links and handles the actual physical transport of the cells.

An ATM network can transport and switch voice, data and video which, seen from the access, use traditional digital interfaces with the same quality of service. This means that a physical connection between any two terminals can be replaced with an equivalent logical connection which is multiplexed with others in a common transmission link. The resource is shared dynamically among all the connections.

Compared with the synchronous time division multiplexing techniques which rigidly link service to resource, the asynchronous technique has the advantage of occupying the transmission link only in proportion to the exact requirement.

The ATM technique completely separates the applications and services transported over a network from the transmission resources used. The ability to construct virtual networks means that the physical network can be shared by many users dynamically and in real time, thereby achieving cost-effective use of infrastructure, for high bit rate services too. Investments at all levels are also future-proofed, because the different applications can be reallocated in time over the same network infrastructure as requirements arise. ATM offers a unique way of coordinating different networks carrying different services into a single physical network.

As digitization and image encoding progress, interactive video services, and more generally multimedia services, are starting to emerge. Their impact on the network will be considerable. Today, ATM is the only transfer technique to offer the high bit rates and flexibility required by these services.

ATM, much more than any other telecommunications technique, is able to meet the current and the future requirements of both operators and users. Compared with other techniques that may compete in certain applications, ATM is special mainly due to its universal nature, both in terms of bit rate and type of information transferred. ATM offers a switching function for all bit rates and this is particularly suitable for high and variable bit rates.

ATM's specific features will make it the preeminent nature vehicle for multimedia services, and especially for varying bit rate video, and will make it one of the essential components of future information superhighways offering new services such as video on demand. In the short term, ATM is also proving of great interests to the operators, because of the flexibility and virtuality that it can introduce into networks, by separating the concept of connection from that of physical resources. This simplifies network management functions and makes optimum use of resources, particularly through statistical multiplexing and the creation of virtual private networks.

Of course, there is still a long way to go before the ATM techniques is in general use, but a revolution is underway which will deeply affect the world of telecommunications, data processing and video. The impact of this upheaval will without any doubt be greater than the advent of digital techniques in analogue networks.

In conclusion, ATM, much more than any other telecommunications technique, is able to meet the current and the future requirements of both operators and users.

New Words and Expressions

intermittent *adj.* 间歇的，断断续续的
scalable *adj.* 可攀登的，可升级的
megabit *n.* 兆位，百万位
gigabit *n.* 吉（咖）比特
irrespective *adj.*（与 of 连用）不顾……的，不考虑……的，不论……的
identifier *n.* 标志[标识，识别]符
simplicity *n.* 简单，简易，朴素，朴实，单纯
convey *v.* 运送，运输
approximation *n.* 近似，近似值
equivalent *adj.* （常与 to 连用）相同的，同等的
proportion *n.* 比例，比率
thereby *adv.* 因此，从而，由此
infrastructure *n.* 基本设施
considerable *adj.* 相当大的，相当多的
telecommunication *n.* 电信，远程通信
variable *adj.* 易变的，不稳定的
preeminent *adj.* 卓越的，杰出的，出类拔萃的
essential *adj.* 必需的，基本的
optimum *adj.* 最好的，最佳的，最有利的
underway *adj.* 起步的，进行中的，航行中的
upheaval *n.* 动乱，剧变

Exercises to the Text

1. Translate the following words and phrases into English.

（1）传输延迟 （2）统计多路复用 （3）虚拟专用网络 （4）交互视频业务 （5）同步时分复用技术

2. Please give a brief definition of the following terms.

(1) ATM (2) B-ISDN (3) Gb/s (4) VC (5) VP (6) TDM

3. Decide whether each of the following statements is true or false according to the text.

(1) ATM is based on the efforts of the ITU-T Broadband Integrated Services Digital Network (B-ISDN) standard.

(2) ATM (Asynchronous Transfer Mode) is both a multiplexing and switching technique.

(3) ATM transfers information in short packets called "cells" with a fixed length of 48 bytes plus five header bytes, irrespective of the underlying type of transmission.

(4) Neither circuit nor packet mode is ATM related to.

(5) An ATM network can be considered, in a first approximation, as being three overlaid functional levels: a services and applications level, an ATM network level and a transmission level.

Text 6: Fiber Optics

Although satellite systems are expected to be the dominant communication medium for long distances during this decade, fiber optics technology is expected to revolutionize the communications industry because of its low cost, high transmission volume, low arrogate, and message security. Fiber optic cables are replacing copper wire as the major communication medium in buildings and cities; major communications companies are currently investing huge sums of money in fiber optics communications networks that can carry digital signals, thus increasing communications and capacity.

In fiber optics communications, signals are converted to light form and fired by laser in bursts through insulated, very thin (1/2000 of an inch) glass or plastic fibers. The pulses of light represent the "on" state in electronic data representation and can occur nearly 1 billion times per second—nearly 1 billion bits can be communicated through a fiber optic cable per second. Equally important, fiber optic cables aren't cumbersome in size: A fiber optic cable (insulated fibers bound together) that is only 1/2-inch thick is capable of supporting nearly 250,000 voice conversation at the same time (soon to be doubled to 500, 000). However, since the data is communicated in the form of pulses of light, specialized communications equipment must be used.

Fiber optic cables are not susceptible to electronic noise and so have much lower error rates than normal telephone wire and cable. In addition, their potential speed for data communications is up to 10,000 times faster than that of microwave and satellite systems. Fiber optics communications are also very resistant to illegal data theft, because taps into it to listen to or change the data being transmitted can be easily detected. In fact, it is currently being used by the U. S. Central Intelligence Agency.

Fiber looks like a common glass cylinder consisting of core and cladding regions. Nowadays, there are three types of fiber optic cable commonly used: single mode, multimode and Plastic Optical Fiber (POF).

Single mode cable is made up of one or a number of quartz fibers with a diameter of 8.3μm to 10μm that has one mode of transmission. Single mode fiber with a relatively narrow diameter, will propagate typically 1310nm or 1550nm. It carries higher bandwidth than multimode fiber, but requires a light source with a narrow spectral width.

Single mode fiber, as is shown in Figure 27, gives you a higher transmission rate and up to 50 times more distance than multimode, but it also costs more. Single mode fiber has a much smaller core than multimode. The small core and single light-wave virtually eliminate any distortion that could result from overlapping light pulses, providing the least signal attenuation and the highest transmission speeds of any fiber cable type.

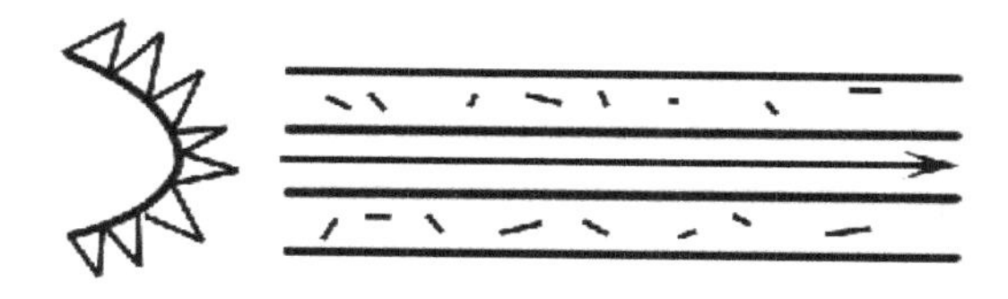

Figure 27 Single mode fiber

By contrast, multimode fiber has a core diameter that is much larger than the wavelength of light transmitted. (The most common size is 62.5μm). Light waves are dispersed into numerous paths, or modes, as they travel through the cable's core typically 850nm or 1300nm. However, in long cable runs (greater than 3000m), multiple paths of light can cause signal distortion at the receiving end, resulting in an unclear and incomplete data transmission.

Multimode fiber can be divided into two types: step index multimode fiber and graded index multimode fiber. Step index multimode was the first fiber design but is too slow for most uses, due to the dispersion caused by the different path lengths of the various modes. Step index fiber is rare-only POF uses a step index design today. Graded index multimode fiber, as the name implies, the refractive index of this fiber gradually decreases from the core out through the cladding to compensate for the different path lengths of the modes. It offers hundreds of times more bandwidth than step index fiber—up to about 2 gigahertz.

POF is a newer plastic-based cable which promises performance similar to glass cable on very short runs, but at a lower cost.

New Words and Expressions

dominant *adj.* 最重要的；有统治权的

revolutionize *v.* 使革命化
arrogate *v.* 非法霸占
volume *n.* 卷，册，（与of连用）体积，容量
inch *n.* 寸，英寸
susceptible *adj.* （与to连用）易受影响的
cylinder *n.* 圆柱体
cladding *n.* 包层，镀层
multimode *n.* [计算机]多模态（光纤的一种传输方式）
quartz *n.* 石英
propagate *v.* 繁殖，增殖
spectral *adj.* 光谱的
eliminate *v.* 除去
attenuation *n.* 变薄，稀薄化，变细，衰减
distortion *n.* 扭曲，变形，曲解，失真
wavelength *n.* [无线电]波长
dispersed *adj.* 分散的，散开的，漫布的
refractive *adj.* 折射的

Exercises to the Text

1. Translate the following words and phrases into English.

（1）光纤 （2）光脉冲 （3）单模光纤 （4）多模光纤 （5）塑料光纤

2. Translate the following paragraphs into Chinese.

(1) Fiber optic cables are not susceptible to electronic noise and so have much lower error rates than normal telephone wire and cable. In addition, their potential speed for data communications is up to 10,000 times faster than that of microwave and satellite systems. Fiber optics communications are also very resistant to illegal data theft, because taps into it to listen to or change the data being transmitted can be easily detected. In fact, it is currently being used by the U. S. Central Intelligence Agency.

(2) By contrast, multimode fiber has a core diameter that is much larger than the wavelength of light transmitted. (The most common size is 62.5μm). Light waves are dispersed into numerous paths, or modes, as they travel through the cable's core typically 850nm or 1300nm. However, in long cable runs (greater than 3000m), multiple paths of light can cause signal distortion at the receiving end, resulting in an unclear and incomplete data transmission.

(3) Multimode fiber can be divided into two types: step index multimode fiber and graded index multimode fiber. Step index multimode was the first fiber design but is too slow for most uses, due to the dispersion caused by the different path lengths of the various modes.

Text 7: Passive Optical Networks (PONs)

This article describes the recent advances made in broadband access network architectures employing Passive Optical Networks (PONs). The potential of PONs to deliver high bandwidths to users in access networks and their advantages over current access technologies have been widely recognized. PONs has made strong progress in terms of standardization and deployment over the past few years.

The access network, also known as the *"first mile"* network, connects the service provider central offices to businesses and residential subscribers. This network is also referred to in the literature as the *subscriber access network,* or the *local loop.* Residential subscribers demand first-mile access solutions that have high bandwidth, offer media-rich Internet services, and are comparable in price with existing networks. Similarly, corporate users demand broadband infrastructure through which they can connect their local-area networks to the Internet backbone.

1. Challenges in Access Networks

Much of the focus and emphasis over the years has been on developing high-capacity backbone networks. Backbone network operators currently provide high-capacity OC-192 (10 Gb/s) links. However, current generation access network technologies such as Digital Subscriber Loop (DSL) provide 1.5 Mb/s of downstream bandwidth and 128 kb/s of upstream bandwidth at best. The access network is, therefore, truly the bottleneck for providing broadband services such as video-on-demand, interactive games, and video conferencing to end users.

In addition, DSL has a limitation that the distance of any DSL subscriber to a central office must be less than 18,000 feet because of signal distortions. Typically, DSL providers do not provide services to distances more than 12,000 feet. Therefore, only an estimated 60% of the residential subscriber base can avail of DSL. Although variations of DSL such as very high bit-rate DSL (VDSL), which can support up to 50 Mb/s of downstream bandwidth, are gradually emerging, these technologies have much more severe distance limitations. For example, the maximum distance which VDSL can be supported over, is limited to 1,500 feet.

The other alternative available for broadband access to end users is through Cable Television (CATV) networks. CATV networks provide Internet services by dedicating some Radio Frequency (RF) channels in coaxial cable for data. However, CATV networks are mainly built for delivering broadcast services, so they don't fit well for distributing access bandwidth. At high load, the network's performance is usually frustrating to end users.

Faster access-network technologies are clearly desired for next-generation broadband applications. The next wave of access networks promises to bring fiber closer to the home. The FTTx model - Fiber to the Home (FTTH), Fiber to the Curb (FTTC), Fiber to the Building (FTTB), etc. - offers the potential for unprecedented access bandwidth to end users. These technologies aim at providing fiber directly to the home, or very near the home, from where technologies such as VDSL can take over. FTTx solutions are mainly based on the Passive

Optical Network (PON). In this article, we shall review major developments in PON in recent years - EPON, APON, GPON and the WDM PON. Finally, we shall review the issues related to deployment of PONs.

2. Passive Optical Network (PON) Architectures

A Passive Optical Network (PON) is a point-to-multipoint optical network. An Optical Line Terminal (OLT) at the central office is connected to many Optical Network Units (ONUs) at remote nodes through one or multiple 1: N optical splitters. The network between the OLT and the ONU is passive, meaning that it doesn't require any power supply. An example of a PON using a single optical splitter is shown in Figure 28. The presence of only passive elements in the network makes it relatively more fault tolerant, and decreases its operational and maintenance costs once the infrastructure has been laid down.

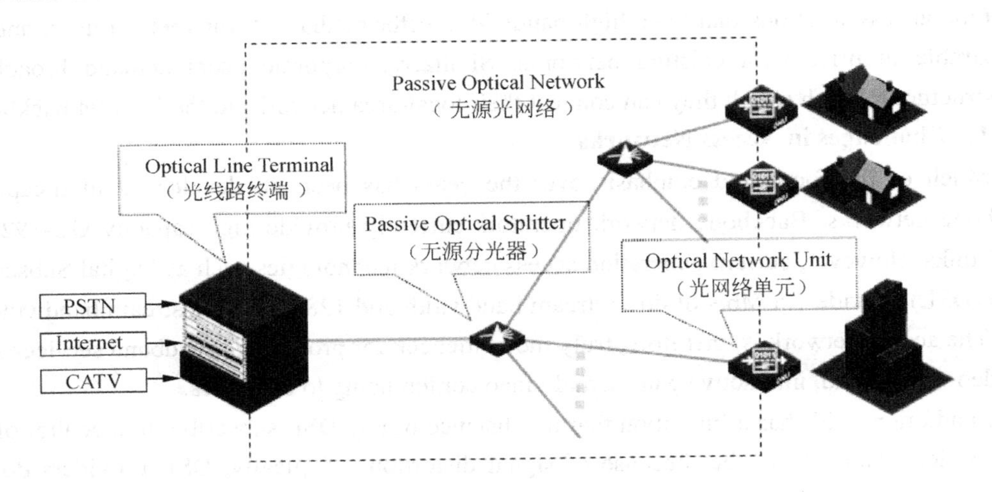

Figure 28 A Passive Optical Network (PON) connecting a central office to residential customers and business establishments.

Passive Optical Networks (PONs) have been considered for the access network for quite some time. A typical PON uses a single wavelength for all downstream transmissions (from OLT to ONUs), and another wavelength for all upstream transmissions (from ONUs to OLT), multiplexed on a single fiber through coarse wavelength-division multiplexing (CWDM).

3. Ethernet PON (EPON) Access Network

Ethernet PON (EPON) is a PON-based network that carries data traffic encapsulated in Ethernet frames (defined in the IEEE 802.3 standard). It uses a standard 8b/10b line coding (in which 8 user bits are encoded as 10 line bits), and it operates at standard Ethernet data rates.

4. Why Ethernet is Gaining Prominence?

The first-generation PON standardized by ITU-T G.983 employed ATM as the medium-access control (MAC) layer protocol. When its standardization effort was started in 1995, the telecom community believed that ATM would be the prevalent technology in backbone networks. ATM had the advantages of streamlining voice and data services while providing operational and performance guarantees. However, since then, Ethernet has grown vastly popular. Ethernet line cards are cheap, and they are widely deployed in LANs today. Since access networks are focused towards end users and LANs, ATM has turned out to be not the best choice to connect to Ethernet-based LANs.

In addition high-speed Gigabit Ethernet deployment is widely accelerating and 10 Gigabit Ethernet products are becoming available. Ethernet is a very efficient MAC protocol to use compared to ATM which imposes a considerable amount of overhead on variable-length Internet Protocol (IP) packets. Newly adopted quality-of-service (QoS) techniques have made Ethernet networks capable of efficiently supporting voice, data, and video. These techniques include full duplex transmission mode, prioritization (802.1p), and virtual LAN (VLAN) tagging (802.1Q). 802. 1p is a specification which allows for prioritization of traffic into different priority classes. 802.1Q defines an architecture for VLANs. Although 802.1Q doesn't directly define any QoS support, it defines a frame-format extension allowing Ethernet frames to carry priority information. EPONs, therefore, have much more promise in future access networks compared to ATM PONs (APONs).

5. EPON Principle of Operation

In the downstream direction (OLT to ONUs), Ethernet frames transmitted by the OLT pass through a 1: N passive splitter and reach each ONU. Typical values of N are between 8 and 64. Packets are broadcast by the OLT and extracted by their destination ONU based on a Logical Link Identifier (LLID), which the ONU is assigned when it registers with the network. Figure 29 shows the downstream traffic in EPON.

In the upstream direction, data frames from any ONU will only reach the OLT and will not reach any other ONU due to the directional properties of a passive optical combiner. Therefore, in the upstream direction, the behavior of EPON is similar to that of a point-to-point architecture. However, unlike in a true point-to-point network, in EPON, data frames from different ONUs transmitted simultaneously may collide. Thus, in the upstream direction, the ONUs need to employ some arbitration mechanism to avoid data collisions and fairly share the channel capacity. A contention-based media-access mechanism (similar to Carrier Sense Multiple Access with Collision Detection (CSMA/CD)) is difficult to implement because ONUs cannot detect a collision in the fiber from the combiner to the OLT due to the directional properties of the combiner. The OLT could detect a collision and inform the ONUs by sending a jam signal; however, propagation delays in PON (the typical distance from the OLT to ONUs is 20 km), greatly reduces the efficiency of such a scheme. To introduce determinism in frame delivery in the upstream direction, different noncontention schemes have been proposed. Figure 29 illustrates an upstream, time-shared, data flow in an EPON.

All ONUs are synchronized to a common time reference and each ONU is allocated a timeslot in which to transmit. Each timeslot is capable of carrying several Ethernet frames. An ONU should buffer frames received from a subscriber until its timeslot arrives. When its timeslot arrives, the ONU would burst all stored frames at full channel speed. If there are no frames in the buffer to fill the entire timeslot, an idle pattern is transmitted.

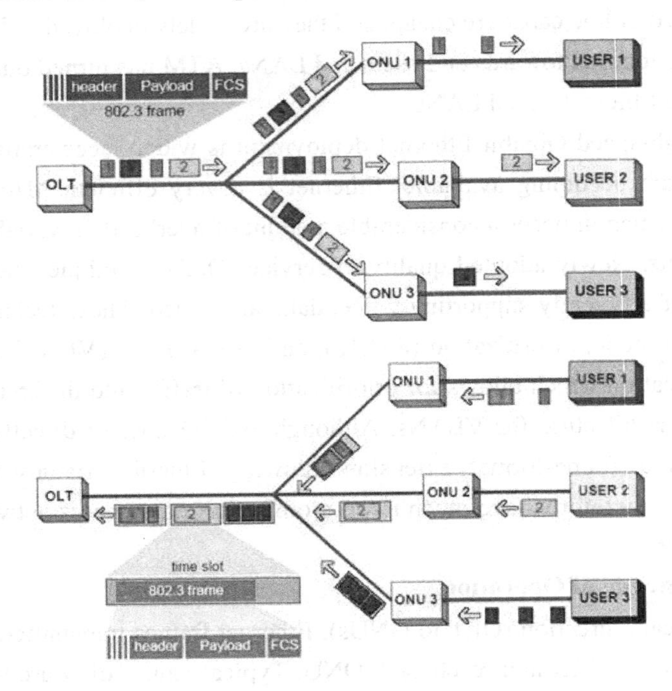

Figure 29 Downstream and Upstream operation in EPON

Thus, timeslot assignment is a very crucial step. The possible timeslot allocation schemes could range from static allocation (fixed time-division multiple access (TDMA)) to a dynamically-adapting scheme based on instantaneous queue size in every ONU (a statistical multiplexing scheme). In the dynamically-adapting scheme, the OLT can play the role of collecting the queue sizes from the ONUs and then issuing time slots. Although this approach may lead to higher signalling overhead between the OLT and the ONUs, the centralized intelligence may lead to more efficient use of bandwidth. More advanced bandwidth-allocation schemes are also possible, including schemes utilizing notions of traffic priority, Quality of Service (QoS), Service-Level Agreements (SLAs), over-subscription ratios, etc.

New Words and Expressions

access *n.* 通路，访问，入门

standardization *n.* 标准化

PON *abbr.* passive optical network 无源光纤网络

Loop *n.* 循环

Infrastructure *n.* 下部构造，基础下部组织

backbone *n.* 脊椎，中枢，骨干，支柱，意志力，勇气，毅力，决心

bottleneck *n.* 瓶颈

distortion *n.* 扭曲，变形，曲解，失真

unprecedented *adj.* 空前的
multipoint *adj.* 多点（式）的，多位置的
node *n.* 节点
maintenance *n.* 维护，保持，生活费用，扶养
multiplex *adj.* 多元的
encapsulate *vt.* 装入胶囊，压缩
protocol *n.* 草案，协议
prevalent *adj.* 普遍的，流行的
accelerate *v.* 加速，促进
overhead *n.* 管理费用，经常费用
duplex *adj.* 双工的，双向的
specification *n.* 详述，规格，说明书，规范
priority *n.* 先，前，优先，优先权
splitter *n.* 分路器，分裂机
collision *n.* 碰撞，冲突
synchronize *v.* （使）同步
frame *n.* 帧，画面，框架
instantaneous *adj.* 瞬间的，即刻的，即时的

Exercises to the Text

1. Translate the following words and phrases into English.

（1）无源光网络 （2）用户接入网 （3）下行带宽 （4）光纤到家 （5）光纤线路终端 （6）光纤网络单元 （7）以太无源光纤网络 （8）介质访问控制 （9）服务质量 （10）国际电信联盟 （11）逻辑链路标识 （12）服务等级协议 （13）数字用户环路 （14）时隙 （15）信道容量

2. Translate the following paragraphs into Chinese.

(1) Thus, timeslot assignment is a very crucial step. The possible timeslot allocation schemes could range from static allocation (fixed time-division multiple access, TDMA) to a dynamically-adapting scheme based on instantaneous queue size in every ONU (a statistical multiplexing scheme).

(2) Passive Optical Networks (PONs) have been considered for the access network for quite some time. A typical PON uses a single wavelength for all downstream transmissions (from OLT to ONUs), and another wavelength for all upstream transmissions (from ONUs to OLT), multiplexed on a single fiber through coarse wavelength-division multiplexing (CWDM).

(3) A Passive Optical Network (PON) is a point-to-multipoint optical network. An Optical Line Terminal (OLT) at the central office is connected to many Optical Network Units (ONUs) at remote nodes through one or multiple 1: N optical splitters. The network between the OLT and the ONU is passive, meaning that it doesn't require any power supply.

(4) In the downstream direction (OLT to ONUs), Ethernet frames transmitted by the OLT

pass through a 1: N passive splitter and reach each ONU. Typical values of N are between 8 and 64. Packets are broadcast by the OLT and extracted by their destination ONU based on a Logical Link Identifier (LLID), which the ONU is assigned when it registers with the network.

Text 8: Television: Basic Principles

1. The Sound and Light Spectrum

Video is a combination of light and sound, both of which are made up of vibrations or frequencies. We are surrounded by various forms of vibrations: visible, tangible, audible, and many other kinds that our senses are unable to perceive. We are in the midst of a wide spectrum which extends from zero to many millions of vibrations per second. The unit we use to measure vibrations per second is Hertz (Hz).

Sound vibrations occur in the lower regions of the spectrum, whereas light vibrations can be found in the higher frequency areas. The sound spectrum ranges from 20 to 20,000 Hertz (Hz). Light vibrations range from 370 trillion to 750 trillion Hz. When referring to light, we speak of wavelengths rather than vibrations.

As a result of the very high frequencies and the speed at which light travels (300, 000 km per second), the wavelength is extremely short, less than one thousandth of a millimeter. The higher the vibration, the shorter the wavelength.

Not all light beams have the same wavelength. The spectrum of visible light ranges from wavelength of 780 nm to a wavelength of 380 nm, as shown in Figure 30. We perceive the various wavelengths as different colors. The longest wavelength (which corresponds to the lowest frequency) is seen by us as the color red followed by the known colors of the rainbow: orange, yellow, green, blue, indigo, and violet which is the shortest wavelength (and highest frequency). White is not a color but the combination of the other colors. Wavelengths which we are unable to perceive (occurring just below the red and just above the violet area), are the infrared and ultraviolet rays, respectively. Nowadays, infrared is used for such applications as remote control devices.

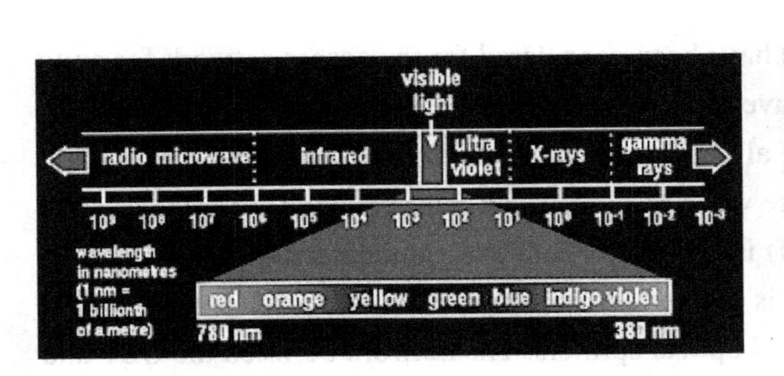

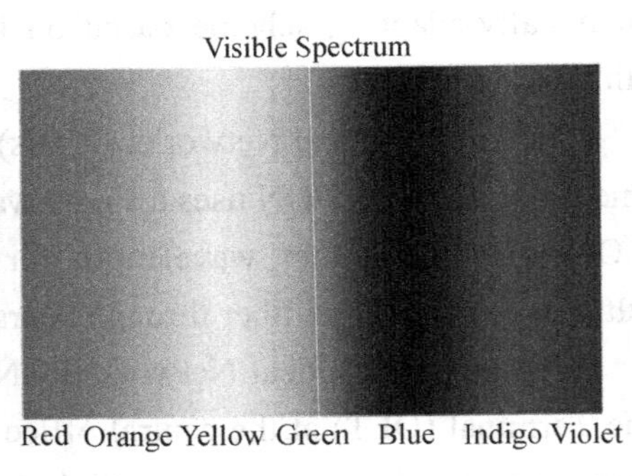

Figure 30 Visible light as part of the electromagnetic spectrum.

Note: visible light is only visible because we can see the source and the objects being illuminated. The light beam itself cannot be seen. The beams of headlights in the mist for instance, can only be seen because the small water drops making up the mist reflect the light.

2. Luminosity

Besides differing in color (frequency), light can also differ in luminosity, or brightness. A table lamp emits less light than a halogen lamp, but even a halogen source cannot be compared with bright sunlight, as far as luminosity is concerned. Luminosity depends on the amount of available light. It can be measured and recorded in a numeric value. In the past, it was expressed in Hefner Candlepower, but nowadays Lux is used to express the amount of luminosity.

Brightness Values:

Candle light at 20 cm	10～15 lx
Street light	10～20 lx
Normal living room lighting	100 lx
Office fluorescent light	300～500 lx
Halogen lamp	750 lx
Sunlight, 1 hour before sunset	1000 lx
Daylight, cloudy sky	5000 lx
Daylight, clear sky	10,000 lx
Bright sunlight	> 20,000 lx

Luminosity is the basic principle of the black-and-white television. All shades between black and white can be created by adjusting the luminosity to specific values.

3. Color Mixing

There are two kinds of color mixing: additive and subtractive color mixing. The mixing of colorants, like paint, is called subtractive mixing. The mixing of colored light is called additive mixing. Color TV is based on the principle of additive color mixing. Primary colors are used to create all the colors that can be found in the color spectrum.

4. Additive Color Mixing

In video, the color spectrum contains three primary colors, namely red, green and blue. By combining these three, all the other colors of the spectrum (including white) can be produced, as shown in Figure 31.

red + blue = magenta
red + green = yellow
blue + green = cyan (turquoise)
green + magenta = white
red + cyan = white
blue + yellow = white
red + blue + green = white

Making colors in this way is based on blending, or adding up colored light, which is why it is called additive color mixing. Combining the three primary colors in specific ratios and known amounts enables us to produce all possible colors.

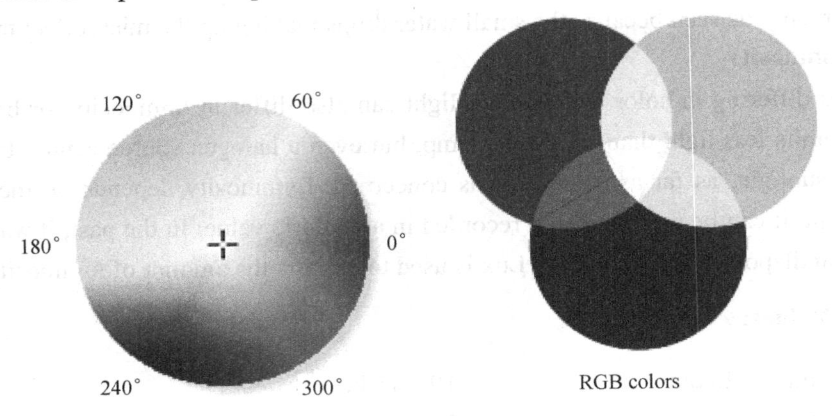

Figure 31 By combining the three primary colors red, green and blue, other colors can be mixed, including white.

White light is derived from a ratio of 30% red, 59% green, and 11% blue. This is also the ratio to which a color TV is set for black-and-white broadcasts. Shades of grey can be created by maintaining the ratio percentages and by varying the luminosity to specific values.

30% red + 59% green + 11% blue = white

5. Light Refraction

Light refraction is the reverse process of color mixing. It shows that white light is a combination of all the colors of the visible light spectrum. To demonstrate refraction a prism is used, which is a piece of glass that is polished in a triangular shape. A light beam travelling through a prism is broken twice in the same direction, causing the light beam to change its original course.

Beams with a long wavelength (the red beams) are refracted less strongly than beams with a short wavelength (the violet beams), causing the colors to fan out. The first fan out is enlarged by the second fan out, resulting in a color band coming out, consisting of the spectrum colors red, orange, yellow, green, blue, indigo, and violet, as shown in Figure 32. There are no clear boundaries between the various colors, but thousands of transitional areas. A rainbow is a perfect example of the principle of light refraction in nature.

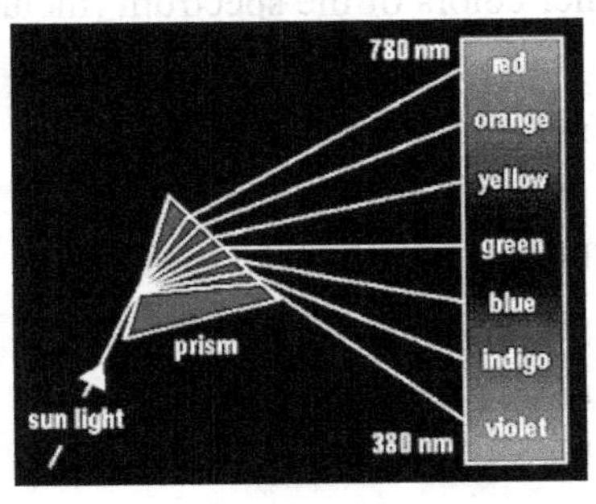

Figure 32 When white light, such as sunlight passes through a prism, it is refracted in the colors of the rainbow.

6. Color Temperature

Color temperature relates to the fact that when an object is heated, it will emit a color that is directly related to the temperature of that object. The higher the color temperature, the more "blue" the light, and the lower the color temperature the more "red" the light. Color temperature of light can be measured in degrees Kelvin (K). Daylight has a color temperature between 6000 and 7000 K. The color temperature of artificial light is much lower: approximately 3000 K. In reality, color temperatures range from 1900 K (candlelight) up to 25,000 K (clear blue sky). Television is set to 6500 K, simulating "standard daylight", as shown in Figure 33.

Lighth source		Degrees Kelvin	Spectrum
sunrise		1000	towards red
candlelight		1900	
light bulb (100W)		2800	
fluorescent lamp		4500	
cloudy sky		6500	
hazy sky		9000	
clear blue sky		25000	towards blue

Figure 33 Various light sources with different color temperatures. Color temperature is expressed in degrees Kelvin.

7. The Human Eye

The eye tends to retain an image for about 80 milliseconds after it has disappeared. Advantage is taken of this in television and cinematography, where a series of still pictures (25 per second) create the illusion of a continuously moving picture. Other characteristics of the human eye are that it is less sensitive to color detail than to black-and-white detail, and that the human eye does not respond equally to all colors. The eye is most sensitive to the yellow/green region, and less in the areas of red and (particularly) blue.

New Words and Expressions

spectrum *n.* 光，光谱，型谱，频谱

vibration *n.* 振动，颤动，摇动，摆动

frequency *n.* 频率，频数，发生次数

tangible *adj.* 切实的，通过触摸可以感知的，确实的，真实的

audible *adj.* 听得见的

perceive *v.* 感知，感到，认识到

Hertz *n.* 赫，赫兹（频率单位：周/秒）

wavelength *n.* 波长

beam *n.* （光线的）束，柱，电波，梁

illuminate *v.* 照亮
luminosity *n.* 发光度
halogen *n.* 卤素
Hefner Candlepower 赫夫纳烛光
Lux *n.* 勒克斯（照明单位）
Magenta *n.* 红紫色，洋红
Cyan *n.* 蓝绿色，青色
Turquoise *n.* 绿宝石，绿松石色，青绿色
Refraction *n.* 折光，折射
Indigo *n.* 靛，靛青色
Violet *n.* 紫罗兰色
prism *n.* 棱镜，棱柱
Kelvin *n.* 绝对温标，开氏温标，简写为K

Exercises to the Text

1. Translate the following words and phrases into English.

（1）光谱 （2）红外遥控装置 （3）亮度 （4）减法混色法 （5）光的折射

2. Translate the following words and phrases into Chinese.

(1) infrared and ultraviolet rays (2) additive mixing (3) Light Refraction (4) light beam (5) color temperature

3. Translate the following paragraphs into Chinese.

(1) Video is a combination of light and sound, both of which are made up of vibrations or frequencies. We are surrounded by various forms of vibrations: visible, tangible, audible, and many other kinds that our senses are unable to perceive.

(2) Not all light beams have the same wavelength. The spectrum of visible light ranges from wavelength of 780 nm to a wavelength of 380 nm. We perceive the various wavelengths as different colors. The longest wavelength (which corresponds to the lowest frequency) is seen by us as the color red followed by the known colors of the rainbow: orange, yellow, green, blue, indigo, and violet which is the shortest wavelength (and highest frequency).

Text 9: Television Receivers

Watching television is North America's favorite pastime, so the manufacture of television receiver sets forms a large portion of today's electronics industry. The systems used for television have been thoroughly standardized, so sets vary little from one manufacturer to another. The system presently used in North America is common to all North American countries, Japan, and a few others. The system common to most European countries differs from the American system mainly in the use of different scanning rates.

Most television broadcasting is now done in color, and although black and white sets may still be obtained, these are usually used for special applications such as security monitoring and microcomputer systems. The color system and all color receivers are made to be compatible with black-and-white transmissions.

1. Cathode-Ray Tube

The cathode-ray tube in the receiver performs the opposite function to the camera tube in the transmitter, converting the electrical signal into an optical image. Figure 34 shows in outline the action of the cathode-ray tube, where, for the moment, the deflecting and focusing arrangements are omitted. The electrical signal, which represents the picture, is applied to the control electrode, which functions in a manner similar to that of the control grid in a tube. The signal causes the potential of the control electrode to vary about a mean level, and as this goes more negative, it reduces the intensity of the electron beam, with a resulting reduction in the brightness of the spot formed on the fluorescent screen. Likewise, as it goes less negative, it increases the brightness of the spot on the screen. The fluorescent screen is made out of a phosphor, which is a material that emits light when bombarded with an electron beam.

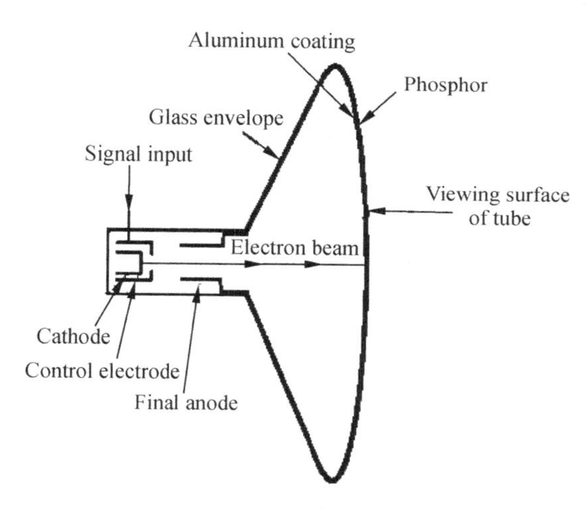

Figure 34 The structure of the cathode-ray picture tube

In modern tubes, the back of the phosphor is coated with a very thin film of aluminum, and this coating is continued down the inside walls of the tube and connected to the final anode. The purpose of the coating is to increase the brightness of the screen (for a given intensity of electron beam) and to ensure that the brightness is even (i. e., no dark patches occur). The coating, being at final anode potential, prevents electrons from the beam from accumulating on the screen, which otherwise would tend to repel the incident beam of electrons, thus reducing brightness. Furthermore, if electrons were allowed to accumulate, they would do so unevenly over the surface of the screen, thus producing dark patches. The aluminum coating improves brightness in another way, by reflecting light forward onto the viewing side, light that would otherwise be lost inside the tube. Some of the electron-beam energy is lost in penetrating the aluminum coating, but since this is kept very thin, the energy lost is more than offset by the gain in brightness.

In both the transmitter camera tube and the receiver picture tube, the electron beam must be made to scan the screen. The beam usually starts at the top left-hand corner of the screen and moves across to the right in an almost horizontal line. There has to be a small downward deflection of the beam as each line is traversed, so that the complete picture is eventually scanned. When the beam reaches the right-hand side, it is very quickly deflected back to the left-hand side again (this is called line flyback), where it starts a new line. The flyback is initiated by a line synchronizing signal called the horizontal sync pulse. This is a special pulse signal introduced at the transmitter which ensures that the receiver flyback occurs at the same time as the transmitter flyback.

When the scanning beam reaches the bottom right-hand corner of the picture, it must be returned to the top left-hand corner again, to repeat the whole scanning procedure. Thus, a second synchronizing signal must be transmitted at this time to ensure that the receiver recommences complete scanning at the same time as the transmitter. This is the field synchronizing signal, called the vertical sync pulse.

2. Interlacing and Picture Scan Repetition Rate

One complete scan of the target area will allow reproduction of one complete picture at the receiver of a television system. A large amount of information must be transmitted during this period, and if the picture is repeated at a high rate, then the bandwidth required for transmission will become excessive. In television, if the picture scan rate is made too low, moving scenes will develop a stop-and-go jerky movement in the same manner that slow-motion moving pictures do. Further, the phosphors used in the receiver picture tubes have a relatively low persistence, allowing the picture to fade out between scans, and the scanning will produce a "flicker" at the picture rate. The picture rate must be sufficiently high so that the normal persistence of the viewer's eyes will override the flicker and merge the picture series into smooth motion. This minimum picture rate has been found to be about 35 to 50 pictures per second.

It has also been found that in television systems, if the scanning rate is near, but not exactly equal to, the supply frequency, voltage pickup from the ac power circuits modulate the scanning circuit amplifiers and cause annoying distortion and jitter in the picture. This interference can be minimized by making the picture scanning rate a multiple of the supply frequency. In the American system, the picture or frame scan rate is 30Hz, which is a submultiple of the supply frequency of 60Hz.

The 30Hz picture repetition rate is too low, and if scanning were done in a straightforward sequential manner, the flicker produced because of picture fading between scans would become objectionable. For this reason, the scanning of each picture frame is divided into two separate fields, and the entire area is scanned twice during a picture period. A total of 525 lines are included in each picture, so that during the first field 262.5 lines are scanned, and during the second field the remaining 262.5 lines are scanned.

At the end of the first 262.5 lines, a vertical retrace occurs, and the second scan begins at the top center of the picture, as shown in Figure 35. As a result, the second field scan lines fall midway between the positions of the first field lines, so that the two fields merge to form a complete scan frame. The effective flicker rate is now 60Hz, well above the threshold, and the frame, or picture repetition rate of 30 per second is low enough to minimize video signal frequencies. European supplies are generally operated at 50Hz, and the picture rate for this system is 25 per second, with a field rate of 50 per second.

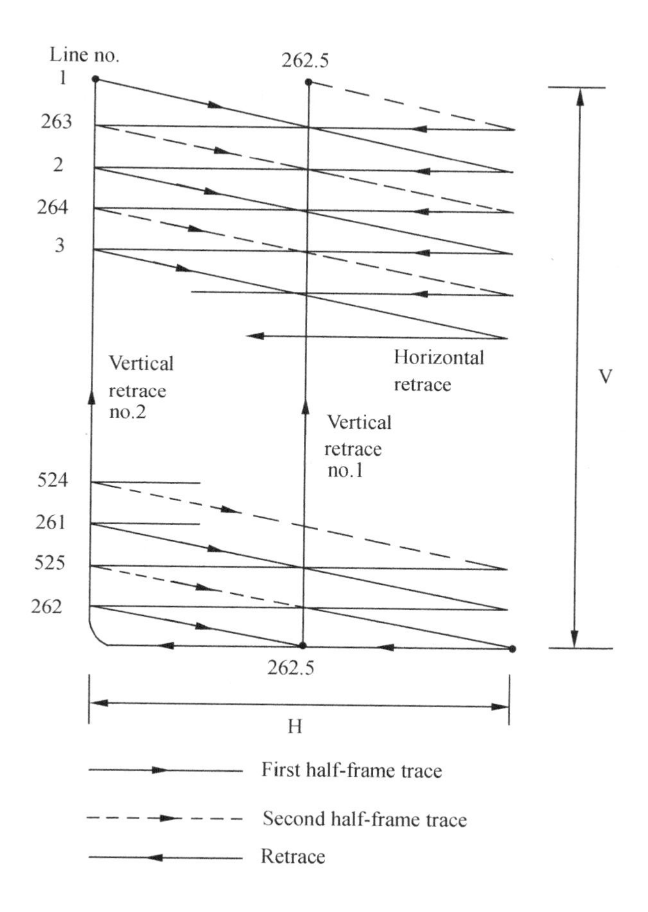

Figure 35 Television frame scan interlacing pattern

Two vertical retraces take place during each complete picture frame period, and for each of these a synchronizing pulse (the vertical sync pulse) is produced in the master camera control unit, which is used to initiate the vertical retracing of both the camera and the receiver scanning beams, so that the two remain in synchronization with each other. The vertical sync frequency, equal to the field frequency, is

$$f_V = 2\text{ P pulses/s},$$

where P is the frame, or picture, repetition rate in pictures/s. The vertical Sync pulse is superimposed on top of a longer "blanking" pulse, which is used to turn off the electron beam of the receiver CRT during the retrace period so that the retrace lines cannot be seen.

3. The Block Diagram of a TV Receiver

Figure 36 shows the block diagram of a typical black-and-white television receiver. Starting from the antenna, which is usually of the "rabbit-ear" type, but can be multielement Yagi or a cable-system (CATV) transmission line, the first section is the tuner assembly. The input signal is coupled into the RF-amplifier stage, which is a tuned class A stage, and then to a mixer-oscillator circuit. For UHF channels, a diode mixer is used with a separate oscillator circuit operating in the frequency-doubling mode. The second harmonic is used to provide the local oscillator signal to the mixer. Tuning is accomplished by means of a turret switch, which places a different set of coils and capacitors in the circuit for each channel. Means are provided for adjusting the oscillator frequency for each channel internally, as well as a vernier trimmer for fine tuning, which can be controlled from the front panel.

The IF used for all receivers and all channels has been standardized to lie in the band 41 to 47MHz, and since the sidebands are flipped over in conversion (for oscillator frequency above signal frequency) the video carrier appears at 45.75MHz, and the sound carrier at 41.25MHz. The bandpass of the video amplifier strip is designed to pass all of these frequencies, with some sloping because of the vestigial sideband. The bandpass characteristics and gain are realized in the IF amplifier string, which can be three to five stages of stagger-tuned amplifiers. AGC is applied to two or more of the IF stages and to the RF-tuner stage. Video detection is accomplished by simple envelope detection at the end of the IF string.

The detected video signal is fed to the input of the audio string, where tuned circuits isolate the audio carrier, which is now at 4.5MHz. The audio IF amplifier stages are designed to pass and limit a band of frequencies about 200 kHz wide around the 4. 5 MHz center frequency. The audio detection circuit is usually a ratio detector, but may be any of the FM detector circuits mentioned previously. The detector is followed by the audio amplifier stages which feed the speaker.

The detected video signal is also passed to the video amplifier, which raises the level of the video signal, removes the sound carrier, and provides the dc bias level for the picture tube cathode. The video signal applied to the picture tube cathode varies the beam intensity, and thus the light intensity produced at the screen.

The detected video signal is also passed to the synchronization circuits, where the video portion is removed by clipping to leave only the sync pulses. The clipped sync pulses are amplified and passed on to trigger the horizontal sweep oscillator. The ramp voltage produced by the oscillator is amplified to drive the horizontal output transformer and the horizontal deflection coil. The large flyback pulse is rectified to provide the high voltage (10 to 20kV) for the picture

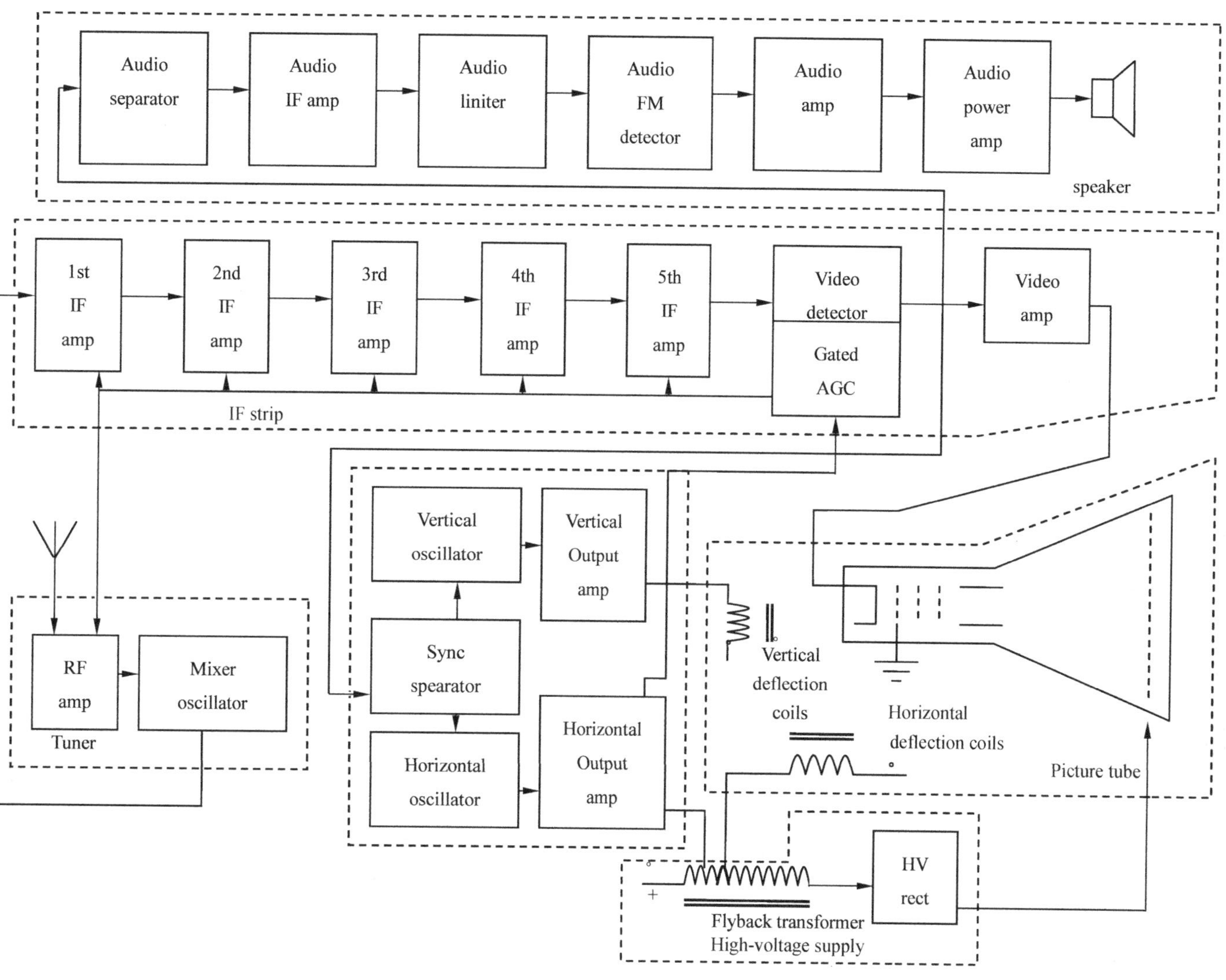

Figure 36 Black-and-white television receiver block diagram

tube target also provides the gating signal to control the AGC circuit.

New Words and Expressions

anode *n.* 阳极
arrangement *n.* 装置，设备，安装，布置
blanking pulse 消隐脉冲
bombard *vt.* 轰击
brightness *n.* 亮度
camera tube 摄像管
cathode-ray tube 阴极射线管
CATV *abbr.* community antenna TV 共用天线电视
clip *vt.* 限幅，消波
coat *n.* &*vt.* 涂，涂层
compatible *adj.* 兼容的
deflect *vt.*&*vi.*（使）偏斜
fade *vt.*&*vi.* 逐渐消失
field *n.* 场
film *n.* 薄膜
fine tuning 细调
flicker *vt.* 闪烁
flip *vt.*&*vi.* 翻转，倒转
flyback *n.* 回扫，回描，逆程扫描
frame *n.* 帧
front panel 前面板
gating signal 选通信号
grid *n.* 栅极
harmonic *n.*& *adj.* 谐波，和声的
incident *adj.* 入射的
intensity *n.* 强度
interlace *v.* 隔行，交错
jerky *adj.* 跳动的，不平稳的
multiple *n.*& *adj.* 倍数，多倍的，多数的
override *vt.* 超越，取代，不顾
penetrate *vt.*& *vi.* 穿透
persistence *n.* 持续，留存
phosphor *n.* 黄磷

picture tube　显像管
repel　*vt.* 排斥
retrace　*vt.* 回扫（描）
scanning rate　扫描速率
spot　*n.* 光点
submultiple　*n.* 约（因）数
superimpose　*v.* 叠加
sweep　*vt.& vi.* 扫描
sync pulse　同步脉冲
traverse　*vt.& vi.* 经过，横过
trigger　*n.* 触发器
tuner assembly　调谐器组件
vernier trimmer　微调电容器
vestigial　*adj.* 残余的

Exercises to the Text

1. Translate the following words and phrases into English.

（1）消隐脉冲　（2）摄像管　（3）阴极射线管　（4）前面板　（5）选通信号　（6）显像管　（7）同步脉冲　（8）调谐器组件　（9）微调电容器

2. Translate the following paragraphs into Chinese.

(1) When the scanning beam reaches the bottom right-hand corner of the picture, it must be returned to the top left-hand corner again, to repeat the whole scanning procedure. Thus, a second synchronizing signal must be transmitted at this time to ensure that the receiver recommences complete scanning at the same time as the transmitter. This is the field synchronizing signal, called the vertical sync pulse.

(2) The detected video signal is also passed to the video amplifier, which raises the level of the video signal, removes the sound carrier, and provides the dc bias level for the picture tube cathode. The video signal applied to the picture tube cathode varies the beam intensity, and thus the light intensity produced at the screen.

(3) The detected video signal is also passed to the synchronization circuits, where the video portion is removed by clipping to leave only the sync pulses. The clipped sync pulses are amplified and passed on to trigger the horizontal sweep oscillator. The ramp voltage produced by the oscillator is amplified to drive the horizontal output transformer and the horizontal deflection coil. The large flyback pulse is rectified to provide the high voltage (10 to 20kV) for the picture tube target also provides the gating signal to control the AGC circuit.

Text 1: About Computers

All computers, from the first room-sized mainframes to today's powerful desktop, laptop and even hand-held PCs, perform the same general operations on information. What changes over time is the information handled, how it is handled, and how quickly and efficiently it can be done.

1. Functions a Computer

A computer is an electronic device that can automatically conduct accurate and fast data manipulation under the control of stored program instructions. It accepts, stores, and processes data and produces output results through output devices like screens and printers.

A computer has four functions: input, process, storage and output.

Input: The input hardware allows you to enter data into the computer. The main input devices used are the keyboard, mouse and other devices such as scanner, as shown in Figure 37, etc.

Figure 37 Keyboard

Processing: The Central Processing Unit (CPU) is the "brain" of your computer. It contains the electronic circuits that cause the computer to follow instructions from ROM (Read-Only Memory) or from a program in RAM (Random Access Memory). By following these instructions information is processed.

Storage: Random Access Memory (RAM) is a short-term memory. It is volatile memory because it is automatically "erased" when the power is turned off or interrupted. The RAM memory is located inside the computer case on the motherboard. A motherboard holds RAM memory, electronic circuits and other computer parts including the central processing unit. Read-Only-Memory (ROM) is not volatile, meaning the memory is still there when power is interrupted or turned off. When the computer is turned back on again, ROM memory is still in storage on the internal hard disk.

Output: Output devices such as monitor or printer shown in Figure 38 and Figure 39 make information you input available for you to view or use.

Figure 38 Monitor

Figure 39 Printer

Computer technology is the combination of electronic technology and calculating technology. Now, it has been developed into a new stage that features the merging of computer and communication, leading to the wonderful Internet world. The computer nowadays possesses rather powers of logical judgment, automatic control and memory capacity. As a result, it can, to some extent, take the place of labors at some occupational posts.

2. The Development of Computers

The first large-scale electronic computer was Electrical Numerical Integrator And Calculator (ENIAC), which became operational in 1946 (See Figure 40). It was originally developed during Word War II to perform complex trajectory calculations, but the war ended before the machine was operational. After the war, it continued to be used, performing calculations for the design of the hydrogen bomb, weather prediction, cosmic-ray analysis, thermal ignition, random numbers, and wind-tunnel design. The ENIAC contained 17,468 vacuum tubes, along with 70,000 resistors, 10,000 capacitors, 1,500 relays, 6,000 manual switches and 5 million soldered joints. It covered 1,800 square feet of floor space, weighed 30 tons, and consumed 160,000 Watts of electrical power, making the lights in Philadelphia go dim each time it was powered up.

First Generation (1940—1956): Vacuum Tubes—Vacuum tubes were used for circuitry and magnetic drums for memory. These computers were somewhat unreliable because the vacuum tubes failed frequently. They were very expensive to operate and in addition to using a great deal of electricity, they generated a lot of heat, which often caused malfunctions. First generation computers relied on machine language to perform operations, and they could only solve one problem at a time. ENIAC, used by the U.S. Bureau of the Census from 1951 to 1962, is an example of first-generation computers.

Figure 40 Electrical Numerical Integrator and Calculator

Second Generation (1956—1963): Transistors—Transistors replaced vacuum tubes and ushered in the second generation of computers. The transistor was invented in 1947, but did not see widespread use in computers until the late 1950s. The first computers of this generation were developed for the atomic energy industry. The transistor was far superior to the vacuum tube, allowing computers become smaller, faster, cheaper, more energy-efficient and more reliable than their first-generation predecessors. Though the transistor still generated a great deal of heat that subjected the computer to damage, it was a vast improvement over the vacuum tube. Second-generation computers moved from cryptic binary machine language to assembly languages, which allowed programmers to specify instructions in words. High-level programming languages were also being developed at this time, such as early versions of COBOL and FORTRAN.

Third Generation (1964—1971): Integrated Circuits—The development of the Integrated Circuit (IC) was the hallmark of the third generation of computers. Transistors were miniaturized and placed on silicon chips, called semiconductors, which drastically increased the speed and efficiency of computers. Computers for the first time became accessible to a mass audience because they were smaller and cheaper than their predecessors.

Fourth Generation (1971—Present): Microprocessors—The microprocessor brought the fourth generation of computers, as thousands of integrated circuits were built onto a single silicon chip. Large-Scale Integrated (LSI) and Very-Large-Scale Integrated (VLSI) circuits were developed that contained hundreds to millions of transistors on one tiny chip. What, in the first generation, filled an entire room could now fit in the palm of the hand. The Intel 4004 chip (See

Figure 41), developed in 1971, located all the components of the computer from the central processing unit and memory to input/output controls on a single chip. In 1981 IBM introduced its first computer for the home user, and in 1984 Apple introduced the Macintosh. Microprocessors moved into many areas of life as more and more everyday products began to use microprocessors. As these small computers became more powerful, they could be linked together to form networks, which eventually led to the development of the Internet. Fourth generation computers also saw the development of Graphic User Interfaces (GUIs), the mouse and handheld devices. In fourth generation, computers' main memory capacity increased and cost decreased, which directly affected the types and usefulness of software that could be used. Software applications like word processing, electronic spreadsheets, database management programs, painting and drawing programs, desktop publishing and so forth become commercially available.

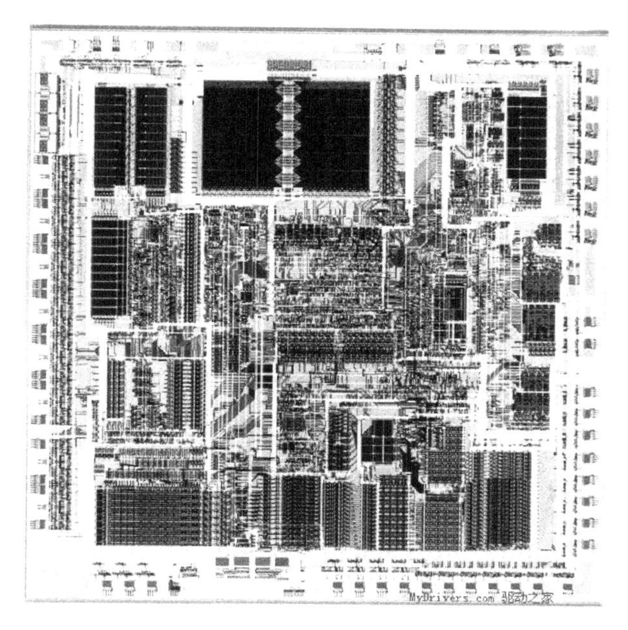

Figure 41 Intel 4004 Chip

Fifth Generation (Present and Beyond): Artificial Intelligence—Fifth generation computing devices, based on artificial intelligence, are still in development, though there are some applications, such as voice recognition, that are being used today. The use of parallel processing and superconductors is helping to make artificial intelligence a reality. Quantum computation and molecular and nanotechnology will radically change the face of computers in years to come. The goal of fifth-generation computing is to develop devices that respond to natural language input and are capable of learning and self-organization.

As science and technique are developing continually, new generations of computer will emerge in the future.

New Words and Expressions

mainframe *n.* 主（计算）机，大型机
hand-held *adj.* 手持式的
automatically *adv.* 自动地，机械地
accurate *adj.* 准确的，精确的
peripheral *adj.* 外围的，外面的
floppy *adj.* 易掉落的，松软的
extract *vt.* 提取
storage *n.* 储存，保管，存储器
central processing unit（CPU） 中央处理器，中央处理装置
erase *vt.* 擦掉，抹去
merge *vt.* 使合并，使结合
trajectory *n.*（射体）轨道，弹道，流轨
hydrogen *n.* [化学]氢，氢气
vacuum tube 电子管
drum *n.* 鼓，鼓状物，鼓膜 *v.* 打鼓
somewhat *adv.* 有点儿，微微
malfunction *vi.* 失灵，发生故障
census *n.* 人口普查，统计 *v.* 实施统计调查
transistor *n.* 晶体管（收音机）
integrated circuits 集成电路
molecular *adj.* 分子的，摩尔的

Exercises to the Text

1. Translate the following words and phrases into English.

（1）台式电脑 （2）膝上型电脑 （3）手提电脑 （4）中央处理器 （5）外围设备 （6）存储设备 （7）随机存储器 （8）电子管 （9）二进制机器语言 （10）高级程序设计语言 （11）集成电路 （12）硅片 （13）操作系统 （14）图形用户界面 （15）人工智能

2. Translate the following paragraphs into Chinese.

(1) A computer is an electronic device that can automatically conduct accurate and fast data manipulation under the control of stored program instructions. It accepts, stores, and processes data and produces output results through output devices like screen and printers.

(2) The Central Processing Unit (CPU) is the “brain” of your computer. It contains the electronic circuits that cause the computer to follow instructions from ROM (Read-Only Memory) or from a program in RAM (Random Access Memory). By following these instructions information is processed.

(3) The first large-scale electronic computer was Electrical Numerical Integrator And Calculator (ENIAC), which became operational in 1946. It was originally developed during

Word War II to perform complex trajectory calculations, but the war ended before the machine was operational.

(4) Transistors replaced vacuum tubes and ushered in the second generation of computers. The transistor was invented in 1947, but did not see widespread use in computers until the late 1950s. The first computers of this generation were developed for the atomic energy industry.

(5) Fifth generation computing devices, based on artificial intelligence, are still in development, though there are some applications, such as voice recognition, that are being used today.

Text 2: Computer Hardware

Hardware refers to the physical equipment that can perform the basic functions contained within the data processing cycle. The hardware may consist of the computer itself plus many auxiliary hardware devices including the keyboard, mouse, monitor, system unit, and other devices. Hardware is controlled by software.

Types of Computer System

This part covers the topic of different types of computer system in use today.

1. Supercomputer

First developed in the 1970s, supercomputers are the fastest amid highest-capacity computers. They are very expensive and their cost ranges from several hundreds of thousands to millions of dollars. Supercomputers are not likely to be visited by general users. They may occupy special air-conditioned rooms and are often used for research. Among their uses are worldwide weather forecasting and analysis of weather phenomena, oil exploration, aircraft design, evaluation of aging nuclear weapons systems, and mathematical research. Unlike microcomputers, which generally have only one central processing unit, Supercomputers have hundreds to thousands of processors and can perform trillions of calculations per second. They are the giants of the computer world, as shown in Figure 42.

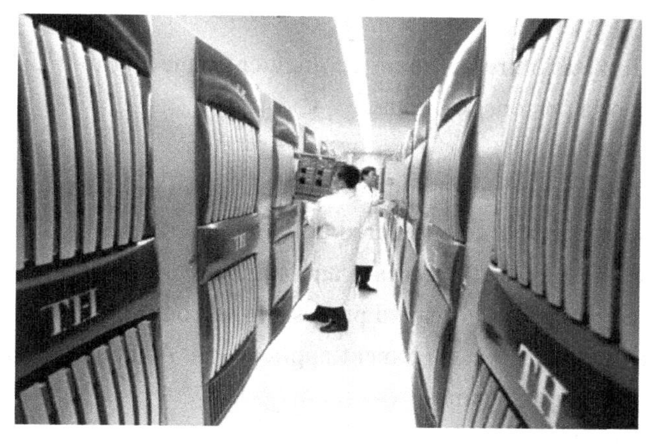

Figure 42 Supercomputer

Though for the common users, a microcomputer is enough, some workplace uses a combination of computers. For example, in a bank there may be a mainframe computer for processing and storage of the complicated data, while the other microcomputers perform some specialized tasks. Furthermore, even microcomputers today are not separated, which are connected to communicate with each other by the wonderful network.

2. Mainframe

This is a very large and expensive computer capable of supporting hundreds, or even thousands of users at a time. In some ways, mainframes are more powerful than supercomputers because they support much more programs at a time. But supercomputers can execute a single program faster than a mainframe.

The main difference between a supercomputer and a mainframe is that a supercomputer puts all its power into executing a few programs as fast as possible, whereas a mainframe uses its power to execute many programs at the same time. And mainframe computers are larger, faster, and more expensive than minicomputers. They are found in banks, insurance companies, airlines, large corporations, and government organizations.

3. Server

A computer or device on a network manages network resources. For example, a file server is a computer and storage device dedicated to storing files. Any user on the network can store files on the server. A print server is a computer that manages one or more printers, and a network server is a computer that manages network traffic. A database server is a computer system that processes database queries.

4. Workstation

a type of computer uses for engineering applications, desktop publishing, software development, and other types of applications that require a common amount of computing power and relatively high quality graphics capabilities.

Like personal computers, most workstations are single-user computers. However, workstations are typically linked together to form a Local Area Network, although they can also be used as stand-alone systems.

In networking, workstation refers to any computer connected to a Local Area Network. It could be a workstation or a personal computer.

5. Personal Computer

A small, relatively inexpensive computer designed for an individual user. In price, personal computer ranges anywhere from a few hundred dollars to thousands of dollars. All are based on the microprocessor technology that enables manufacturers to put an entire CPU on one chip. Businesses use personal computers for word processing, accounting, desktop publishing, and for running spreadsheet and database management applications. At home, the most popular use for personal computers is for playing games.

6. Embedded-Special Computers

A specialized computer system is part of a larger system or machine. Often an embedded system is housed on a single microprocessor board with the programs stored in ROM. Some embedded systems include an operating system, but many are so specialized that the entire logic can be made as a single program.

Components of a Computer

A computer consists of a variety of hardware components that work together with software to perform calculations, organize data and communicate with other computers. These components falls into five basic categories: system units, input/output devices, storage devices, and communications devices.

System unit, sometimes called a chassis, is a box-like case made from metal or plastic that protects the internal electronic components of the computer from damage. The circuitry in the system unit usually is part of or is connected to a circuit board called the motherboard. Two main components on the motherboard are the central processing unit and memory (primary memory or main memory).

1. Central Processing Unit

Central processing unit, sometimes called the processor, is the electronic device that interprets and carries out the basic instructions that operate the computer. In a microcomputer system, the CPU is the heart of the computer. Now the Pentium chip or processor, made by Intel, is the most common CPU. Other example is the CPU made by Motorola which is used in Apple computers.

Information from an input device or from the computer's memory is communicated via the bus to the central processing unit (CPU), which is the part of computer that translates commands and runs programs. The CPU is a microprocessor chip—that is, a single piece of silicon containing millions of tiny, microscopically wired electrical components. Information is stored in a CPU memory location called a register. Registers can be thought of as the CPU's tiny scratchpad, temporarily storing instructions or data. When a program is run called the program counter keeps track of which program instruction comes next by maintaining the memory location of the next program instruction to be executed. The CPU's control unit coordinates and times the CPU's functions, and it uses the program counter to locate and retrieve the next instruction from memory.

In a typical sequence, the CPU locates the next instruction in the appropriate memory device. The instruction then travels along the bus from the computer's memory to the CPU, where it is stored in a special instruction register. The current instruction is analyzed by a decoder, which determines what the instruction will do. Any data the instruction needs are retrieved via the bus and placed in the CPU's registers. The CPU executes the instruction, and the results are stored in another register or copied to specific memory locations via a bus. This entire sequence of steps is called an instruction cycle. Frequently, several instructions may be in process simultaneously, each at a different stage in its instruction cycle. This is called pipeline processing.

CPU has two important parts: the Control Unit and the Arithmetic/Logic Unit. The tow components are connected by a kind of electronic road called a bus. (A bus also connects these components with other parts of the computer.)

2. Memory

Memory is a temporary holding place for data and instructions. There are two basic references to memory in computer, logical and physical. Logical memory is patterns in the way memory is accessed and stored, which could be pictured as a flow chart, while physical memory is the actual hardware. System memory is used by the operating system as its main workspace or desktop. Read Only Memory (ROM) provides the instructions, which the computer uses each time it boots. Cache is used by the CPU for very short term quick storage, like a shelf above a desk. Random Access Memory (RAM) is the main location for most operations performed by the computer as directed by the CPU, and user input.

3. Input Devices

Input devices are the peripherals that allow users to enter data, programs, commands, and user responses into a computer and transform them into a suitable form for processing. The input devices may vary depending on the user's particular application and requirements. Some input devices are: keyboard, mouse, input pen, touch screen, microphone, scanner, digital camera, and bar code reader. In the following we will introduce some input devices in more detail.

1) Keyboards

Computer keyboards are one of the most basic system components in your computer. They are used to type information into the computer. They were originally designed to imitate electric typewrites so that typists could learn to use computers more easily. The demands of computer technology, however, have led to keys and keyboard layouts never before seen on a typewriter. While a typewriter has only about 50 keys, a computer keyboard may have 100 or more. And the standard keyboard has 101 keys. Some special ones are designed to make typing easier.

A computer's keys are generally divided into four clusters: alphanumeric keys, function keys, cursor keys and the numeric keypad. Alphanumeric keys include letters, numbers, and punctuation marks. They are arranged much like the keys on a typewriter. Function keys are labeled Fl, F2, and so on up to F12 or F15. They can be used for giving common commands such as "Print" or "Quit program". The precise purpose of any function key varies from one program to another. Cursor keys are used to move the cursor around on the screen. The cursor is the little blinking symbol that indicates where things will happen next on the screen. When you are typing, the cursor always blinks just to the right of the last character you typed. Cursor keys include the arrow keys for moving up, down, left, and right, as well as the PgDn, PgUp, Home, and End keys. The numeric keypad includes the mathematical keys found on a standard calculator.

2) Mouse

Most modern computers today are run using a mouse controlled pointer. Generally if the mouse has two buttons, the left one is used to select objects and text and the right one is used to access menus. A mouse with a third buttons can be used by special software programs. The mouse usually has a cable that is connected to the computer's system unit. On the bottom side of the mouse is a ball that translates the mouse movement into digital signals. Some brands of mouse, such as the Microsoft mouse, have a wheel between the left and right mouse buttons with which one can scroll through the contents of a file, a mouse shown in Figure 43.

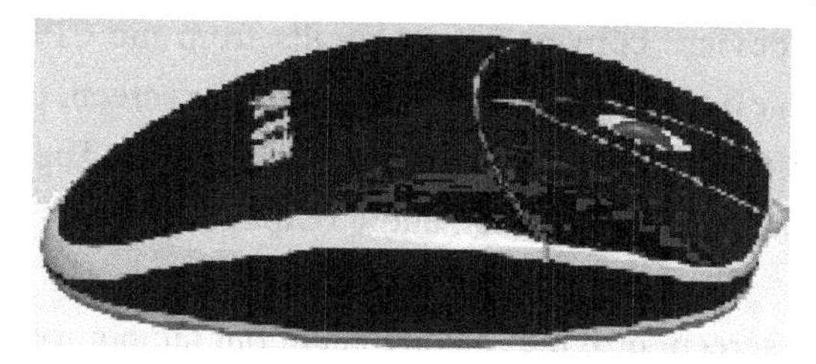

Figure 43 Mouse

Mice are popular because it is easier to point than type, and because the arrow keys don't work very well for drawing pictures or moving things on the screen. A mouse consists of a ball mounted under a plastic housing with one or more buttons on top. As you move the mouse around the tabletop, sensors inside register the rolling of the ball and move the cursor around the screen to match. There are three basic ways of giving commands with the mouse. First, you can click the button to identify something— perhaps to indicate which part of a drawing you want to change. Second, you can drag the mouse; that is, you can hold the button down while you move the mouse. Dragging can be used to move a drawing across the screen. The third way to give a command is to double-click the mouse's button by pressing it twice within about a half second. Double-clicking is used to select things on the screen.

The simplest pointing tool of all, of course, is the finger. In fact, touch screens are widely used in department store advertising displays, information kiosks, lottery game machines, and other places where users are not expected to have much familiarity with computers. On these machines you simply touch the part of the display screen you want to select, just as you might push a vending machine's button. Depending on the sensing method used by the touch screen, your finger might interrupt a network of infrared rays protected across the screen's surface. This would tell the computer where you pointed.

3) Scanner

Scanners use laser light and reflected light to translate hardcopy images of text, drawings, photos, and the like into digital form. The images can then be processed by a computer, displayed on a monitor, stored on a storage device, or communicated to another computer. Internet and Web users scan photos into their computer systems to send to online friends or to post on Web pages.

Output, the result produced by the central processing unit, is a computer's whole reason for being. Output is usable information; that is, raw input data that has been processed by the computer into information. The most common forms of output are text, graphics, audio, and video. Text output, for example, may consist of characters that create words, sentences, and paragraphs. A graphic may be a digital representation of non-text information such as drawings, charts, and photographs.

4. **Output Devices**

Output devices are instruments of interpretation and communication between humans and computer systems. These devices convert output results from the CPU into a form that can be used by people. The popular output devices may be a monitor screen, printer, headsets, speakers, data projectors, facsimile machines, and multifunction devices. The following pages discuss three kinds of output devices: monitor, printer, audio output device.

1) Monitor

A monitor, or display screen, provides a convenient but temporary way to view information. The earliest computer monitors were simply converted television sets. While ordinary TVs are still used for some video games, most computer programs today demand higher-quality monitors.

For a monitor, two important characteristics are size and clarity. A monitor's size is measured diagonally and is still quoted in inches. Popular sizes are 15 inches (38 cm) and 17 inches (43 cm). Larger monitors make working at a computer easier on the eyes and are essential for use in DeskTop Publishing (DTP) and Computer Aided Design (CAD) work. A monitor's clarity is indicated by its resolution, which is measured in pixels. A pixel (short for picture clement) is the smallest unit on the screen that can be turned on and off or made different shades. The greater the resolution (the more pixels) the more the monitors require a higher resolution.

The monitor shows information on the screen when you type. This is called outputting information. When the computer needs more information it will display a message on the screen, usually through a dialog box. There are two kinds of computer monitor: CRT monitor shown in Figure 44 and LCD monitor shown in Figure 38.

Figure 44　CRT Monitor

2) Printer

The printer takes the information on your screen and transfers it to paper or a hard copy. There are many different types of printers with various levels of quality. They are categorized in three types: ink-jet printer, laser printer, and thermal printer.

Different kinds of computer printers use surprisingly different technologies. Some printers squirt ink, some apply heat to sensitive paper, others hammer inked ribbons, and still others create images with lasers. Laser printers work by reflecting a laser beam from a rapidly rotating octagonal mirror onto a light-sensitive roller. Ink-jet printers work by squirting tiny droplets of liquid ink at the paper. Dot-matrix primer features a movable print head containing a row of tiny pins. The pins push an inked ribbon against the paper, producing a matrix (or pattern) of dots. As the print head moves back and forth across the page, the dots can form either letters or graphics.

3) Audio Output Device

Audio output devices are a component of the computer that are used to play music, vocalize translations from one language to another, and communicate information from the computer system to user. The most widely used audio output devices are speakers and headsets (See Figure 45 and Figure 46). These devices are connected to a sound card in the system unit. The sound card is used to capture and play back recorded sounds. Most speakers have tone and volume controls to adjust settings. When using speakers, anyone within listening distance can hear the output. With the headset, only the user who is using it can hear the output.

Figure 45　Speaker

Figure 46　Headset

Storage devices, also called secondary storage devices, refer to backing up or keeping digital data in a secure place for future use. It provides additional storage separate from memory. Storage devices differ from memory, in that they can hold these items permanently. Storage devices are needed for large volumes of data and also for data that must be retained after the computer is turned off. Types of storage media include magnetic disks, hard disks, optical disks, magnetic tapes, Zip disks and PC cards.

Communication devices are used to facilitate the connections between computers and between groups of connected computers called networks. Such connections allow the sharing of resources, including hardware, software and data. The communication component of a computer system vastly extends the computer's range and utility. A modem is a communication device that enables computers to communicate via telephone lines or cable. Modems are available as both internal and external devices.

New Words and Expressions

Forecast　*vt.* 预测；预报（天气）　*n.* 预测；预报

exploration　*n.* 搜寻，考察，探究，探索，考查，调查

microprocessor　*n.* 微处理器

Computer Aided Design (CAD)　计算机辅助设计

microcomputer *n.* 微型计算机
workstation *n.* 工作区，工作站
minicomputer *n.* 小型计算机；小型电脑
supercomputer *n.* 超级计算机，巨型计算机
additional *adj.* 增加的，添加的；附加的，追加的，外加的，额外的
cache *n.* [计算机]高速缓冲内存，高速缓冲存储器
primary memory 主存储器
boot *n.* 引导；自引；启动；引导程序
secondary storage 辅助存储器，二级存储器

Exercises to the Text

1. Translate the following words and phrases into English.

（1）超级计算机 （2）存储容量 （3）大型机 （4）文件服务器 （5）打印机服务器 （6）数据库服务器 （7）嵌入式专用计算机 （8）喷墨打印机 （9）激光打印机 （10）热敏打印机

2. Translate the following paragraphs into Chinese.

(1) First developed in the 1970s, supercomputers are the fastest amid highest-capacity computers. They are very expensive and their cost ranges from several hundreds of thousands to millions of dollars. Supercomputers are not likely to be visited by general users.

(2) The main difference between a supercomputer and a mainframe is that a supercomputer puts all its power into executing a few programs as fast as possible, whereas a mainframe uses its power to execute many programs at the same time. And mainframe computers are larger, faster, and more expensive than minicomputers. They are found in banks, insurance companies, airlines, large corporations, and government organizations.

(3) Memory is a temporary holding place for data and instructions. There are two basic references to memory in computer, logical and physical. Logical memory is patterns in the way memory is accessed and stored, which could be pictured as a flow chart, while physical memory is the actual hardware.

Text 3: Computer Software

Computer software, also called a computer program or a program, is a series of instructions that tells the hardware of a computer what to do and how to do it. Computer software is so called in contrast to computer hardware, which encompasses the physical interconnections and devices required to store and execute (or run) the software.

In computers, software is loaded into RAM and executed in the central processing unit. At the lowest level, software consists of a machine language specific to an individual processor. A machine language consists of groups of binary values signifying processor instructions (object code), which change the state of the computer from its preceding state. Software is an ordered sequence of instructions for changing the state of the computer hardware in a particular sequence. It is usually written in high-level programming languages that are easier and more efficient for humans to use (closer to natural language) than machine language. High-level languages are compiled or interpreted into machine language object code. Software may also be written in an assembly language, essentially, a mnemonic representation of a machine language using a natural language alphabet. Assembly language must be assembled into object code via an assembler.

The term "software" was first used in this sense by John W. Tukey in 1958. In computer science and software engineering, computer software is all computer programs. The concept of reading different sequences of instructions into the memory of a device to control computations was invented by Charles Babbage as a part of his difference engine. There are two basic types of software: system software and application software.

System software includes the operating system which couples the computer's hardware with the application software. The purpose of the operating system is to provide an environment in which application software executes in a convenient and efficient manner. And the operating system's tasks fall into six broad categories:

- Processor management—Breaking the tasks down into manageable chunks and prioritizing them before sending to the CPU.
- Memory management—Coordinating the flow of data in and out of RAM and determining when virtual memory is necessary.
- Device management—Providing an interface between each device connected to the computer, the CPU and applications.
- Storage management—Directing where data will be stored permanently on hard drives and other forms of storage.
- Application Interface—Providing a standard communications and data exchange between software programs and the computer.
- User Interface—Providing a way for you to communicate and interact with the computer.

In addition to the operating system, system software includes the utility program, the device driver and the language translator.

If a computer program is not system software then it is application software. Application software includes middleware, which couples the system software with the user interface. Application software also includes utility programs that help users solve application problems, like the need for sorting.

System Software

System software is a subclass of software. It consists of a collection of programs that control the operations of a computer and its devices. It serves as the interface between the user and the computer hardware. System software is not a single program. It is a collection or a system of programs that handle hundreds of technical details with little or no user intervention.

1. Operating Systems

Operating Systems (OS) are series of programs that control and run the operation of the computer hardware and provide an interface through which the user communicates with the computer.

OS stands for operating system. It is the main software that coordinates computer resources, provides an interface between users and the computer, and runs applications in a safe, efficient and abstract way. For example, an OS ensures safe access to a printer by allowing only one application program to send data directly to the printer at any one time. An OS encourages efficient use of the CPU by suspending programs that are waiting for I/O operations to complete to make way for programs that can use the CPU more productively. An OS also provides convenient abstractions (such as files rather than disk locations), which isolate application programmers and users from the details of the underlying hardware.

Notice that:

- The operating system kernel is in direct control of the underlying hardware. The kernel provides low-level device, memory and processor management functions (e.g. dealing with interruptions from hardware devices, sharing the processor among multiple programs, allocating memory for programs etc.).
- Basic hardware-independent kernel services are exposed to higher-level programs through a library of system calls (e.g. services to create a file, begin execution of a program, or open a logical network connection to another computer).
- Application programs (such as word processors, spreadsheets) and system utility programs (simple but useful application programs that come with the operating system, e.g. programs which find text inside a group of files) make use of system calls. Applications and system utilities are launched using a shell (a textual command line interface) or a graphical user interface that provides direct user interaction.

Operating systems can be distinguished from one another by the system calls, system utilities and user interface they provide, as well as by the resource scheduling policies implemented by the kernel.

2. Utilities Programs

Utilities programs, also known as service programs, perform specific tasks related to managing computer resources. Typical Utilities include such programs as shells, text editors, compilers, and (sometimes) the file system.

A Shell is a program, which serves as the primary interface between the user and the operating system. The shell is a "command interpreter", i.e. it prompts the user to enter commands for tasks which the user wants done, reads and interprets what the user enters, and directs the OS to perform the requested task. Such commands may call for the execution of another utility such as a text editor or compiler or a user program or application, the manipulation of the file system, or some system operation such as logging in or out. There are many variations of shells available from relatively simple command line interpreters such as

DOS or more powerful command line interpreters such as the Bourne Shell, sh, or C Shell, csh in the Unix environment to more complex, but easy to use graphical user interfaces such as Macintosh or Windows. Users should become familiar with the particular shell(s) available on the computer they are using, as it will be the primary means of access to the facilities of the machine.

A text editor is a program for entering programs and data and storing them in the computer. This information is organized as a unit called a file, similar to a file in an office filing cabinet, and stored on the disk. There are many text editors available, such as vi and emacs on Unix or Linux systems. Users should familiarize themselves with those available on their system.

In today's computing environment, most programming is done in High-Level Languages (HLL) such as C. However, the computer hardware cannot understand these languages directly. Instead, the CPU executes programs coded in a lower level language called the machine language. A utility called a compiler is a program, which translates the HLL program into a form understandable to the hardware.

Finally, another important utility, or task of the operating system is to manage the file system for users. A file system is a collection of files in which a user keeps programs, data, text material, graphical images, etc. The file system provides a means for the user to organize files, giving them names and gathering them into directories or folders and to manage their file storage. Typical operations, which may be done with files, include creating new files, destroying, renaming, and copying files.

Most operating systems also include several utility programs that perform specific tasks related to managing a computer, its devices, or its programs. The following are commonly used utility programs:

- File management programs make it easier to manage your files. In DOS's prime, it didn't take much to improve on the text-only type-it-all-yourself methods that DOS provided. Many programs were written to help the user find files, create and organize directories, copy, move, and rename files. The newer graphical interfaces that come with operating systems like Windows 95 have reduced the need for alternate file management programs.
- A disk defragmenter is a utility program that locates and eliminates unnecessary fragments and rearranges files and unused disk space to optimize operations.
- Memory management software handles the locations where programs put their current data within RAM. They move certain memory-resident items out of the way. This can effectively increase the memory available by gathering all the unused pieces together in one spot, thereby creating a useable amount.
- A Backup program restores selected files or an entire hard drive onto another storage device. It is a secure method of saving any data that may otherwise be lost accidentally. The software will compress the data to take up the least amount of space.
- Data recovery programs are used to recover deleted or damaged (corrupted) files.
- Data compression programs squeeze out the slack space generated by the formatting

schemes.

- Anti-virus programs are other must-have programs. They monitor the computer for the presence of viruses, which are nasty programs that copy themselves to spread to other computers. Viruses can be extremely destructive to your files.
- A Screen saver is a utility that causes the monitor's screen to display a moving image on a blank screen if no keyboard activity occurs for a specified period.

3. Device Drivers

Device drivers are specialized programs designed to allow particular input or output devices to communicate with the rest of the computer system. It is a specially written program, which understands the operation or the device it interfaces with, such as a printer, video card, sound card modem, adapter or CD ROM drive. Rather than access the device directly, a device driver translates commands from the operating system or user into commands understood by the device it interfaces with.

Whenever a new device is added to a computer system, a new device driver must be installed before the device can be used. Windows supplies hundreds of different device drivers with its system software.

4. Language Translators

Language translators convert the programming instructions into a language that computers understand and process.

A language translator is software that translates a program written by a programmer in a language such as C or C++ into machine language, which the computer can understand. All system software and applications software must be translated into machine language for execution by the computer. There are three types of language translators: assembler, compiler and interpreter.

- An assembler is a program that translates the assembly language source program into machine language.
- A compiler is software that looks at an entire high-level program before translating it into machine language. The programming instructions of a high-level language are called the source code. The compiler translates the source code into machine language, which in this case is called the object code. With compilers, the object code can be saved and run later. This means that the object code doesn't have to be recompiled. A compiler is likely to perform many or all of the following operations: lexical analysis, preprocessing, semantic analysis, code generation, and code optimization.
- Interpreter is software that converts high-level language statements into machine language one at a time, in succession. Unlike compilers, which must look at an entire program before converting it into machine language, interpreters provide the programmer with immediate feedback regarding the accuracy of their coded instructions.

Application Software

Application software is another subclass of software. It consists of programs that perform specific tasks for users. Application software can be used as a productivity/business tool; to assist with graphics and multimedia projects; to support home, personal, and educational activities; and to facilitate communications. Specific application software products, called software packages, are available from software vendors. Although application software is also available as shareware, freeware, and public-domain software, these usually have fewer capabilities than retail software packages. In general, software is divided into a number of categories.

- Business software allows users to perform tasks related to running a business, such as paying accounts, keeping track of products, sending money and writing reports and letters. Examples of business software are Microsoft Works and Lotus Notes.
- Education software is designed to teach or educate users. These include encyclopedias, reference books and instructional programs. Examples of education software are Encyclopedia Britannica and Microsoft Magic School Bus.
- Entertainment/Games software is designed for users to have fun with. Its purpose is to keep users entertained. This includes games software. Example of entertainment software is Microsoft Age of Empires.
- Utility software is designed for users to perform routine tasks associated with the storage and manipulation of the user's information. This includes software such as schedulers, clocks, media players and communication tools. Examples of utility software are McAfee Virus Scan and Arcada backup software.

Although software applications differ in their use of specific commands and functions, most of them have some common features. A user interface is the portion of the application with which the user works. The user interface controls how you enter data or instructions and how information is displayed on the computer screen. Many of today's software programs have a GUI. A graphical user interface combines text, graphics, and other visual images to make software easier to use. Now, most applications have an interface that displays information in windows. A window is simply a rectangular area that can contain a document, program, or message. More than one window can be opened and displayed on the computer screen at one time.

The majority of software packages have menus to present commands. Usually, menus are displayed in a menu bare at the top of the screen. When one of the menu items is selected, a pull-down menu appears. This is a list of options or commands associated with the selected menu. One of the commands on the menu bar is Help. It provides assess to a variety of Help features. These features include a table of contents, a keyword index and a search feature to help you locate the information you need. Below the menu bar, there are toolbars. Toolbars contain buttons and menus that provide quick access to commonly used commands. The standard toolbar and the formatting toolbar are common to most applications.

Most users, whether at home or in business, are drawn to task-oriented software, sometimes called productivity software, which can help them to work faster and make their lives easier. The collective set of business tasks is limited, and the number of general paths towards performing these tasks is limited, too. Thus, the tasks and the software solutions fall, for the most part, into just a few categories, which can be found in most business environments. These major categories are word processing (including desktop publishing), spreadsheets, database management, graphics, and communications. A brief description of each category follows:

1. Word Processing/Desktop Publishing

The most widely used personal computer software is word processing software. This software allows users to create, edit, format, store, and print text and graphics in one document. In this definition, the three words edit, format, and store, reveal difference between word processing and plain typing. Since memo or document can be stored on disk, it can be retrieved, changed, reprinted or manipulated at another time. Therefore, word processing can be a great timesaver. Unchanged parts of the stored document do not need to be retyped, and the whole revised document can he reprinted as if new. Popular word processing softwares are Word, WPS and so on. They have some common features.

- Word wrap and the Enter key: One basic word processing feature is word wrap. When you finish a line, a word processor decides for you and automatically moves the insertion point to the next lines. To begin a new paragraph or leave a blank line, you press the Enter key.
- Search and replace: A search or find command allows you to locate any character, word, or phrase in your document. When you search, the insertion point moves to the first place the item appears. If you want, the program will continue to search for all other locations where the item appears. The replace command automatically replaces the word you search for with another word. The search and replace commands are useful for finding and fixing errors.
- Cut, copy, and paste: With a word processor, you select the portion of text to be moved by highlighting it. Using either the menu or button bar, choose the command to cut the selected text. The selected text disappears from your screen. Then move the insertion point to the new location and choose the paste command to reinsert the text into the document. In a similar manner, you can copy selected portions of text from one location to another.

As the number of features in word processing packages has grown, word processing has crossed the border into desktop publishing territory. Desktop publishing packages are usually better than word processing packages at meeting high-level publishing needs, especially when it comes to typesetting and color reproduction. Many magazines and newspapers today rely on desktop publishing software. Businesses use it to produce professional-looking newsletters, reports and brochures-both to improve internal communication and to make a better impression on the outside world.

2. Electronic Spreadsheets

Spreadsheets, made up of columns and rows, have been used as business tools for centuries.

A manual spreadsheet can be tedious to prepare and, when there are changes, a considerable amount of calculation may need to be redone. An electronic spreadsheet is still a spreadsheet, but the computer does the work. It is an electronic worksheet used to organize and manipulate numbers and display options for analysis. Electronic spreadsheets are used by financial analysts, accountants, contractors, and others concerned with manipulating numeric data. Electronic spreadsheets allow you to try out various "what-if " kinds of possibilities. That is a powerful feature. You can manipulate numbers by using stored formulas and calculate different outcomes.

An electronic spreadsheet has several parts. The worksheet area of the electronic spreadsheet has letters for column headings across the top. It also has numbers for row headings down the left side. The intersection of a column and row is called a cell. The cell holds a single unit of information. The position of a cell is called the cell address. For example, "A1" is the cell address of the first position on an electronic spreadsheet, the topmost and leftmost position. A cell pointer—also known as the cell selector—indicates where data is to be entered or changed in the electronic spreadsheet. The cell pointer can be moved around in much the same way that you move the insertion pointer in a word processing program. Excel is the most common spreadsheet software. It has some common features of spreadsheet programs.

- Format: Label is often used to identify information in a worksheet. It is usually a word or symbol. A number in cell is called a value. Labels and values can be displayed or formatted in different ways. A label can be centered in the cell or positioned to the left or right. A value can be displayed to show decimal places, dollars, or percent. The number of decimal positions can be altered, and the width of columns can be changed.
- Formulas: One of the benefits of the electronic spreadsheets is that you can manipulate data through the use of formulas. Formulas are instructions for calculations. They make connections between numbers in particular cells.
- Functions: Functions are built-in formulas that perform calculations automatically.
- Recalculation: Recalculation or what-if analysis is one of the most important features of the electronic spreadsheets. If you change one or more numbers in your spreadsheet, all related formulas will recalculate automatically. Thus you can substitute one value for another in the cells affected by your formula and recalculate the results. For more complex problems, recalculation enables you to store long, complicated formulas and many changing values and quickly produce alternatives.

3. Graphics

It might seem wasteful to show graphics to business people when standard computer printouts are readily available. However, graphics, maps and charts can help people compare data and spot trends more easily, and make decisions more quickly. In addition, visual information is usually more compelling than a page full of numbers.

Adobe's Photoshop and Microsoft's PowerPoint application software are used for graphics. They can be used in two ways: for doing original drawings, and for creating visual aids to support to an oral presentation.

4. Communications

From the viewpoint of an employee with a personal computer at home, communications means that the user can hook a phone up to their home computer and communicate with the computer at the office, or gain access to data stored in someone else's computer in another location. Microsoft's Internet Explorer application software is used for doing email, World Wide Web browsing, and participating in Internet discussion groups.

New Words and Expressions

Assembly *n.* 集合，集会；议会
permanently *adv.* 永久地；长期不变地
translator *n.* 翻译者；[无线电]译码机；变换器；传送器；转发器
kernel *n.*（果实的）核，仁；内核，核心，要点
variation *n.* 变量，变度，偏差
immediate *adj.* 直接的；最接近的；即时的，立即的
vendor *n.* 自动售货机
encyclopedia *n.* 百科全书；专科全书
graphical *adj.* 书写的，绘画的；印刷的，雕刻的
manipulate *v.* 操纵，利用；[计算机]操作
eliminate *v.* 除去，排除，剔除；[计算机]消除
Linux *n.* 一个自由操作系统，一种可免费使用的类 UNIX 操作系统
Variation *n.* 变化，变动，变更；变量；变度；变化率
Portion *n.* 一部分；一定数量；区；段

Exercises to the Text

1. Translate the following words and phrases into English.

（1）计算机软件 （2）调制解调器 （3）系统软件 （4）应用软件 （5）用户界面 （6）实用程序 （7）设备驱动程序 （8）源代码 （9）教学软件 （10）菜单栏 （11）数据库管理 （12）文字处理软件 （13）高级排版 （14）金融分析师 （15）万维网

2. Translate the following paragraphs into Chinese.

(1) A Shell is a program, which serves as the primary interface between the user and the operating system. The shell is a "command interpreter", i.e. it prompts the user to enter commands for tasks which the user wants done, reads and interprets what the user enters, and directs the OS to perform the requested task.

(2) A text editor is a program for entering programs and data and storing them in the computer. This information is organized as a unit called a file, similar to a file in an office filing cabinet, and stored on the disk.

(3) In today's computing environment, most programming is done in High-Level Languages (HLL) such as C. However, the computer hardware cannot understand these languages directly.

Instead, the CPU executes programs coded in a lower level language called the machine language. A utility called a compiler is a program, which translates the HLL program into a form understandable to the hardware.

(4) Whenever a new device is added to a computer system, a new device driver must be installed before the device can be used. Windows supplies hundreds of different device drivers with its system software.

(5) An electronic spreadsheet has several parts. The worksheet area of the electronic spreadsheet has letters for column headings across the top. It also has numbers for row headings down the left side. The intersection of a column and row is called a cell. The cell holds a single unit of information. The position of a cell is called the cell address.

Text 4: Operating System

The most important program on any computer is the operating system or OS. The OS is a large program made up of many smaller programs. It controls how the CPU communicates with other hardware components. It also makes computers easier to operate by people who don't understand programming languages. In other words, operating systems make computer users friendly. It is the basic software that controls a computer.

Operating systems are normally unique to their manufacturers and the hardware in which they are run. Generally, when a new computer system is installed, operational software suitable to that hardware is purchased. Users want reliable operational software that can effectively support their processing activities. Though operational software varies with manufacturers, it has similar characteristics. Modern hardware, because of its sophistication, requires that operating systems meet certain specific standards. For example, considering the present state of the field, an operating system must support some form of online processing.

Functions of Operating Systems

Regardless of the size of computer, most operating systems provide similar functions. The following parts discuss functions common to operating systems.

1. Process Management and Coordination

The CPU executes a large number of programs. While its main concern is the execution of user programs, the CPU is also needed for other system activities. These activities are called processes. A process is a program in execution. Typically, a batch job is a process. A time-shared user program is a process. A system task, such as spooling, is also a process.

In general, a process will need certain resources such as the CPU, memory, files, I/O devices, etc. to accomplish its task. These resources are given to the process when it is created. In addition to the various physical and logical resources that a process obtains when it is created, some initialization data (input) may be passed along. For example, a process whose function is to display on the screen of a terminal the status of a file, such as Fl, will receive as input the name of the file Fl and execute the appropriate program to obtain the desired information.

Emphasized that a program by itself is not a process; a program is a passive entity, while a process is an active entity. Two processes may be associated with the same program; they are nevertheless considered two separate execution sequences. A process is the unit of work in a system. Such a system consists of a collection of processes, some of which are operating system processes that execute system code, and the rest being user processes that execute user code. All of those processes can potentially execute concurrently.

The operating system is responsible for the following activities in connection with processes managed:

- The creation and deletion of both user and system processes.
- The suspension and resumption of processes.
- The provision of mechanisms for process synchronization.
- The provision of mechanisms for deadlock handling.

2. Memory Management

Memory is central to the operation of a modern computer system. It is a large array of words or bytes, each with its own address. The CPU fetches from and stores in memory. In order for a program to be executed, it must be mapped to absolute addresses and loaded into memory. Interaction is achieved through a sequence of reads or writes of specific memory addresses.

In order to improve both the utilization of the CPU and the speed of the computer's response to its users, several processes must be kept in memory. There are many different algorithms depending on the particular situation. Selection of a memory management scheme for a specific system depends upon many factors, but especially upon the hardware design of the system. Each algorithm requires its own hardware support.

The operating system is responsible for the following activities in connection with memory management:

- Keeping track of which parts of memory are currently being used and by whom.
- Deciding which processes are to be loaded into memory when memory space becomes available.
- Allocating and deallocating memory space as needed.

3. Secondary Storage Management

The main purpose of a computer system is to execute programs. These programs, together with the data they access, must be in main memory during execution. Since the main memory is too small to permanently accommodate all data and programs, the computer system must provide secondary storage to backup main memory. Most modem computer systems use disks as the primary on-line storage of information, of both programs and data. Most programs, like compilers, assemblers, sort routines, editors, formatters, and so on, are scored on the disk until loaded into memory, and then use the disk as both the source and destination of their processing. Hence the proper management of disk storage is of central importance to a computer system.

The operating system is responsible for the following activities in connection with disk management:

- Free space management.
- Storage allocation.
- Disk scheduling.

4. File System Management

File management is one of the most visible services of an operating system. Computers can store information in several different physical forms; magnetic tape, disk, and drum are the most common. Each of these devices has its own characteristics and physical organization.

For convenient use of the computer system, the operating system provides a uniform logical view of information storage. The operating system abstracts from the physical properties of its storage devices to a logical storage unit, the file. In general a file is a sequence of bits, bytes, lines or records whose meaning is defined by its creator and user. Files are mapped by the operating system onto physical devices. Commonly, files represent programs (both source and object forms) and data. Data files may be numeric, alphabetic or alphanumeric. Files may be free-form, such as text files, or may be rigidly formatted.

The operating system implements the abstract concept of the file by managing mass storage devices, such as types and disks. Also, files are normally organized into directories for ease of access. Finally, when multiple users have access to files, it may be desirable to control by whom and in what ways files may be accessed.

The operating system is responsible for the following activities in connection with file management:

- The creation and deletion of files.
- The creation and deletion of directories.
- The support of manipulating files and directories.
- The mapping of files onto disk storage.
- Backup of files on stable (non-volatile) storage.

5. Input and Output Management

One of the purposes of an operating system is to hide the peculiarities of specific hardware devices from the user. For example, in Unix, the peculiarities of I/O devices are hidden from the bulk of the operating system itself by the I/O system. The I/O system consists of:

- A buffer caching system.
- A general device driver code.
- Drivers for specific hardware devices.

Only the device driver knows the peculiarities of a specific device.

6. Coordinating Network Communications

It used to be that network communications functions were handled by a separate Network Operating System. Now, however, some communications functions are included in the regular OS, to meet personal network connection needs. Business network administration still requires a separate Network Operating System, such as Netware or Windows NT.

7. Protection and Security

The various processes in an operating system must be protected from each other's activities. For that purpose, various mechanisms can be used to ensure that files, memory segments, CPU and other resources can be operated on only by those processes that have gained proper authorization from the operating system. For example, memory addressing hardware ensures that a process can only be executed within its own address space. The timer ensures that no process can gain control of the CPU without relinquishing it.

8. Providing User Interface

Users interact with software through its user interface. A user interface controls how you enter data and instructions and how information is displayed on the screen. There are two types of user interfaces, called command line and graphical. Many operating systems use a combination of these two interfaces to define how users interact with their computers.

Types of Operating Systems

Within the board family of operating systems, there are generally four types, categorized on the types of computers they control and the sort of applications they support.

1. Real-time operating system

Real-time operating systems are used to control machinery, scientific instruments and industrial systems. A very important part of a real-time operating system is managing the resources of the: computer so that a particular operation executes in precisely the same amount of time every time it occurs.

2. Single-user, single-task operating system

As the name implies, this operating system is designed for one user to effectively do one thing at a time, The Palm OS for Palm handheld computers is a good example of a modem single-user, single-task operating system.

3. Single-user, multi-tasking operating system

This is the type of operating system most people use on their desktop and laptop computers today. Microsoft's Windows and Apple's Mac OS platforms are both examples of single-user, multi-tasking operating systems that will let a single user for a Windows to edit a text in a word processor while downloading a file from the Internet while printing the text of an E-mail message.

4. Multi-user operating system

A multi-user operating system allows many different users to take advantage of one computer's resources simultaneously. The operating system must make sure that the requirements of the various users are balanced and that each of the programs they are using has sufficient and separate resources so that a problem with one user doesn't affect the entire community of users. UNIX, VMS and mainframe operating system, such as MVS, are examples of multi-user operating systems.

How Operating Systems Work

The operating system is the system software that manages and controls the activities of the computer. It supervises the operation of the CPU, controls input, output, and storage activities; and provides various support services. It can be visualized as the chief manager of the computer system. The operating system determines which computer resources will be used for solving which problems and the order in which they will be used.

Operating systems control different computer processes, such as running a spreadsheet program or accessing information from the computer's memory. One important process is interpreting commands, enabling the user to communicate with the computer. Some command interpreters are text-oriented, requiring commands to be typed in or to be selected via function keys on a keyboard. Other command interpreters use graphics and let the user communicate by pointing and clicking on an icon, an on-screen picture that represents a specific command. Beginners generally find graphically oriented interpreters easier to use, but many experienced computer users prefer text-oriented command interpreters because they are more powerful.

Operating systems can use a technique known as virtual memory to run processes that require more main memory than is actually available. To implement this technique, space on the hard drive is used to mimic the extra memory needed. Accessing the hard drive is more time-consuming than accessing main memory, however, so performance of the computer slows.

When a computer is turned on, it searches for instructions in its memory. These instructions tell the computer how to start up. Usually, one of the first sets of these instructions is a special program called the operating system, which is the software that makes the computer work. It prompts the user (or other machines) for input and commands, reports the results of these commands and other operations, stores and manages data, and controls the sequence of the software and hardware actions. When the user requests that a program run, the operating system loads the program in the computer's memory and runs the program. Popular operating systems, such as Microsoft Windows and the Macintosh system (Mac OS), have graphical user interfaces (GUIs)—that use tiny pictures, or icons, to represent various files and commands. To access these files or commands, the user clicks the mouse on the icon or presses a combination of keys on the keyboard. Some operating systems allow the user to carry out these tasks via voice, touch, or other input methods.

The operating system's small bootstrap program is stored in ROM and supplies the instructions needed to load the operating system's core into memory when the system boots. This core part of the operating system, called the kernel, provides the most essential operating system services, such as memory management and file access. The kernel stays in memory all the time your computer is on. Other parts of the operating system, such as customization utilities, are loaded into memory as they are need.

At first operating systems were designed to help applications interact with the computer hardware. While this is still the case, the importance of the operating system has grown to the point where the operating system defines the machine. Most users engaged in the Mac, PC and Unix battle are arguing about the operating systems on these machines, not the hardware platform itself. The operating system provides a layer of abstraction between the user and the bare machine. Users and applications do not see the hardware directly, but view it through the operating system. This abstraction can be used to hide certain hardware details from users and applications. Thus, the users do not see changes in the hardware.

Common Operating Systems

There are many types of operating systems. They include, from most recent to oldest:

Windows XP is the most popular operating system from Microsoft. The release of XP means that all the desktop versions are now built on the Windows NT/2000 codebase. For anyone who runs Windows 3.1, 95, 98 or ME, it is strongly recommended. As an upgrade from Windows 2000/NT 4, it is an attractive rather than compelling option, as XP is built on the NT/2000 codebase.

Windows CE is for small devices like palmtop and handheld computers. Late versions of a number of major applications are available to run on these devices. You can link your small computer to a regular one to synchronize documents and data.

Windows 2000 is an upgrade of Windows NT, which works well both at home and as a workstation at a business. Windows 2000 includes technologies, which allow hardware to he automatically detected, along with other enhancements over Windows NT.

Windows ME is an upgraded version of Windows 98 but it has been historically plagued with programming errors which may be frustrating for home users.

Windows 98 is one of the operating systems from Microsoft. It was previously known by the code name Memphis and at one time was referred to as Windows 97. It is an advanced operating system designed for today's very powerful microcomputer. Windows 98 has some major advantages compared to Windows 95. These include high performance, Internet integration, ease of use, and audio and video capability.

Windows NT is an operating system designed to run on a wide range of powerful computers and microcomputers. It is a very sophisticated and powerful operating system. Developed by Microsoft, Windows NT is not considered a replacement for Windows. Rather, it is an advanced alternative designed for very powerful microcomputers and networks. Windows NT has two major advantages when compare to Windows:

① Multiprocessing—It is similar to multitasking except that the applications are run independently at the same time. For instance, you could be printing a word processing document and using a database management program at the same time. With multitasking, the speed at which the document is printed is affected by the demands of the database management program. With multiprocessing, the demands of the database management program do not affect the printing of the document.

② Networking—In many business environments, workers often use computer to communicate with one another and to share software using a network. This is made possible and controlled by special system software. Windows NT has network capabilities and security checks built into the operating system. This makes network installation and use relatively easy.

Windows 95 is the first version of Windows after the older Windows 3.x versions. It offers a better interface and better library functions for programs.

MS-DOS (Microsoft Disk Operating System) was very popular in the 1980s and early 1990s. IBM's version of DOS is called PC-DOS and Microsoft's version is called MS-DOS. The two OS are identical except for subtle differences.

Unix was originally developed by AT&T. It has been a popular OS for more than two decades because of its multi-user, multi-tasking environment, stability, portability and powerful networking capabilities. Unix continues to be popular on powerful workstations. Many computer old-timers love Unix and its command line interface. But all those commands are not easy to remember for newcomers. X-Windows is a graphical interface for Unix that some think is even easier to work with than Windows 98.

Linux is an operating system similar to Unix that is becoming more and more popular. It is an open-source program created by Linus Torvalds at the University of Finland, starting in 1991. Open source means that the underlying computer code is freely available to everyone. Programmers can work directly with the code and add features.

Mac OS is designed to run on Macintosh computers. It offers a high-quality graphical user interface and is very easy to use. While its market share is much less than that of Windows, it is a very powerful and easy-to-use operating system. Apple Macintosh System 7.5 designed for Apple computers using Motorola's PowerPC microprocessor, is a significant milestone for Apple. It is a very powerful operating system like Windows NT and OS/2. System 7.5 has network capabilities and can read Windows and OS/2 files. It has several advantages:

① Ease of use. The graphical user interface has made the Macintosh popular with many new comers to computing. This is because it is easy to learn.

② Quality graphics. Macintosh has established a high standard for graphics processing. This is a principal reason why the Macintosh is popular for desktop publishing. Users are easily able to merge pictorial and text materials to produce nearly professional-looking newsletters, advertisements, and the like.

③ Consistent interfaces. Macintosh applications have a consistent graphics interface. Across all applications, the user is provided with similar screen displays, menus, and operations.

④ Multitasking. Like Windows, Windows NT, and OS/2, the Macintosh System enables you to do multitasking. That is, several programs can run at the same time.

⑤ Communications between programs. The Macintosh system allows applications programs to share data and commands with other applications programs.

OS/2 stands for Operating System/2, was originally developed jointly by IBM and Microsoft. Like Windows 98 and Window NT, OS/2 Wrap is designed for very powerful microcomputers and has several advance features.

① Minimum system configuration: Like Windows NT, OS/2 requires significant memory and hard disk space. However, OS/2 requires slightly less.

② Windows application: Like Windows NT, OS/2 does not have a large number of application programs written especially for it. OS/2 can also run Windows programs, but it runs these programs slightly faster than Windows NT.

③ Common user interface: Microcomputer application programs written specifically for Windows NT, as well as for OS/2, have consistent graphics interfaces. Across applications, the user is provided with similar screen displays, menus, and operations. Additionally, OS/2 offers a consistent interface with mainframe computers, minicomputers, and microcomputers.

Windows Vista is the latest generation of Microsoft's operating system. It provides several enhancements over its direct predecessor, Windows XP, Service Pack 2. As soon as you start up your computer and log into Windows, you will notice a difference. A new Start menu, new desktop backgrounds, and even the new Sidebar that docks on the right side of your screen will tell you that you are experiencing something new. Then go ahead and click the new Start menu button and navigate the menus to launch a program. No longer does your screen fill with multiple layers of menus that reach to the edge of your screen and back. Instead, each time you drill down in a menu, it overlays the previous menu to make it easier to find the program you want.

New Words and Expressions

drill *n.* 操练，训练，演习 *vt.* 练习
laptop *n.* 膝上计算机
mainframe computer 大型计算机
desktop search 桌面搜索
communicate *vt. &vi.* 传递，传播
component *adj.* 组成的，构成（全物）的 *n.* 部分，成分
manufacturer *n.*[u] 制造，生产 [pl] 机器制造物，制造品，产品
sophistication *n.* 高深，奥妙，复杂
manipulate *vt.* 熟练地使用；操纵 控制 *n.* 操纵，操作，利用
concurrently *adv.* 一致地；同时发生地，并存地
synchronization *n.* 同步
utilization *n.* 利用，被利用
peculiarity *n.* 古怪，怪癖；特性，特色
authorization *n.* 授权，批准
simultaneously *adv.* 同时地
configuration *n.* 某物的构造，结构，布局，形状，外观
automatically *adv.* 自动地

Exercises to the Text

1. Translate the following words and phrases into English.

（1）实时操作系统 （2）联机处理 （3）实用程序 （4）进程管理与协调 （5）辅助存储器管理 （6）磁盘调度 （7）逻辑存储单元 （8）用户界面 （9）单任务操作系统（10）多用户操作系统 （11）命令解释器 （12）功能键 （13）虚拟内存 （14）多道处理

2. Translate the following paragraphs into Chinese.

(1) The operating system is the system software that manages and controls the activities of the computer. It supervises the operation of the CPU, controls input, output, and storage

activities; and provides various support services. It can be visualized as the chief manager of the computer system.

(2) When a computer is turned on, it searches for instructions in its memory. These instructions tell the computer how to start up. Usually, one of the first sets of these instructions is a special program called the operating system, which is the software that makes the computer work.

(3) The CPU executes a large number of programs. While its main concern is the execution of user programs, the CPU is also needed for other system activities. These activities are called processes. A process is a program in execution. Typically, a batch job is a process. A time-shared user program is a process. A system task, such as spooling, is also a process.

(4) In order to improve both the utilization of the CPU and the speed of the computer's response to its users, several processes must be kept in memory. There are many different algorithms depending on the particular situation. Selection of a memory management scheme for a specific system depends upon many factors, but especially upon the hardware design of the system.

(5) File management is one of the most visible services of an operating system. Computers can store information in several different physical forms; magnetic tape, disk, and drum are the most common. Each of these devices has its own characteristics and physical organization.

Text 5: Creating a LAMP server with the Raspberry Pi

To configure a Raspberry Pi as a webserver is similar to the guide to using Xubuntu as a LAMP webserver, but adds some of the things that need to be handled differently for the Raspberry Pi. A LAMP server is one of the most common configuration for webservers which standard for:

- **L**inux—operating system.
- **A**pache—webserver (http) software.
- **M**ysql—database server.
- **P**HP or **Perl**—programming languages.

All this configuration is done at the command line. This may not be quite as easy as clicking a few icons, but it has many advantages, including the ability to remotely manage and install the server. It also means that the computer can spend more of it's time server up web pages and less processor time drawing a GUI, which is after all the whole point of a webserver.

1. Why use the Pi?

I think there are some good reasons for doing this.

Learn Web programming The aim of the Raspberry Pi Foundation is to teach programming to children. Learning to program web based applications can be a useful skill to learn. It's debatable whether it's better to learn to program a desktop application before web programming or vice versa, but it's certainly something that is a useful skill to learn.

As an interface The Raspberry Pi is useful as a device for collecting information from various sensors. A web server can be a good way of accessing that information.

Dedicated network device You could use it as a dedicated network service for the home.

As a test or development server When creating a web application it is useful to have a dedicated server to test the application on. Ideally this should be identical hardware and software to the production, but if that is not possible then the Raspberry Pi could be an inexpensive alternative.

2. Making the Raspberry Pi secure

The first priority is to make the Raspberry Pi a little more secure. The image includes a default username and password, which once connected to the Internet would allow anyone to login and have free roam of the device.

To change the password for the pi user after logging in issue "passwd" and follow the prompts for changing the password.

3. Performance tuning the operating system

Performance tuning is something that you would normally leave until later, but in the case of the Raspberry Pi there is an single option that can be done to improve performance for servers. By configuring it here we can let it get picked up by the reboot later saving us from having to reboot the server.

The Raspberry Pi has 256MB (or 512MB for later versions) of RAM. This RAM is however shared between the graphics and main system memory. By default 64MB is allocated to graphics.

4. Setting up networking

The next step is to give the Raspberry Pi an static IP address. By default the Raspberry Pi will request a dynamic IP address. This however may change in future which would make it hard to connect to the webserver. Instead we provide it with an address that doesn't change. And this address can be used on the local network, but not on the Internet.

5. Enabling SSH

SSH (Secure Shell) is a network protocol that allows you to login and control the computer through the command line remotely. As the name suggests it is secure as it encrypts communication across the network (so that others cannot see your password etc).

6. Making the server available on the Internet

Next we are going to configure the router to allow ssh logins and web traffics through its firewall to the Raspberry Pi.

You did remember to change the default password for the pi username didn't you! If you haven't already changed the default password then do it now otherwise anyone will be able to login to your Raspberry Pi.

As a home user the ip address used on your local network is a private address range that will not work over the Internet. Instead your ISP will provide a single dynamic IP address which is used by the router. To allow traffic to flow from the Internet to your Raspberry Pi needs the IP address of the Pi to be made to look as though it is from the router. This is a process called

Network Address Translation (NAT).

The ports that need to be allowed through are port 80 (http) and if you would like to be able to login to the computer from the Internet then port 22 (ssh).

The final stage is to have a DNS entry point at your router's IP address.

7. Install apache webserver

The Apache webserver is available to download from the Debian repositories. This can be done through the apt tools.

First have you refreshed the software repositories? If not run sudo apt-get update to make sure that it knows about any new packages / versions available.

Apache is installed by entering the following

```
sudo apt-get install apache2
```

8. Install MySQL

The MySQL database server is also available through the Debian repositories and installed as

```
sudo apt-get install mysql-server
```

During the install there is a prompt request for a password. The password is for the mysql root user.

9. Install PHP

Perl is installed as part of the operating system so I will just be adding PHP.

10. Setup complete

Once the setup is complete you can access the web page by pointing your browser to the router IP address or DNS entry.

You should get a page back stating that it works, but that there is no content loaded.

To test that the webserver and PHP are working correctly then delete the file /var/www/html/index.html * and create a file /var/www/html/index.php with the contents of this page.

New Words and Expressions

configuration *n.* 配置；布局，构造

command *n.* 指令；控制力

processor *n.* 加工；数据处理机；（计算机的）中央处理器

vice versa *adv.* 反之亦然；反过来也一样

interface *n.* 接口；交界面

dedicated *adj.* 专用的

alternative *adj.* 替代的；另类的；备选的；其他的；*n.* 可供选择的事物

roam *vt.* 漫游；漫步 *n.* 漫步，漫游

performance *n.* 性能

dynamic *adj.* 动态的；不断变化的，充满变数的

protocol *n.*（数据传递的）协议

encrypt *v.* 加密，将……译成密码

traffic　*n.* 交通，运输量；通信量
repository　*n.* 仓库；贮藏室
entry　*n.* 进入，入场；入口处，门口

Exercises to the Text

1. Translate the following words and phrases into English.

（1）Web 服务器　（2）命令行　（3）操作系统　（4）数据库服务器　（5）编程语言　（6）图形用户界面　（7）桌面应用程序　（8）性能调优　（9）网络地址转换

2. Translate the following paragraphs into Chinese.

(1) All this configuration is done at the command line. This may not be quite as easy as clicking a few icons, but it has many advantages, including the ability to remotely manage and install the server. It also means that the computer can spend more of it's time server up web pages and less processor time drawing a GUI, which is after all the whole point of a webserver.

(2) The aim of the Raspberry Pi Foundation is to teach programming to children. Learning to program web based applications can be a useful skill to learn. It's debatable whether it's better to learn to program a desktop application before web programming or vice versa, but it's certainly something that is a useful skill to learn.

(3) The first priority is to make the Raspberry Pi a little more secure. The image includes a default username and password, which once connected to the Internet would allow anyone to login and have free roam of the device.

Part 5 Programming Languages

Text 1: About Programming Languages

Just as there are many different languages in use throughout the world for humans to communicate with, there are different languages that are used to communicate with a computer. Programming languages, in computer science, are the artificial languages used to write a sequence of instructions (a computer program) that can be run by a computer. They are standardized communication technique for expressing instructions to a computer. Similar to natural languages, such as English, programming languages have a vocabulary, grammar, and syntax. However, natural languages are not suited for programming computers because they are ambiguous, meaning that their vocabulary and grammatical structure may be interpreted in multiple ways. The languages used to program computers must have simple logical structures, and the rules for their grammar, spelling, and punctuation must be precise. They enable a programmer to precisely specify what data a computer will act upon, how the data will be stored or transmitted and precisely what actions to take under various circumstances.

1. The Development of Programming Languages

Programming languages have been under development for years and will remain so for many years to come. Perhaps the languages of tomorrow will be more natural with the invention of quantum and biological computers. The following are detailed descriptions for five generation languages.

2. First Generation: Machine Languages

It has been shown that a program instruction comprises a particular combination of binary digits. Different makes and types of computers use different "codes" of binary digits to represent instructions. Such codes are referred to as machine or instruction codes. At this level, a computer has a basic repertoire of instructions that it can perform, known as the instruction set. Typically this instruction set includes:

- Basic arithmetic operations,
- Comparators of various kinds (e.g. equality operators and so on),
- Facilities to deal with sequences of characters,
- Input/output operators.

In the early days of computer programming, all programs had to be written in machine code. For example, a short (3 instruction) program might look like this:

0111 0001 0000 1111

1001 1101 1011 0001

1110 0001 0011 1110

An instruction in machine language generally tells the computer four things: (1) where to find one or two numbers or simple pieces of data in the main computer memory (Random Access Memory, or RAM), (2) a simple operation to perform, such as adding the two numbers together, (3) where in the main memory to put the result of this simple operation and (4) where to find the next instruction to perform. While all executable programs are eventually read by the computer in machine language, they are not all programmed in machine language. It is extremely difficult to program directly in machine language because the instructions are sequences of 1s and 0s.

Machine code has several significant disadvantages associated with it:

- It is not intuitively obvious what a machine code instruction does simply from its encoding. Consequently it is very difficult to read and write machine code.
- The writing of machine code is extremely time consuming and error prone.
- Many different machine codes exist (one for each make and type of computer).

These disadvantages all serve to severely limit the applications to which computers can be applied when using machine code.

3．Second Generation: Assembler Languages

Assembler languages were initially developed to address the disadvantages associated with machine code programming. They used symbolic codes instead of lists of binary instructions. Consequently programming became more “friendly”. An example of assembly code is given below:

```
MOV AX 01
MOV BX 02
ADD AX BX
```

In assembler language each line of the program corresponds to one instruction in machine code. For a program written in an assembler language to be executable, it must be translated into machine code using a translating program called an assembler.

Assembly languages share certain features with machine languages. For instance, it is possible to manipulate specific bits in both assembly and machine languages. Programmers use assembly languages when it is important to minimize the time it takes to run a program, because the translation from assembly language to machine language is relatively simple. Assembly languages are also used when some part of the computer has to be controlled directly, such as individual dots on a monitor or the flow of individual characters to a printer.

Although use of assembly languages offers some advantages, there are still a number of significant disadvantages associated with their use: (1) each model of computer has its own assembly language associated with it, (2) assembler programming still requires great attention to detail and hence remains both time consuming and tedious, and (3) the risk of program error is not significantly reduced.

Note that there are some computer applications, such as interfacing with peripherals, where an assembler language is still a necessity.

4. Third Generation: High Level Languages

From the early 1950s onwards high-level languages were developed with the express aim of providing the means whereby computer programs could be written more efficiently and in a less error prone manner. High-level languages are relatively sophisticated sets of statements utilizing words and syntax from human language. They are more similar to normal human language than assembly or machine languages and are therefore easier to use for writing complicated programs. These programming languages allow larger and more complicated programs to be developed faster. The advantages offered are as follows.

- Programs written in a high-level language are more adapted to human modes of expression than to the computer's set of instructions. Programs are expressed in "half-English", and arithmetic calculations are written in a way familiar to mathematics.
- Programmers can concentrate more closely on the problem to be solved rather than the mass of detail required for machine code or assembly language programming.
- High-level languages are not necessarily dedicated to a particular type or model of computer, a feature known as portability.

Ultimately, a computer can only operate on programs defined using machine code. Consequently, a program written in a high-level language such as Java cannot be run directly. To execute a computer program written in a high level language it must be either compiled or interpreted. For this reason, programs written in a high-level language may take longer to execute and use up more memory than programs written in an assembly language.

1) Compiler

A compiler is software that looks at an entire high-level program before translating it into machine language. The programming instructions of a high-level language are called the source code. The central task of a compiler is to translate source code written in a high level language into a machine form, which in this case is called the object code. The advantage is that the machine executable form runs much faster than if it were interpreted (see below). The disadvantage is that different machines and operating systems have different machine codes associated with them. Generally, compilers have two functions:

- Checking for syntactic errors of source code
- Translating source code into object code

Examples of high-level languages using compilers are C and C++.

2) Interpreter

In the case of interpretation, each line of the program is decoded and "interpreted" by a special program known as an interpreter. Different interpreters are required for different languages (and different machines). Interpretation occurs every time a line in a program is executed. This means that a line which occurs many times must be interpreted on each occasion. This wastes computer time, and causes programs to run relatively slowly.

However, the repeated examination of the source program by an interpreter allows interpretation to be more flexible than when using a compiler. Interpreters also provide a faster and easier way of testing small programs or fragments of programs. Further, in the context of error detection, because the interpreter works directly with the source code, errors can be reported accurately with reference to line numbers (this is not the case when programs are compiled). Another, more questionable advantage, is that parts of a program which are not executed need not be interpreted.

5. Fourth Generation: Very High Level Languages

Compared with third generation languages, 4GLs, for forth generation languages, are much more user-oriented and allow programmers to develop programs with fewer commands, although they require more computing power. 4GLs are called nonprocedural because a programmer and even users can write programs that only tell the computer what they want done, without specifying all the procedures for doing it. 4GLs consist of report generators, query languages, application generators, and interactive database management system languages.

- Report generator also called a report writer, is a program for end-users that are used to produce a report. Report generators were the precursor to today's query language.
- Query language is an easy-to-use language for retrieving data from a database management system.
- Application generator is a programmer's tool that generates applications programs from descriptions of the problem rather than by traditional programming. The benefit is that the programmer does not need to specify how the data should be processed.

4GLs may not entirely replace third generation languages because they are usually focused on specific tasks and therefore offer fewer options. Still, they improve productivity because programs are easy to write.

6. Fifth Generation: Artificial Intelligence

Fifth-generation language is artificial intelligence, making computers behave like humans. The term was coined in 1956 by John McCarthy at the Massachusetts Institute of Technology.

New Words and Expressions

ambiguous *adj.* 含糊不清的，模棱两可的
punctuation *n.* 标点；标点法；标点符号的使用
statement *n.* 语句
precursor *n.* 先驱者，前导，先进者
debug *vt.* 驱除（某处的）害虫；排除程序等中的错误
portable *adj.* 可移植的；便携的
quantum *n.* 量，额；定量，定额；份；总量
assembly language 汇编语言

Exercises to the Text

1. Translate the following words and phrases into English.

（1）程序设计语言 （2）计算机科学 （3）人工语言 （4）自然语言 （5）电子设备

（6）面向对象编程 （7）机器语言 （8）汇编语言 （9）高级语言 （10）目标代码

2. Translate the following paragraphs into Chinese.

(1) Similar to natural languages, such as English, programming languages have a vocabulary, grammar, and syntax. However, natural languages are not suited for programming computers because they are ambiguous, meaning that their vocabulary and grammatical structure may be interpreted in multiple ways.

(2) The languages used to program computers must have simple logical structures, and the rules for their grammar, spelling, and punctuation must be precise. They enable a programmer to precisely specify what data a computer will act upon, how the data will be stored or transmitted and precisely what actions to take under various circumstances.

(3) An instruction in machine language generally tells the computer four things: ①where to find one or tow numbers or simple pieces of data in the main computer memory (Random Access Memory, or RAM), ②a simple operation to perform, such as adding the two numbers together, ③where in the main memory to put the result of this simple operation and ④where to find the next instruction to perform.

(4) Assembly languages share certain features with machine languages. For instance, it is possible to manipulate specific bits in both assembly and machine languages. Programmers use assembly languages when it is important to minimize the time it takes to run a program, because the translation from assembly language to machine language is relatively simple. Assembly languages are also used when some part of the computer has to be controlled directly, such as individual dots on a monitor or the flow of individual characters to a printer.

(5) High-level languages are relatively sophisticated sets of statements utilizing words and syntax from human language. They are more similar to normal human language than assembly or machine languages and are therefore easier to use for writing complicated programs. These programming languages allow larger and more complicated programs to be developed faster.

Text 2: C

C was developed in the early 1970s, and it was developed by Dennis Ritchie as a systems programming language for UNIX. C has grown into a very popular language now. C might best be described as a “medium level language”. Like a true high level language, there is a one-to-many relationship between a C statement and the machine language instructions it is compiled into. However, unlike most high level languages, C let you easily do chores (such as bit and pointer manipulation) additionally performed by assembly languages. Therefore, C is an especially good tool to use for developing operating systems (such as the UNIX operating system), or other system software.

1. What is C

The C programming language is a popular and widely used programming language for creating computer programs. It is one of thousands of programming languages currently in use. C has been around for several decades and has won widespread acceptance because it gives programmers maximum control and efficiency. If you are programmer, or if you are interested in becoming a programmer, there are a couple of benefits you gain from learning C:

- You will be able to read and write code for a large number of platforms—everything from microcontrollers to the most advanced scientific systems can be written in C, and many modern operating systems are written in C.
- The jump to the object oriented C++ language becomes much easier. C++ is an extension of C, and it is nearly impossible to learn C++ without learning C first.

Lexically, C is more cryptic than other languages. In fact, C is an easy language to learn. It is a bit more cryptic in its style than some other languages, but you get beyond that fairly quickly. For example, brackets are often used to obviate the need for keywords. However, underlined characters are allowed in identifiers, which can make them more understandable.

There are a number of monadic and addict (called binary) operators. Some have unexpected precedence. Brackets may be ignored by the compiler, with occasionally surprising results. There are shift operations. Overflows on integer arithmetic may be ignored. There are some composite symbols with special meanings: for example "&&" means "and then" and "||" means "or else".

There are several integer types of different sizes and there are floating point numbers, * pointers (C talks of indirection), arrays and structures. C is not strongly typed: for example, some compilers do not insert run-time checks on array subscripts, etc. Type conversion is permissive. Address arithmetic can be performed on pointers; Null is demoted by a zero value.

C has procedures and functions. There are few features (apart from procedures and functions) to support modularization, however; separate (strictly independent) compilation is allowed.

C is what is called a compiled language. This means that once you write your C program, you must run it through a C compiler to turn your program into an executable that the computer can run (execute). The C program is the human-readable form, while the executable that comes out of the compiler is the machine-readable and executable form. What this means is that to write and run a C program, you must have access to a C compiler. If you are using a UNIX machine (for example, if you are writing CGI scripts in C on your host's UNIX computer, or if you are a student working on a lab's UNIX machine), the C compiler is available for free. It is called either "cc" or "gcc" and is available on the command line. If you are a student, then the school will likely provide you with a compiler—find out what the school is using and learn about it. If you are working at home on a Windows machine, you are going to need to download a free C compiler or purchase a commercial compiler. A widely used commercial compiler is Microsoft's Visual C++ environment (it compiles both C and C++ programs). Unfortunately, this program costs several hundred dollars. If you do not have hundreds of dollars to spend on a commercial compiler, then you can use one of the free compilers available on the Web.

We will start at the beginning with an extremely simple C program and build up from there. I will assume that you are using the UNIX command line and gcc as your environment for these examples; if you are not, all of the code will still work fine—you will simply need to understand

and use whatever compiler you have available.

2. The Simplest C Program

The best way to get started with C is to write, compile, and execute a simple program. Now, let's start with the simplest possible C program and use it both to understand the basics of C and the C compilation process. Type the following program into a standard text editor. Then save the program to a file named samp.c. If you leave off .c, you will probably get some sort of error when you compile it, so make sure you remember the .c. Also, make sure that your editor does not automatically append some extra characters (such as .txt) to the name of the file. Here's the first program:

```
#include<stdio.h>
int main()
{
  printf("This is output from my first program!\n");
  return 0;
}
```

When executed, this program instructs the computer to print out the line "This is output from my first program!"—then the program quits. You can't get much simpler than that!

Position: When you enter this program, position #include so that the pound sign is in column 1 (the far left side). Otherwise, the spacing and indentation can be any way you like it. The spacing and indentation shown above is a good example to follow.

To compile this code, take the following steps:

① On a UNIX machine, type gcc samp.c -o samp (if gcc does not work, try cc). This line invokes the C compiler called gcc, asks it to compile samp.c and asks it to place the executable file it creates under the name samp. To run the program, type samp.

② On a DOS or Windows machine using DJGPP, at an MS-DOS prompt type gcc samp.c -o samp.exe. This line invokes the C compiler called gcc, asks it to compile samp.c and asks it to place the executable file it creates under the name samp.exe. To run the program, type samp.

③ If you are working with some other compiler or development system, read and follow the directions for the compiler you are using to compile and execute the program.

You should see the output "This is output from my first program!" when you run the program. If you mistyped the program, neither will it compile, nor will it run. You should edit it again and see where you went wrong in your typing. Fix the error and try again.

Let's walk through this program and start to see what the different lines are doing.

- This C program starts with #include<stdio.h>. This line includes the "standard I/O library" into your program. The standard I/O library lets you read input from the keyboard(called "standard in"), write output to the screen (called "standard out"), process text files stored on the disk, and so on. It is an extremely useful library. C has a large number of standard libraries like stdio, including string, time and math libraries. A library is simply a package of code that someone else has written to make your life easier.

- The line int main() declares the main function. Every C program must have a function named main somewhere in the code. At run time, program execution starts at the first line of the main function.
- In C, the "{" and "}" symbols mark the beginning and end of a block of code. In this case, the block of code making up the main function contains two lines.
- The printf statement in C allows you to send output to standard out (for us, the screen). The portion in quotes is called the format string and describes how the data is to be formatted when printed. The format string can contain string literals, symbols for carriage returns (\n), and operators as placeholders for variables. If you are using UNIX, you can type man 3 printf to get complete documentation for the printf function. If not, see the documentation included with your compiler for details about the printf function.
- The return 0; line causes the function to return an error code of 0 (not error) to the shell that started execution.

3. Variables

As a programmer, you will frequently want your program to "remember" a value. For example, if your program requests a value from the user, or if it calculates a value, you will want to remember it somewhere so you can use it later. The way your program remembers things is by using variables. For example:

```
int b;
```

This line says, "I want to create a space called b that is able to hold one integer value." A variable has a name (in this case, b) and a type (in this case, int, an integer). You can store a value in b by saying something like:

```
b=5;
```

You can use the value in b by saying something like:

```
printf("%d",b);
```

In C, there are several standard types for variables:

- int—integer (whole number) values.
- float—floating point values.
- char—single character values (such as "m" or "Z").

4. Printf

The printf statement allows you to send output to standard out. For us, standard out is generally the screen. Here is another program that will help you learn more about printf:

```
#include<stdio.h>
int main()
{
   int a,b,c;
   a=5;
```

```
    b=7;
    c=a+b;
    printf("%d+%d=%d\n",a,b,c);
    return 0;
}
```

Type this program into a file and save it as add.c. Compile it with the line gcc add.c –o add and then run it by typing add (or ./add). You will see the line "5+7=12" as output.

Here is an explanation of the different lines in this program:

- The line int a,b,c; declares three integer variables named a,b and c.
- The next line initializes the variable named a to the value 5.
- The next line sets b to 7.
- The next line adds a and b and "assigns" the result to c. The computer adds the value in a (5) to the value in b (7) to form the result 12, and then places that new value (12) into the variable c. The variable c is assigned the value 12. For this reason, the = in this line is called "the assignment operator."
- The printf statement then prints the line "5+7=12." The %d placeholders in the printf statement act as placeholders for values. There are three %d placeholders, and at the end of the printf line there are the names for three variables: a,b and c. C matches up the first %d with a and substitutes 5 there. It matches the second %d with b and substitutes 7. It matches the third %d with c and substitutes 12.

Then it prints the completed line to the screen: 5+7=12. The +, the = and the spacing are a part of the format line and get embedded automatically between the %d operators as specified by the programmer.

New Words and Expressions

leverage *n.* 杠杆作用，影响力
binary *adj.* 双重的，双的；二进制的
subscript *adj.* 写在下面的 *n.* 脚注，下标，下角数码
modularization *n.* 模块化
executable *adj.* 可执行的，可实行的，可以做成的
integer *n.* 整数；完整的东西
automatically *adv.* 自动地；机械地

Exercises to the Text

1. Translate the following words and phrases into English.

（1）系统编程语言 （2）中级语言 （3）高级语言 （4）机器语言 （5）汇编语言 （6）系统软件 （7）免费编译器 （8）间距和缩进 （9）标准输入输出库 （10）标准类型

2. Translate the following paragraphs into Chinese.

(1) C has been around for several decades and has won widespread acceptance because it gives programmers maximum control and efficiency. If you are programmer, or if you are

interested in becoming a programmer, there are a couple of benefits you gain from learning C.

(2) There are some composite symbols with special meanings: for example "&&" means "and then" and "||" means "or else". "==" is used for equality to avoid confusion with "=" in assignments, and "!=" is used for inequality.

(3) C is what is called a compiled language. This means that once you write your C program, you must run it through a C compiler to turn your program into an executable that the computer can run (execute). The C program is the human-readable form, while the executable that comes out of the compiler is the machine-readable and executable form. What this means is that to write and run a C program, you must have access to a C compiler.

(4) The standard I/O library lets you read input from the keyboard(called "standard in"), write output to the screen (called "standard out"), process text files stored on the disk, and so on. It is an extremely useful library. C has a large number of standard libraries like stdio, including string, time and math libraries. A library is simply a package of code that someone else has written to make your life easier.

Text 3: C++

C++ is a general-purpose programming language with high-level and low-level capabilities. It is a statically typed, free-form, multi-paradigm, usually compiled language supporting procedural programming, data abstraction, object-oriented programming, and generic programming.

1. Origins of the C++ Language

The C++ programming languages can be thought of as the C Programming language with classes (and other modern features added). The C programming language was developed by Dennis Ritchie of AT&T Bell Laboratories in the 1970s. It was first used for writing and maintaining the UNIX operating system. (Up until that time, UNIX systems programs were written either in assembly language or in language called B, a language developed by Ken Thompson, the originator of UNIX.) C is a general-purpose language that can be used for writing any sort of program, but its success and popularity are closely tied to the UNIX operating system. If you wanted to maintain your UNIX system, you need to use C. C and UNIX fit together so well that soon not just systems programs but almost all commercial programs that ran under UNIX were written in the C language. C became so popular that versions of the language were written for other popular operating systems; its use is thus not limited to computers that use UNIX. However, despite its popularity, C was not without its shortcomings.

The C language is peculiar because it is a high-level language with many of the features of a low-lever language. C is somewhere in between the two extremes of a very high-level language and a low-level language, and therein lies both its strengths and its weakness. Like (low-level) assembly language, C language programs can directly manipulate the computer's memory. On the other hand, C has the features of a high-level language, which makes it easier to read and write than assembly language. This makes C an excellent choice for writing systems programs, but for other programs (and in some sense even for systems programs) C is not as easy to understand as other languages; also, it does not have as many automatic checks as some other high-level languages.

To overcome these and other shortcomings of C, Bjarne Stroustrup of AT&T Bell Laboratories developed C++ in the early 1980s. Stroustrup designed C++ to be a better C. Most of C is a subset of C++, and so most C programs are also C++ programs. (The reverse is not true; many C++ programs are definitely not C programs.) The basic syntax and semantics of C and C++ are the same. If you are familiar with C, you can program in C++ immediately. C++ has the same types, operators, and other facilities defined in C that usually correspond directly to computer architecture. Unlike C, C++ has facilities for classes and so can be used for object-oriented programming.

2. C++ and Object-Oriented Programming

Object-oriented programming is a programming technique that allows you to view concepts as a variety of objects. By using objects, you can represent the tasks that are to be performed, their interaction, and any given conditions that must be observed. A data structure often forms the basis of an object; thus, in C or C++, the struct type can form an elementary object. Communicating with objects can be done through the use of messages. Using messages is similar to calling a function in a procedure-oriented program. When an object receives a message, methods contained within the object respond. Methods are similar to the functions of procedure-oriented programming. However, methods are part of an object.

The main characteristics of Object-oriented programming (OOP) are encapsulation, inheritance, and polymorphism. Encapsulation is a form of information hiding or abstraction. Inheritance has to do with writing reusable code. Polymorphism refers to a way that a single name can have multiple meanings in the context of inheritance. C++ accommodates OOP by providing classes, a kind of data type combining both data and algorithms. C++ is not what some authorities would call a "pure OOP language." C++ tempers its OOP features with concerns for efficiency and what some might call "practicality." This combination has mad C++ currently the most widely used OOP language, although not all of its usage strictly follows the OOP philosophy.

3. The Character of C++

C++ allows the programmer to create classes, which are somewhat similar to C structures. In C++, it can be assigned methods, functions associated to it, of various prototypes, which can access and operate within the class, somewhat like C functions often operate on a supplied handler pointer. The C++ class is an extension of the C language structure. Because the only difference between a structure and a class is that structure members have public access by default and a class member has private access by default, you can use the keywords class or struct to define equivalent classes.

The C++ class is an extension of the C and C++ struct type and forms the required abstract data type for object-oriented programming. The class can contain closely related items that share attributes. Stated more formally, an object is simply an instance of a class.

Ultimately, there should emerge class libraries containing many object types. You could use instances of those object types to piece together program code.

Typically, an object's description is part of a C++ class and includes a description of the object's internal structure, how the object relates with other objects, and some form of protection that isolates the functional details of the object from outside the class. The C++ class structure does all of this.

In a C++ class, you control functional details of the object by using private, public, or protected descriptors. In object-oriented programming, the public section is typically used for the interface information (methods) that makes the class reusable across applications. If data or methods are contained in the public section, they are available outside the class. The private section of a class limits the availability of data or methods to the class itself. A protected section containing data or methods is limited to the class and any derived subclasses.

C++'s connection to the C language gives it a more traditional look than newer object-oriented languages, yet it has more powerful abstraction mechanisms than many other currently popular languages. C++ has a template facility that allows for full and direct implementation of algorithm abstraction. C++ templates allow you to code using parameters for types. The newest C++ standard, and most C++ compilers, allow multiple namespaces to accommodate more reuse of class and function names. The exception handling facilities in C++ are similar to what you would find in other programming languages. Memory management in C++ is similar to that in C. The programmer must allocate his or her own memory and handle his or her own garbage collection. Most compilers will allow you to do C-style memory management in C++, since C is essentially a subset of C++. However, C++ also has its own syntax for a C++ style of memory management, and you are advised to use the C++ style of memory management when coding in C++.

Inheritance in object-oriented programming allows a class to inherit properties from a class of objects. The parent class serves as a pattern for the derived class and can be altered in several ways. If an object inherits its attributes from multiple parents, it is called multiple inheritance. Inheritance is an important concept since it allows reuse of a class definition without requiring major code changes. Inheritance encourages the reuse of code since child classes are extensions of parent classes.

Another important object-oriented concept that relates to the class hierarchy is that common messages can be sent to the parent class objects and all derived subclass objects. In formal terms, this is called polymorphism.

Polymorphism allows each subclass object to respond to the message format in a manner appropriate to its definition. Imagine a class hierarchy for gathering data. The parent class might be responsible for gathering the name, social security number, occupation, and number of years of employment for an individual. You could then use child classes to decide what additional information would be added based on occupation. In one case a supervisory position might include yearly salary, while in another case a sales position might include an hourly rate and commission information. Thus, the parent class gathers general information common to all child classes while the child classes gather additional information relating to specific job descriptions. Polymorphism allows a common data-gathering message to be sent to each class. Both the parent and child classes respond in an appropriate manner to the message.

Polymorphism gives objects the ability to responds to messages from routines when the object's exact type is not known. In C++ this ability is a result of late binding. With late binding, the addresses are determined dynamically at run time, rather than statically at compile time, as in traditional compiled languages. This static method is often called early binding. Function names are replaced with memory addresses. You accomplish late binding by using virtual functions. Virtual functions are defined in the parent class when subsequent derived classes will overload the function by redefining the function's implementation.

Virtual functions utilize a table for address information. The table is initialized at run time by using a constructor. A constructor is invokes whenever an object of its class is created. The job of the constructor here is to link the virtual function with the table of address information. During the compile operation, the address of the virtual function is not known; rather, it is given the position in the table of addresses that will contain the address for the function.

4. The C++ Program

All procedure-like entities are called functions in C++. Things that are called procedures, methods, functions, or subprograms in other languages are all called functions in C++. A C++ program is basically just a function called main; when you run a program, the run-time system automatically invokes the function named main. Other C++ terminology is pretty much the same as most other programming languages. Here's the C++ program.

```
1  #include<iostream>
2  using namespace std;
3  int main()
4  {
5    int numberOfLanguage;
6    cout<<"Hello reader.\n"
7        <<"Welcome to C++.\n";
8    cout<<"How many programming languages have you used?";
9    cin>>numberOfLanguages;
10    if(numberOfLanguages<1)
11      cout<<"Read the preface. You may prefer\n"
12        <<"a more elementary book by the same author.\n";
13    else
14      cout<<"Enjoy the book.\n";
15  return 0;
16  }
```

A C++ program is really a function definition for a function named main. When the program is run, the function named main is invoked. The body of the function main is enclosed in braces, {}. When the program is run, the statements in the braces are executed. Here are two possible screen displays that might be generated when a user runs the program.

DIALOGUE 1

```
Hello reader.
Welcome to C++.
How many programming languages have you used? 0← User types in 0 on the keyboard.
Read the preface. You may prefer a more elementary book by the same author.
```

DIALOGUE 2

```
Hello reader.
Welcome to C++.
How many programming languages have you used? 1← User types in 1 on the keyboard.
Enjoy the book
```

Variable declarations in C++ are similar to what they are in other programming languages. The fifth line declares the variable numberOfLanguages. The type int is one of the C++ types for whole numbers (integers).

If you have not programmed in C++ before, then the use of cin and cout for console I/O is likely to be new to you. But the general idea can be observed in this sample program. For example, consider the eighth line and the ninth lines. The eighth line outputs the text within the quotation marks to the screen. The ninth line reads in a number that the user enters at the keyboard and sets the value of the variable numberOfLanguages to this number.

The eleventh line and the twelfth output two strings instead of just one string. The symbolism \n is the new line character, which instructs the computer to start a new line of output.

New Words and Expressions

architecture *n.* 建筑学
capability *n.* 能力
enhance *v.* 改善，提高，增进
semantics *n.* 语义学，语义论
concept *n.* 概念，观念，思想
interaction *n.* 交互作用，相互作用
extension *n.* 伸展，扩大，延长，延期，电话分机
equivalent *adj.* 等价的，相等的，同意义的 *n.* 等价物，相等物
description *n.* 描述，描写，说明书，类型
subclass *n.* 子类，亚类
hierarchy *n.* 层级，等级制度
inherit *vt.* 继承，遗传而得 *vi.* 成为继承人
inheritance *n.* 遗产，继承，遗传
hourly *adv.* 频繁地，随时，每小时地 *adj.* 每小时的，以钟点计算的，频繁的

Exercises to the Text

1. Translate the following words and phrases into English.

（1）编程语言 （2）面向对象程序设计 （3）抽象机制 （4）代码重用 （5）接口信息 （6）地址信息表 （7）虚函数 （8）附加信息 （9）函数定义 （10）变量声明

2. Translate the following paragraphs into Chinese.

(1) C++ is a general-purpose programming language with high-level and low-level capabilities. It is a statically typed, free-form, multi-paradigm, usually compiled language supporting procedural programming, data abstraction, object-oriented programming, and generic programming.

(2) C is a general-purpose language that can be used for writing any sort of program, but its success and popularity are closely tied to the UNIX operating system. If you wanted to maintain your UNIX system, you need to use C. C and UNIX fit together so well that soon not just systems programs but almost all commercial programs that ran under UNIX were written in the C language.

(3) Object-oriented programming is a programming technique that allows you to view concepts as a variety of objects. By using objects, you can represent the tasks that are to be performed, their interaction, and any given conditions that must be observed.

(4) Polymorphism allows each subclass object to respond to the message format in a manner appropriate to its definition. Imagine a class hierarchy for gathering data. The parent class might be responsible for gathering the name, social security number, occupation, and number of years of employment for an individual. You could then use child classes to decide what additional information would be added based on occupation.

(5) Virtual functions utilize a table for address information. The table is initialized at run time by using a constructor. A constructor is invokes whenever an object of its class is created. The job of the constructor here is to link the virtual function with the table of address information.

Text 4: Java

Java-has become enormously popular. Java's rapid rise and wide acceptance can be traced to its design and programming feature, particularly its promise that you can write a program once and run anywhere. As stated in the Java-language white paper by Sun, Java simple, object-oriented, distributed, interpreted, robust, secure, architecture-neutral, portable, high-performance, multithreaded, and dynamic.

What Is Java?

Java is a major departure from the HTML coding that makes up most Web pages. Sitting atop markup languages such as HTML and XML, Java is an object-oriented, network-friendly high-level programming language that allows programmers to build applications that can run on almost any operating system. With Java, big applications programs can be broken into mini-applications, or "applets", that can be downloaded off the Internet and run on any computer. Moreover, Java enables a Web page to deliver, along with visual content, applets that when downloaded can make Web pages interactive.

Some microcomputers include special Java microprocessors designed to run Java software directly. However, Java is not compatible with many existing microprocessors, such as those from Intel and Motorola. For this reason, these users need to use a small "interpreter" program, called a Java Virtual Machine that translates a Java program into a language that any computer or operating system can understand. They also need a Java-capable browser in order to view Java special effects on the Web.

Java development programs are available for programmers. In addition, Java software packages—such as ActionLine, Activator Pro, AppletAce, and Mojo-give nonprogrammers the ability to add multimedia effects to their Web pages, by producing applets that any Java-equipped browser can view. Such packages can be used by anyone who understands multimedia file formats and is willing to experiment with menu options.

Java is the latest, flashiest object-oriented language. One especially powerful feature of OOP (object-oriented programming) languages is a property which is known as inheritance. Inheritance allows an object to take on the characteristics and functions of other objects to which it is functionally connected. Programmers connect objects by grouping them together in different classes and by grouping the classes into hierarchies. These classes and hierarchies allow programmers to define the characteristics and functions of objects without needing to repeat source code. Thus, using OOP languages can greatly reduce the time it takes for a programmer to write an application, and also reduce the size of the program. OOP languages are flexible and adaptable, so programs or parts of programs can be used for more than one task. Programs written with OOP languages are generally shorter in length and contain fewer bugs, or mistakes, than those written with non-OOP languages.

Java has taken the software world by storm due to its close ties with the Internet and Web browsers. It is designed as a portable language that can run on any Web-enabled computer via that computer's Web browser. As such, it offers great promise as the standard Internet and Intranet programming language.

It has the syntax of C++, making it easy (or difficult) to learn, depending on your experience. But it has improved on C++ in some important areas. For one thing, it has no pointers, which are low-level programming constructs that make for error-prone programs. It has garbage collection, a feature that frees the programmer from explicitly allocating and de-allocating memory. It also runs on a virtual machine, which is software built into a Web browser that executes the same standard compiled Java bytecodes no matter the type of computer.

Java development tools are being rapidly deployed, and are available from such major software companies as IBM, Microsoft, and Symantec.

The Characteristics of Java

1. Java Is Simple

No language is simple, but Java is a bit easier than the popular object-oriented programming language C++, which was the dominant software-development language before Java. Java is partially modeled on C++, but greatly simplified and improved. For instance, pointers and multiple inheritances often make programming complicated. Java replaces the multiple inheritances in C++ with a simple language construct called an interface, and eliminates pointers.

Java uses automatic memory allocation and garbage collection, whereas C++ requires the programmer to allocate memory and collect garbage. Also, the number of language constructs is small for such a powerful language. The clean syntax makes Java programs easy to write and read. Some people refer to Java as "C++--" because it is like C++ but with more functionality and fewer negative aspects.

2. Java Is Object-Oriented

Java is inherently object-oriented. Although many object-oriented languages began strictly as procedural languages, Java was designed from the start to be object-oriented. Object-oriented programming (OOP) is a popular programming approach that is replacing traditional procedural programming techniques.

Software systems developed using procedural programming languages are based on the paradigm of procedures. Object-oriented programming models the real world in terms of objects. Everything in the world can be modeled as an object. A circle is an object, a person is an object, and a Windows icon is an object. Even a loan can be perceived as an object. A Java program is object-oriented because programming in Java is centered on creating objects, manipulating objects, and making objects work together.

One of the central issues in software development is how to reuse code. Object-oriented programming provides great flexibility, modularity, clarity, and reusability through encapsulation, inheritance, and polymorphism. For years, object-oriented technology was perceived as elitist, requiring a substantial investment in training and infrastructure. Java has helped object-oriented technology enter the mainstream of computing. Its simple, clean syntax makes programs easy to write and read. Java programs are quite expressive in terms of designing and developing applications.

3. Java Is Distributed

Distributed computing involves several computers working together on a network. Java is designed to make distributed computing easy. Since networking capability is inherently integrated into Java, writing network program is like sending and receiving data to and from a file.

4. Java Is Interpreted

You need an interpreter to run Java programs. The programs are compiled into the Java Virtual Machine code called bytecode. The bytecode is machine-independent and can run on any machine that has a Java interpreter, which is part of the Java Virtual Machine (JVM).

Most compilers, including C++ compilers, translate programs in a high-level language to machine code. The code can only run on the native machine. If you run the program on other machines, it has to be recompiled on the native machine. For instance, if you compile a C++ program in Windows, the executable code generated by the compiler can only run on the Windows platform. With Java, you compile the source code once, and the bytecode generated by a Java compiler can run on any platform with a Java interpreter. The Java interpreter translates the bytecode into the machine language of the target machine.

5. Java Is Robust

Robust means reliable. No programming language can ensure complete reliability. Java puts a lot of emphasis on early checking for possible errors, because Java compilers can detect many problems that would first show up at execution time in other languages. Java has eliminated certain types of error-prone programming constructs found in other languages. It does not support pointers, for example, thereby eliminating the possibility of overwriting memory and corrupting data.

Java has a runtime exception-handling feature to provide programming support for robustness. Java forces the programmer to write the code to deal with exceptions. Java can catch and respond to an exceptional situation so that the program can continue its normal execution and terminate gracefully when a runtime error occurs.

6. Java Is Secure

As an Internet programming language, Java is used in a networked and distributed environment. If you download a Java applet (a special kind of program) and run it on your computer, it will not damage your system because Java implements several security mechanisms to protect your system against harm caused by stray programs. The security is based on the premise that nothing should be trusted.

7. Java Is Architecture-Neutral

Java is interpreted. This feature enables Java to be architecture-neutral, or to use an alternative term, platform-independent. With a Java Virtual Machine (JVM), you can write one program that will run on any platform.

Java's initial success stemmed from its Web-programming capability. You can run Java applets from a Web browser, but Java is for more than just writing Web applets. You can also run standalone Java applications directly from operating systems, using a Java interpreter. Today, software vendors usually develop multiple versions of the same product to run on different platforms. Using Java, developers need to write only one version that can run on every platform.

8. Java Is Portable

Because Java is architecture-neutral, Java programs are portable. They can be run on any platform, without being recompiled. Moreover, there are no platform-specific features in the Java language. In some languages, such as Ada, the largest integer varies on different platforms. But in Java, the range of the integer is the same on every platform, as is the behavior of arithmetic. The fixed range of the numbers makes the program portable.

The Java environment is portable to new hardware and operating systems. In fact, the Java compiler itself is written in Java.

9. Java's Performance

Java's performance is sometimes criticized. The execution of the bytecode is never as fast as it would be with a compiled language, such as C++. Because Java is interpreted, the bytecode is not directly executed by the system, but is run through the interpreter. However, its speed is more than adequate for most interactive applications, where the CPU is often idle, waiting for input or for data from other sources.

CPU speed has increased dramatically in the past few years, and this trend will continue. There are many ways to improve performance. Users of the earlier Sun Java Virtual Machine certainly noticed that Java was slow. However, the new JVM is significantly faster. The new JVM uses the technology known as just-in-time compilation. It compiles bytecode into native machine code, stores the native code, and reinvokes the native code when its bytecode is executed. Sun recently developed the Java HotSpot Performance Engine, which includes a compiler for optimizing the frequently used code. The HotSpot Performance Engine can be plugged into a JVM to dramatically boost its performance.

10. Java Is Multithreaded

Multithreading is a program's capability to perform several tasks simultaneously. Multithread programming is smoothly integrated in Java, whereas in other languages you have to call procedures specific to the operating system to enable multithreading. Multithreading is particularly useful in graphical user interface (GUI) and network programming. In GUI programming, there are many things going on at the same time. A user can listen to an audio recording while surfing a Web page. In network programming, a server can serve multiple clients at the same time. Multithreading is a necessity in multimedia and network programming.

11. Java Is Dynamic

Java was designed to adapt to an evolving environment. New class can be loaded on the fly without recompilation. There is no need for developers to create, and for users to install, major new software versions. New features can be incorporated transparently as needed.

What Can Java Technology Do?

The most common types of programs written the Java programming are applets and application. If you've surfed the Web, you're probably already familiar with applets. An applet is a program that adheres to certain conventions that allow it to run within a Java-enabled. At the beginning of this trail is an applet that displays an animation of the Java technology mascot, Duke, waving at you.

However, the Java programming language is not just for writing cute, entertaining applets for the Web. The general-purpose, high-level Java programming language is also a powerful software platform. Using the generous API, you can write many types of programs.

An application is a standalone program that runs directly on the Java platform. A special kind of application known as a server serves and supports clients on a network. Examples of servers are web servers, proxy servers, mail servers, and print servers. Another specialized program is a Servlet. A Servlet can almost be thought of as an applet that runs on the server side. Java Servlets are a popular choice for building interactive web applications, replacing the use of CGI scripts. Servlets are similar to applets in that they are runtime extensions of applications. Instead of working in browsers, though, Servlets run within Java Web servers, configuring or tailoring the server.

How does the API support all these kinds of programs? It does so with packages of software components that provide a wide range of functionality. Every full implementation of the Java platform gives you the following features:

- The essentials—Objects, strings, threads, numbers, input and output data structures, system properties, date and time, and so on.
- Applets—The set of conventions used by applets.
- Networking—URLs, TCP (Transmission Control Protocol), UDP (User Datagram Protocol) sockets, and IP (Internet Protocol) addresses.
- Internationalization—Help for writing programs that can be localized for users worldwide. Programs can automatically adapt to specific locales and be displayed in the appropriate language.
- Security—Both low level and high level, including electronic signatures, public and private key management, access control, and certificates.
- Software components—Known as JavaBeans, can plug; into existing component architectures.
- Object serialization—Allows lightweight persistence and communication via Remote Method Invocation (RMI).
- Java Database Connectivity (JDBC)—Provides uniform access to a wide range of relational databases.

The Java platform also has APIs for 2D and 3D graphics, accessibility, servers, collaboration, telephony, speech, animation, and more.

Note: The Java 2 SDK, Standard Edition v. 1.3. The Java 2 Runtime Environment (JRE) consists of the virtual machine, the Java platform core classes, and supporting files. The Java 2 SDX includes the JRE and development tools such as compilers and debuggers.

New Words and Expressions

incorporated *adj.* 组成公司的，合成一体的

enormously *adv.* 巨大地，庞大地，非常地，在极大程度上

architecture *n.* 建筑学 建筑式样，建筑风格

multithreaded *adj.* 多线程的，多重线串的

interactive *adj.* 相互作用的，交互式的

microcomputer *n.* 微型计算机，微电脑
inheritance *n.* 继承（性）
memory allocation 存储器分配
garbage *n.* 垃圾，无用信息（或单元、数据）
garbage collection 垃圾收集，无用（存储）单元收集
functionality *n.* 功能性
bytecode *n.* 字节码，字节代码，位元码
inherent *adj.* 内在的；固有的
allocation *n.* 分配，配置，安置
paradigm *n.* 范例，样式；范式
modularity *n.* 模块性，模块化
clarity *n.* 明晰，明确（性）
reusability *n.* 可重用性，可重复使用性
encapsulation *n.* 封装，密封；压缩
polymorphism *n.* 多态性，多形性
elitist *adj.* 杰出人物的；上等的，高级的
infrastructure *n.* 基础结构，基础设施
mainstream *n.* 主流
expressive *adj.* 表现的；富于表现力的
integrate *v.* （使）成为一体（或结合、合并）
compiler *n.* 编辑者，汇编者，编译器，编译程序
platform *n.* 月台，站台，讲台，平台
arithmetic *n.* 算术，算法
bytecode *n.* 字节码，字节代码
prone *adj.* 有……倾向的，易于……的
overwrite *v.* 改写，重写
corrupt *v.* 毁坏，损坏，破坏
runtime *n.* 运行时刻，运行（时）期
exception *n.* 异常（事件），例外
terminate *v.* （使）终止
stray *n.* 迷路的；杂散的；寄生的
premise *n.* 假设；前提
Web browser 万维网浏览器
standalone *adj.* （相对于操作系统而言）独立的
integer *n.* 整数
optimizing *n.* 优化，最佳化 *adj.* 最佳的
multithreading *n.* 多线程，多线索
extension *n.* 扩展，扩张，延长
internationalization *n.* 国际化
animation *n.* 活泼，生气，卡通片绘制

Exercises to the Text

1. Translate the following words and phrases into English.

（1）存储器分配 （2）面向对象程序设计 （3）机器码 （4）万维网浏览器 （5）可执行代码 （6）运行期异常处理 （7）Java 虚拟机 （8）即时编译 （9）可移植应用软件 （10）本机代码

2. Translate the following paragraphs into Chinese.

(1) Java is a major departure from the HTML coding that makes up most Web pages. Sitting atop markup languages such as HTML and XML, Java is an object-oriented, network-friendly high-level programming language that allows programmers to build applications that can run on almost any operating system.

(2) Java has taken the software world by storm due to its close ties with the Internet and Web browsers. It is designed as a portable language that can run on any Web-enabled computer via that computer's Web browser. As such, it offers great promise as the standard Internet and Intranet programming language.

(3) No language is simple, but Java is a bit easier than the popular object-oriented programming language C++, which was the dominant software-development language before Java. Java is partially modeled on C++, but greatly simplified and improved. For instance, pointers and multiple inheritances often make programming complicated. Java replaces the multiple inheritances in C++ with a simple language construct called an interface, and eliminates pointers.

(4) Multithreading is particularly useful in graphical user interface (GUI) and network programming. In GUI programming, there are many things going on at the same time. A user can listen to an audio recording while surfing a Web page. In network programming, a server can serve multiple clients at the same time. Multithreading is a necessity in multimedia and network programming.

(5) Java was designed to adapt to an evolving environment. New class can be loaded on the fly without recompilation. There is no need for developers to create, and for users to install, major new software versions. New features can be incorporated transparently as needed.

Text 5: ActionScript Basics

Where to Place ActionScript Code?

If you have a new ActionScript project, do you know where to put the code for it to execute properly? The answer is, place ActionScript code in the constructor and additional methods of the class.

In ActionScript 1.0 and 2.0, you had many choices as to where to place your code: on the timeline, on buttons and movie clips, on the timeline of movie clips, in external .as files referenced with #include, or as external class files. ActionScript 3.0 is completely class-based, so all code must be placed in methods of your project's classes.

When you create a new ActionScript project, the main class is automatically created, and opened in the Code view. It should look something like this:

```
package {
    import flash.display.Sprite;
    public class ExampleApplication extends Sprite
    {
        public function ExampleApplication()
        {
        }
    }
}
```

Even if you are familiar with classes in ActionScript 2.0, there are some new things here.

The first thing you'll notice is the word package at the top of the code listing. Packages are used to group classes of associated functionality together. In ActionScript 2.0, packages were inferred through the directory structure used to hold the class files. In ActionScript 3.0, however, you must explicitly specify packages. For example, you could have a package of utility classes. This would be declared like so:

```
package com.as3cb.utils {
}
```

If you don't specify a package name, your class is created in the default, top-level package. You should still include the package keyword and braces.

Next, place any import statements. Importing a class makes that class available to the code in the file and sets up a shortcut so you don't have to type the full package name every time you want to refer to that class. For example, you can use the following import statement:

```
import com.as3cb.utils.StringUtils;
```

Thereafter you can refer to the StringUtils class directly without typing the rest of the path. As shown in the earlier example, you will need to import the Sprite class from the flash.display package, as the default class extends the Sprite class.

Next up is the main class, ExampleApplication. You might notice the keyword public in front of the class definition. Although you can't have private classes within a package, you should label the class public. Note that the main class extends Sprite. Also, a .swf itself is a type of sprite or movie clip, which is why you can load a .swf into another .swf and largely treat it as if it were just another nested sprite or movie clip. This main class represents the .swf as a whole, so it should extend the Sprite class or any class that extends the Sprite class (such as MovieClip).

Finally, there is a public function (or method, in class terminology) with the same name as the class itself. This makes it a constructor. A class's constructor is automatically run as soon as an instance of the class is created. In this case, it is executed as soon as the .swf is loaded into the

Flash player. So where do you put your code to get it to execute? Generally, you start out by putting some code in the constructor method. Here's a very simple example that just draws a bunch of random lines to the screen:

```
package {
    import flash.display.Sprite;
    public class ExampleApplication extends Sprite {
        public function ExampleApplication() {
            graphics.lineStyle(1, 0, 1);
            for(var i:int=0;i<100;i++) {
                graphics.lineTo(Math.random() * 400, Math.random() * 400);
            }
        }
    }
}
```

Save and run the application. Your browser should open the resulting HTML file and display the .swf with 100 random lines in it. As you can see, the constructor was executed as soon as the file was loaded into the player.

In practice, you usually want to keep code in the constructor to a bare minimum. Ideally the constructor would just contain a call to another method that initializes the application.

For beginners, now that you know where to enter code, here is quick primer on terminology. These definitions are briefly stated and intended to orient people who have never programmed before.

- Variables

Variables are convenient placeholders for data in your code, and you can name them anything you'd like, provided the name isn't already reserved by ActionScript and the name starts with a letter, underscore, or dollar sign (but not a number). The help files installed with Flex Builder 2 contain a list of reserved words. Variables are convenient for holding interim information, such as a sum of numbers, or to refer to something, such as a text field or sprite. Variables are declared with the var keyword the first time they are used in a script. You can assign a value to a variable using an equal sign (=), which is also known as the assignment operator. If a variable is declared outside a class method, it is a class variable. Class variables, or properties, can have access modifiers, public, private, protected, or internal. A private variable can only be accessed from within the class itself, whereas public variables can be accessed by objects of another class. Protected variables can be accessed from an instance of the class or an instance of any subclass, and internal variables can be accessed by any class within the same package. If no access modifier is specified, it defaults to internal.

- Functions

Functions are blocks of code that do something. You can call or invoke a function (that is, execute it) by using its name. When a function is part of a class, it is referred to as a method of the class. Methods can use all the same modifiers as properties.

- Scope

A variable's scope describes when and where the variable can be manipulated by the code in a movie. Scope defines a variable's life span and its accessibility to other blocks of code in a script. Scope determines how long a variable exists and from where in the code you can set or retrieve the variable's value. A function's scope determines where and when the function is accessible to other blocks of code.

- Event handler

A handler is a function or method that is executed in response to some event such as a mouseclick, a keystroke, or the movement of the playhead in the timeline.

- Objects and classes

An object is something you can manipulate programmatically in ActionScript, such as a sprite. There are other types of objects, such as those used to manipulate colors, dates, and text fields. Objects are instances of classes, which means that a class is a template for creating objects and an object is a particular instance of that class. If you get confused, think of it in biological terms: you can consider yourself an object (instance) that belongs to the general class known as humans.

- Methods

A method is a function associated with an object that operates on the object. For example, a text field object's replaceSelectedText( ) method can be used to replace the selected text in the field.

- Properties

A property is an attribute of an object, which can be read and/or set. For example, a sprite's horizontal location is specified by its x property, which can be both tested and set. On the other hand, a text field's length property, which indicates the number of characters in the field, can be tested but cannot be set directly (it can be affected indirectly, however, by adding or removing text from the field).

- Statements

ActionScript commands are entered as a series of one or more statements. A statement might tell the playhead to jump to a particular frame, or it might change the size of a sprite. Most ActionScript statements are terminated with a semicolon (;).

- Comments

Comments are notes within code that are intended for other humans and ignored by Flash. In ActionScript, single-line comments begin with // and terminate automatically at the end of the current line. Multiline comments begin with /* and are terminated with */.

- Interpreter

The ActionScript interpreter is that portion of the Flash Player that examines your code and attempts to understand and execute it. Following ActionScript's strict rules of grammar ensures that the interpreter can easily understand your code. If the interpreter encounters an error, it often fails silently, simply refusing to execute the code rather than generating a specific error message.

- Handling Events

If you want to have some code repeatedly execute, add a listener to the enterFrame event and assign a method as a handler.

In ActionScript 2.0 handling the enterFrame event was quite simple. You just had to create a timeline function called onEnterFrame and it was automatically called each time a new frame began. In ActionScript 3.0, you have much more control over the various events in a .swf, but a little more work is required to access them.

If you are familiar with the EventDispatcher class from ActionScript 2.0, you should be right at home with ActionScript 3.0's method of handling events. In fact, EventDispatcher has graduated from being an externally defined class to being the base class for all interactive objects, such as sprites.

To respond to the enterFrame event, you have to tell your application to listen for that event and specify which method you want to be called when the event occurs. This is done with the addEventListener method, which is defined as follows:

```
addEventListener(type:String, listener:Function)
```

The type parameter is the type of event you want to listen to. In this case, it would be the string, "enterFrame". However, using string literals like that opens your code to errors that the compiler cannot catch. If you accidentally typed "enterFrame", for example, your application would simply listen for an "enterFrame" event. To guard against this, it is recommended that you use the static properties of the Event class. You should already have the Event class imported, so you can call the addEventListener method as follows:

```
addEventListener(Event.ENTER_FRAME, onEnterFrame);
```

Now if you accidentally typed Event.ENTER_FRAME, the compiler would complain that such a property did not exist.

The second parameter, onEnterFrame, refers to another method in the class. Note, that in ActionScript 3.0, there is no requirement that this method be named onEnterFrame. However, naming event handling methods on plus the event name is a common convention. This method gets passed an instance of the Event class when it is called. Therefore, you'll need to import that class and define the method so it accepts an event object:

```
import flash.events.Event;
private function onEnterFrame(event:Event) {
}
```

The event object contains information regarding the event that may be useful in handling it. Even if you don't use it, you should still set your handler up to accept it. If you are familiar with the ActionScript 2.0 version of EventDispatcher, you'll see a difference in implementation here. In the earlier version, there was an issue with the scope of the function used to handle the event, which often required the use of the Delegate class to correct. In ActionScript 3.0, the scope of the handling method remains the class of which it is a method, so there is no necessity to use Delegate to correct scope issues.

Here is a simple application that draws successive random lines, using all the concepts discussed in this recipe:

```
package {
    import flash.display.Sprite;
    import flash.events.Event;
        public class ExampleApplication extends Sprite {
            public function ExampleApplication() {
            graphics.lineStyle(1, 0, 1);
            addEventListener(Event.ENTER_FRAME, onEnterFrame);
        }
        private function onEnterFrame(event:Event):void {
            graphics.lineTo(Math.random() * 400, Math.random() * 400);
        }
    }
}
```

New Words and Expressions

terminology *n.* 专门用语，术语
interim *adj.* 暂时的，临时的，间歇的 *n.* 过渡时期
invoke *vt.* 援引，援用；行使（权利等）
modifier *n.* [语言学]修饰语
retrieve *vt.* 取回，挽回，弥补，恢复，补偿，回忆，检索
manipulate *vt.* 熟练控制（操作）
semicolon *n.* 分号
successive *adj.* 连续的，相继的
regarding *prep.*（表示论及）关于；至于；就……而论
version *n.* 版本，形式
terminate *vt.&vi.* 结束；使终结
execute *vt.* 实行，实施，执行；完成，实现，履行
underscore *vt.* 在……下画线；强调
nested *adj.* 嵌套的
explicitly *adv.* 明白地，明确地

Exercises to the Text

1. Translate the following words and phrases into English.

（1）变量 （2）函数 （3）作用域 （4）事件处理器 （5）对象和类 （6）方法 （7）属性 （8）语句 （9）注释 （10）解释器

2. Translate the following paragraphs into Chinese.

(1) Variables are convenient placeholders for data in your code, and you can name them anything you'd like, provided the name isn't already reserved by ActionScript and the name starts with a letter, underscore, or dollar sign (but not a number).

(2) A property is an attribute of an object, which can be read and/or set. For example, a sprite's horizontal location is specified by its x property, which can be both tested and set. On the other hand, a text field's length property, which indicates the number of characters in the field, can be tested but cannot be set directly.

(3) In the earlier version, there was an issue with the scope of the function used to handle the event, which often required the use of the Delegate class to correct. In ActionScript 3.0, the scope of the handling method remains the class of which it is a method, so there is no necessity to use Delegate.

(4) If you are familiar with the EventDispatcher class from ActionScript 2.0, you should be right at home with ActionScript 3.0's method of handling events. In fact, EventDispatcher has graduated from being an externally defined class to being the base class for all interactive objects, such as sprites.

Part 6 Computer Network

Text 1: About Computer Networks

A network, in computer science, is a group of computers and associated devices that are connected by communications facilities. A network can involve permanent connections, such as cables, or temporary connections made through telephone or other communications links. A network can be as small as a local area network consisting of a few computers, printers, and other devices, or it can consist of many small and large computers distributed over a vast geographic area.

Computer networks are a central part of the information age. Small or large, they exist to provide computer users with the means of communicating and transferring information electronically.

Networking Basics

A network is a set of devices (often referred to as nodes) connected by a media link. A node can be a computer, printer, or any other device capable of sending and / or receiving data generated by other nodes on the network. The links connecting the devices are often called communications channels. A network requires a Network Operating System (NOS) to manage network resources. It may be a completely self-contained operating system, such as NetWare, or it may require an existing operating system, such as Windows NT in order to function. The following paragraphs explain the advantages of using a network.

- Facilitating Communications: Using a network, people can communicate efficiently and easily via e-mail, instant messaging, chat rooms, telephone, video-telephone calls, and: video-conferencing. Sometimes these communications occur within a business's network. Other times, they occur globally over the Internet.
- Sharing Hardware: In a networked environment, each computer on a network can access and use hardware on the network. Suppose several people require the use of a laser printer. If the personal computers are on a network together with a laser printer, they will all be able to share it. Business and home users' network share their hardware for one main reason, which is cost reduction. It may be too costly to provide each user with the same piece of hardware, such as a printer.
- Sharing Data and Information: In a networked environment, any authorized computer user can access data and information stored on other computers on the network. The capability of providing access to and storage of data and information on shared storage devices is an important feature of many networks.

- Sharing Software: Users connected to a network can access software on the network. To support multiple user access of software, most software vendors sell network versions of their software.

Types of Computer Networks

Computer network can be used for numerous services, both for computers and for individuals. For companies, networks of personal computers using shared servers often provide flexibility and a good price performance ratio. For individuals, networks offer access to a variety of information and entertainment resources.

From the transmission technology aspect, networks are classified as Local Area Networks (LANs), Wide Area Networks (WANs), and Metropolitan Area Network (MAN). LANs are usually located in single buildings or campuses, and handle interoffice communications. WANs cover a large geographical area and connect cities and countries. MANs are used to link office building within city.

1. Local Area Network

LAN is an acronym for local area network. It is a group of computers and other devices dispersed over a relatively limited area and connected by a communications link. A LAN can connect all of the workstations, peripherals, terminals, and other devices in a single building. LAN makes it possible for using computer technology to efficiently share such things as files and printers, and to make possible communications such as e-mail. They combine data, communications, computing, and file servers together.

LANs are designed to do the following:

- operate within a limited geographic area,
- allow many users to access high-bandwidth media,
- provide full-time connectivity to local services, and
- connect physically adjacent devices.

2. Metropolitan Area Network

MAN is an acronym for metropolitan area network. It is a high speed network that can carry voice, data and images up to 200 Mb/s or faster over distances of up to 75 km. Based on the network architecture, the transmission speed can be higher for shorter distances. A MAN is smaller than a wide area network but generally operates at a higher speed. The range is usually within a mile-perhaps one office, one building. Cellular phone systems expand the flexibility of MANs by allowing links to car phones and portable phones.

All these networks may consist of various combinations of computers, storage devices, and communications devices.

3. Wide Area Network

WAN is an acronym for wide area network. It is a geographically widespread network, which relies on communications capabilities to link the various network segments. A WAN can be one large network, or it can consist of a number of linked LANs.

Due to the increased use of computers in departments and businesses, it soon became apparent that even LANs were not sufficient. In a LAN system, each department or business was a kind of electronic island. What was needed was a way for information to move efficiently and quickly from one business to another, thus the development of WANs.

The popular adoption of the personal computer (PC) and the local area network (LAN) during the 1980s has led to the capacity to access information on a distant database; download an application from overseas; send a message to a friend in a different country; and share files with a colleague — all from a personal computer.

The networks that allow all this to be done so easily are sophisticated and complex entities. They rely for their effectiveness on many cooperating components. The design and deployment of the worldwide computer network can be viewed as one of the great technological wonders of recent decades.

Network Topology

A network topology is the configuration, or physical arrangement, of devices in a communication network. It can be laid out in different ways. The five basic topologies are bus, star, ring, tree, and hybrid.

1. Bus Topology

A bus topology uses a single backbone segment (length of cable) that all the hosts connect to directly. The bus network works like a bus system at rush hour, with various buses pausing in different bus zones to pick up passengers. In a bus network, all communications devices are connected to a common cable, called a bus, using co-ax.STP or UTP. There is no central computer or server, and data transmission is bi-directional at a rate of about one to ten Mb/s. Each communications device transmits electronic messages to other devices. If some of those messages collide, the device waits and tries to retransmit.

The advantage of a bus network is that it is relatively inexpensive to install. The disadvantage is that, if the bus fails, the entire network fails.

2. Star Network

In a star network, all microcomputers and other communications devices are connected to a central hub, such as a file server or host computer, usually via (UTP). Electronic messages are sent through the central hub to their destinations at rates of one to 100 Mb/s. The central hub monitors the flow of traffic.

The advantage of a star network is that, if a connection is broken between any communications device and the hub, the rest of the devices on the network will continue operating. The primary disadvantage is that a hub failure is catastrophic.

3. Ring Network

In a ring network, all microcomputers and other communication devices are connected to a continuous loop. Electronic messages are passed around the ring in one direction, with each node serving as a repeater, until they reach the right destination. There is no central host computer or server. Rings generally are UTR, STR, or fiber-optic cable.

Typically, token ring network, in which a bit pattern (called a "token") determines which user on the network I can send information. The advantage of a ring network is that messages flow in only one direction. Thus, there is no danger of collisions. The disadvantages are the current speed limit and the relatively high cost.

4. Tree Network

In a tree network, it combines characteristics of linear bus and star topologies. It consists of groups of star-configured workstations connected to a linear bus backbone cable. Tree topologies allow for the expansion of an existing network.

5. Hybrid Network

Hybrid networks are combinations of star ring, and bus networks. For example, a small college campus might use a bus network to connect buildings while using star and ring networks within certain buildings.

Intranets

Recognizing the efficiency and power of the Internet, many organizations apply Internet and Web technologies to their own internal networks. An Intranet is an internal network that uses Internet technologies. Intranets generally make company information accessible to employees and facilitate working in groups. Simple intranet applications include electronic publishing of organizational materials such as telephone directories, event calendars, procedure manuals, employee benefits information, and job postings. An intranet typically also includes a connection to the Internet. More sophisticated uses of intranets include groupware applications such as project management, chat rooms, newsgroups, group scheduling, and videoconferencing.

An intranet essentially is a small version of the Internet that exists within an organization. It uses TCP/IP technologies, has a Web server, supports multimedia Web pages coded in HTML, and is accessible via a Web browser such as Microsoft Internet Explorer or Netscape Navigator. Users can post and update information on the intranet by creating and posting a Web page, using a method similar to that used on the Internet.

Sometimes a company uses an extranet, which allows customers or suppliers to access part of its intranet. Federal Express, for example, allows customers to access their intranet to print air bills, schedule pickups, and even track shipped packages as they travel to their destination.

Network Operating System

Unlike operating systems, such as DOS and Windows, which are designed for single users to control one computer, Network Operating Systems (NOS) coordinate the activities of multiple computers across a network. The network operating system acts as a director to keep the network running smoothly.

The two major types of network operating systems are:

1. Peer-to-Peer

Peer-to-peer network operating systems allow users to share resources and files located on their computers and to access shared resources found on other computers. However, they do not have a file server or a centralized management source. In a peer-to-peer network, all computers are considered equal; they all have the same abilities to use the resources available on the network. Peer-to-peer networks are designed primarily for small to medium local area networks. AppleShare and Windows for Workgroups are examples of programs that can function as peer-to-peer network operating systems.

Advantages of a peer-to-peer network:

- Less initial expense—No need for a dedicated server.
- Setup easily—An operating system (such as Windows XP) already in place may only need to be reconfigured for peer-to-peer operations.

Disadvantages of a peer-to-peer network:

- Decentralized—No central repository for files and applications.
- Security—Does not provide the security available on a client/server network.

2. Client/Server

Client/server network operating systems allow the network to centralize functions and applications in one or more dedicated file servers. The file servers become the heart of the system, providing access to resources and providing security. Individual workstations (clients) have access to the resources available on the file servers. The network operating system provides the mechanism to integrate all the components of the network and allow multiple users to simultaneously share the same resources irrespective of physical location. Novell Netware and Windows 2000 Server are examples of client/server network operating systems.

Advantages of a client/server network:

- Centralized—Resources and data security are controlled through the server.
- Scalability—Any or all elements can be replaced individually as needs increase.
- Flexibility—New technology can be easily integrated into system.
- Interoperability—All components (client/network/server) work together.
- Accessibility—Server can be accessed remotely and across multiple platforms.

Disadvantages of a client/server network:

- Expense—Requires initial investment in dedicated server.
- Maintenance—Large networks will require a staff to ensure efficient operation.
- Dependence—When server goes down, operations will cease across the network.

New Words and Expressions

associate *v.* 联合；把……联系在一起
facility *n.* 设备
cable *n.* 电缆；电报
temporary *adj.* 暂时的
distribute *v.* 分发；发行
vast *adj.* 巨大的
node *n.* 节点，中心点
facilitate *v.* 使……便利
via *prep.* 经由，经过

conference　n. 会议
laser　n. 激光；激光器
reduction　v. 缩减；降低；减小
authorize　v. 授权
multiple　adj. 多个的；多重的
vendor　n. 摊贩；卖家
ratio　n. 比率
metropolitan　n. 大城市；首都
interoffice　adj.（同一组织的）各部门间的；各办公室之间的
acronym　n. 首字母缩写词
disperse　v.（使）散开；（使）分散
peripheral　adj. 外围的；边缘的
cellular　adj. 细胞的
portable　adj. 手提的
sufficient　adj. 充分的
sophisticate　n. 久经世故的人　vt. 使复杂
component　n. 组成部分；成分
topology　n. 拓扑图
configuration　n. 布局；配置
hybrid　n.（动植物）杂种；混合型
collide　v. 冲突
hub　n. 中心；枢纽
catastrophic　n. 大灾难
loop　n. 环形；圈
fiber-optic　n. 视觉纤维
linear　adj. 线的；直线的
intranet　n. 内联网
manual　adj. 手工的；人力的
browser　n. 浏览者
federal　adj. 联邦的
initial　adj. 最初的
dedicate　v. 奉献；贡献
client　n. 委托人
scalability　n. 可伸缩性；可量测性
interoperability　n. 互用性，[计算机]让软硬件在多种机器上能有意义地沟通
maintenance　n. 维持
instant messaging　（网上的）即时通信（服务）

Exercises to the Text

1. Translate the following words and phrases into English.

（1）通信设备　（2）计算机网络　（3）即时信息　（4）局域网　（5）城域网　（6）广

域网 （7）网络拓扑 （8）数据传输 （9）中央集线器 （10）网络操作系统

2. Translate the following paragraphs into Chinese.

(1) A network, in computer science, is a group of computers and associated devices that are connected by communications facilities. A network can involve permanent connections, such as cables, or temporary connections made through telephone or other communications links. A network can be as small as a local area network consisting of a few computers, printers, and other devices, or it can consist of many small and large computers distributed over a vast geographic area.

(2) In a networked environment, each computer on a network can access and use hardware on the network. Suppose several people require the use of a laser printer. If the personal computers are on a network together with a laser printer, they will all be able to share it. Business and home users' network share their hardware for one main reason, which is cost reduction. It may be too costly to provide each user with the same piece of hardware, such as a printer.

(3) LAN is an acronym for local area network. It is a group of computers and other devices dispersed over a relatively limited area and connected by a communications link. A LAN can connect all of the workstations, peripherals, terminals, and other devices in a single building. LAN makes it possible for using computer technology to efficiently share such things as files and printers, and to make possible communications such as e-mail. They combine data, communications, computing, and file servers together.

(4) MAN is an acronym for metropolitan area network. It is a high speed network that can carry voice, data and images up to 200Mb/s or faster over distances of up to 75km. Based on the network architecture, the transmission speed can be higher for shorter distances. A MAN is smaller than a wide area network but generally operates at a higher speed. The range is usually within a mile-perhaps one office, one building. Cellular phone systems expand the flexibility of MANs by allowing links to car phones and portable phones.

(5) Recognizing the efficiency and power of the Internet, many organizations apply Internet and Web technologies to their own internal networks. An Intranet is an internal network that uses Internet technologies. Intranets generally make company information accessible to employees and facilitate working in groups. Simple intranet applications include electronic publishing of organizational materials such as telephone directories, event calendars, procedure manuals, employee benefits information, and job postings. An intranet typically also includes a connection to the Internet.

Text 2: Application of Computer Networks

Computer networks have become an indispensable part of business, industry and entertainment in the short time they have been around. Some of the network applications in different fields are as follows.

1. Marketing and Sales

Computer networks are used extensively in both marketing and sales organizations. Marketing professionals use them to collect, exchange and analyze data relating to customer needs and product development cycles. Sales applications include teleshopping, which uses order-entry computers or telephones connected to an order-processing network, on-line reservation services for hotels, airlines, and so on.

2. Manufacturing

Computer networks are used today in many aspects of manufacturing, including the manufacturing process itself. Two applications that use networks to provide essential services are Computer-Assisted Design (CAD) and Computer-Assisted Manufacturing (CAM), both of which allow multiple users to work on a project simultaneously.

3. Searching with Directories

Directories are usually a good choice if you want information about a particular category, but have less of an exact subject in mind. A directory also uses a database, but that is usually screened by a human editor so it is much smaller, though often more accurate. For example, while a spider may classify a page about "computer chips" under the keyword "chips" together with information about potato chips, a human editor wouldn't. One of the largest directories—the Open Directory Project—claims to have indexed over 2 million Web pages using over 30,000 volunteer editors.

To use a directory located on a search site, categories are selected instead of typing keywords. After selecting a main category, a list of more specific subcategories for the selected main category is displayed. Eventually, after selecting one or more subcategories, a list of appropriate Web pages is displayed.

4. Payment Services

You need a valid checking account or national credit card (such as a VISA or MasterCard) and an e-mail address. You first need to set up an account with an online payment service, such as PayPal, Billpoint, etc. When you buy an item from a participating merchant or individual, you type your user name and password (on a secure Web page) to authorize the payment service to pay the merchant or individual a specific amount. The payment service then either charges your credit card or deducts the appropriate amount from your checking or payment service account; you usually receive confirmation of the transaction via e-mail.

5. Online TV and Movies

Online multimedia is still in the early stages, so what we have access to online today is likely just a drop in the bucket of what will be available a few years from now. The most common type of online TV and movie activity at the present time is viewing news clips, movie trailers, music videos, taped interviews, and similar short, prerecorded videos. These clips can be found on specialty sites dedicated to providing multimedia Web content; links to video clips are also widely found on news and entertainment sites.

Though some video files are played after they have been completely downloaded, because of the size of the files (and their corresponding long download times), it is becoming very common for multimedia files to use a streaming approach, where a small piece of the file is downloaded and buffered (temporarily stored on your hard drive), then the music or video begins playing while the remainder of the file downloads simultaneously. Video clips typically play using RealPlayer, QuickTime, or Windows Media Player. Depending on the speed of your Internet connection and PC, some video files may be choppy, if the playing needs to stop for a moment to wait for more of the file to be downloaded and buffered.

In addition to the types of video mentioned so far, some sites host short original films that can be played, usually for free. There are also Web sites that offer old TV shows that can be viewed on demand, as well as live TV broadcast over the Internet, though these applications are fairly limited at the current time. Television or movies delivered through your Internet connection are referred to as Internet TV, which is still in the infancy stages. However, it is expected to eventually become a very common Internet application, once broadband Internet connections become more widespread. Possibilities include receiving live broadcasts, movies on demand (paying for video downloads instead of physically renting videos at a video store), and interactive TV. This latter application—television synchronized with online activities—has been available on a very limited basis for several years.

6. Electronic Messaging

Probably the most widely used network application is electronic mail (e-mail). Now, we will introduce how to use the e-mails.

1) Introduction

Electronic mail, or e-mail, is an electronic system for sending and receiving messages and files over a computer network. It is one of the first applications to appear on the Internet and it remains the most widely used. Compared to the phone or paper-based documents, though, e-mail is still relatively new in the workplace.

If you are connected to the Internet, you can send an e-mail message to anyone who has an Internet e-mail address, regardless of whether or not they are on the Internet at the time you send the message. E-mail messages are routed from the sender's PC to his or her ISP, and then through the Internet to the receiver's ISP. When the receiver logs on to the Internet and requests his or her mail, the message is sent to that PC. To indicate what e-mail should be retrieved, the mail server being used and the appropriate e-mail address are specified in the browser's Options or Preferences screen. Some browsers allow multiple e-mail accounts (such as both a personal and school account) to be set up at one time. Others support only one, so the settings must be changed to check a different e-mail account when necessary.

Web-based e-mail—such as HotMail and Yahoo! Mail—works a little differently. Instead of specifying the e-mail settings in your browser, with Web-based e-mail you type your user name and password in the designated boxes on the mail service's Web site. This feature makes Web-based e-mail more flexible, since a user's e-mail can be accessed from any computer without changing the browser settings. Instead of being sent to your PC, Web-based e-mail is typically stored on the e-mail server and just viewed through the mail service's Web site. Web-based e-mail is a popular option for travelers, students, and other users who switch between computers frequently.

When you are using e-mail, here are a few guidelines to keep in mind:

- E-mail is increasingly used for professional purposes—Not long ago, e-mail was considered a secondary form of communication. Today, e-mail is a principal form of communication in most workplaces, so people expect e-mail messages to be professional and nonfrivolous.
- E-mail is a form of public communication—Your readers can purposely or mistakenly send your e-mail messages to countless others. So, you should not say things with e-mail that you would not say openly to your supervisors, coworkers, or clients.
- E-mail is increasingly formal—In the past, readers forgave typos, misspellings, and bloopers in e-mail messages, especially when e-mail was new and difficult to use. Today, readers expect e-mails to be more formal, reflecting the quality expected in other forms of communication.
- E-mail standards and conventions are still being formed—How e-mail should be used in the workplace is still being worked out. People hold widely different views about the appropriate (and inappropriate) use of e-mail. So, you need to pay close attention to how e-mail is used in your company and your readers' companies.

You should also keep in mind that legal constraints shape how e-mail is used in the workplace. E-mail, like any other written document, is protected by copyright law. So, you need to be careful not to use e-mails in any way that might violate copyright law. For example, if you receive e-mail from a client, you cannot immediately post it to your company's website without that client's permission.

Also, lawyers and courts treat e-mail as written communication, equivalent to a memo or letter. For example, much of the antitrust case against Microsoft in the late 1990s was built on recovered e-mail messages in which Bill Gates and other executives chatted informally about aggressively competing with other companies.

Legally, any e-mail you send via the employer's computer network belongs to the employer. So, your employers are within their rights to read your e-mail without your knowledge or permission. Also, deleted e-mails can be retrieved from the company's servers, and they can be used in a legal case.

2) Basic Features of E-Mail

(1) Head.

The header has lines for To and Subject. Usually, there are also lines like cc, bcc, and Attachments, which allow you to expand the capabilities of the message.

To line—Here is where you type the e-mail address of the person to whom you are sending the e-mail. You can put multiple addresses on this line, allowing you to send your message to many people.

cc and bcc lines—They are used to copy the message to people who are not the primary readers, like your supervisors or others who might be interested in your conversation. The cc line shows your message's recipient that others are receiving copies of the message too. The bcc line ("blind cc") allows you to copy your messages to others without anyone else knowing.

Subject line—It signals the topic of the e-mail. Usually a small phrase is used. If the message is a response to a prior message, e-mail programs usually automatically insert a "Re:" into the subject line. If the message is being forwarded, a "Fwd:" is inserted in the subject line.

Attachments line—It signals whether there are any additional files, pictures, or programs attached to the e-mail message. An attached document retains its original formatting and can be downloaded right to the reader's computer.

(2) Message Area.

After the header, the message area is where you can type your comments to your readers. It should have a clear introduction, body, and conclusion.

Introduction—The introduction should minimally define the subject, state your purpose, and state your main point. Also, if you want the reader to do something, you should mention it up front, not at the end of the e-mail.

Body—The body should provide the information needed to prove or support your e-mail purpose.

Conclusion—The conclusion should restate the main point and look to the future. Most readers of e-mail never reach the conclusion, so you should tell them any action items early in the message and then restate them in the conclusion.

(3) Signature.

E-mail programs usually let you create a signature file that automatically puts a signature at the end of your messages. Signature files can be both simple and complex. They allow you to personalize your message and add in additional contact information. By creating a signature file, you can avoid typing your name, title, phone number, etc. at the end of each message you write.

(4) Attachments.

Attachments are files, pictures, or programs that readers can download to their own computer.

Sending attachments—If you would like to add an attachment to your e-mail message, click on the button that says "Attach Document" or "Attachment" in your e-mail software program. Most programs will then open a textbox that allows you to find and select the file you want to attach.

Receiving attachments—If someone sends you an attachment, your e-mail program will use an icon to signal that a file is attached to the e-mail program will use an icon to signal that a file is attached to the e-mail message. Click on that icon. Most e-mail programs will then allow you to save the document to your hard disk. From there, you can open the file.

3) Managing E-mail

When you send e-mail, copies of the messages that you send are stored in a folder named

Sent, Sent Items, or something similar so that you can read them or resend them, if necessary. These messages remain there, and your retrieved e-mail messages remain in your Inbox folder, until you delete them or move them into a different folder (you can create new folders in either program using the mail program's File menu and then drag messages into the desired folder). Once an e-mail message is deleted, it is moved into a special folder for deleted items (called Trash in Netscape Mail and Deleted Items in Outlook Express)—you should clean out these folders periodically by selecting Empty Trash from Netscape's File menu or Empty "Deleted Items" Folder from Outlook Express' Edit menu to free up space on your hard drive.

New Words and Expressions

indispensable *adj.* 必需的，必不可少的
extensive *adj.* 广大的；广泛的
reservation *n.* 预定；预留
manufacturing *n.* 制造业
simultaneously *adv.* 同时地；同步地
category *n.* 种类
accurate *adj.* 精确的
classify *v.* 把……分类
valid *adj.* 有效的
authorize *vt.* 授权；批准
deduct *v.* 扣除；减去
transaction *n.* 事务；事项
bucket *n.* 水桶
trailer *n.* 预告片
dedicate *v.* 奉献
buffer *n.* 缓冲器
choppy *adj.* 波涛汹涌的
broadcast *n.* 无线电；电视节目
infancy *n.* 初级阶段
broadband *adj.* 多频率的
synchronize *v.* 使同步
indicate *v.* 表示
retrieve *v.* 重新获取
specify *v.* 规定；详述
flexible *adj.* 灵活的
frivolous *adj.* 轻薄的；轻浮的
supervisor *n.* 监督者
typo *n.* 打字排印错误
blooper *n.* 大挫折

constraint *n.* 约束；强制
violate *v.* 侵犯；冒犯
antitrust *adj.* 反垄断的
executive *adj.* 执行的
recipient *n.* 接受者
retain *v.* 保留；保持
item *n.* 条款
signature *n.* 签名
Netscape *n.* 网景公司

Exercises to the Text

1. Translate the following words and phrases into English.

（1）营销和销售 （2）计算机辅助设计 （3）计算机辅助制造 （4）网页 （5）电子邮件的地址 （6）网络电视 （7）电子邮件 （8）邮件服务器 （9）版权法 （10）发送附件

2. Translate the following paragraphs into Chinese.

(1) Computer networks are used extensively in both marketing and sales organizations. Marketing professionals use them to collect, exchange and analyze data relating to customer needs and product development cycles. Sales applications include teleshopping, which uses order-entry computers or telephones connected to an order-processing network, on-line reservation services for hotels, airlines, and so on.

(2) Computer networks are used today in many aspects of manufacturing, including the manufacturing process itself. Two applications that use networks to provide essential services are Computer-Assisted Design (CAD) and Computer-Assisted Manufacturing (CAM), both of which allow multiple users to work on a project simultaneously.

(3) Online multimedia is still in the early stages, so what we have access to online today is likely just a drop in the bucket of what will be available a few years from now. The most common type of online TV and movie activity at the present time is viewing news clips, movie trailers, music videos, taped interviews, and similar short, prerecorded videos.

(4) You should also keep in mind that legal constraints shape how e-mail is used in the workplace. E-mail, like any other written document, is protected by copyright law. So, you need to be careful not to use e-mails in any way that might violate copyright law. For example, if you receive e-mail from a client, you cannot immediately post it to your company's website without that client's permission.

(5) Legally, any e-mail you send via the employer's computer network belongs to the employer. So, your employers are within their rights to read your e-mail without your knowledge or permission. Also, deleted e-mails can be retrieved from the company's servers, and they can be used in a legal case.

Text 3: Network Security

Security is a broad topic and covers a multitude of sins. Most security problems are intentionally caused by malicious people trying to gain some benefit or harm someone.

Network security problems can be divided roughly into four intertwined areas: secrecy, authentication, nonrepudiation, and integrity control. Secrecy, also called confidentiality, has to do with keeping information out of the hands of unauthorized users. This is what usually comes to mind when people think about network security. Authentication deals with determining whom you are talking to before revealing sensitive information or entering into a business deal. Nonrepudiation deals with signatures: How do you prove that your customer really placed an electronic order for ten million left-handed doohickeys at 89 cents each when he later claims the price was 69 cents? Or maybe he claims he never placed any order. Finally, how can you be sure that a message you received was really the one sent and not something that a malicious adversary modified in transit or concocted?

And all these issues (secrecy, authentication, nonrepudiation, and integrity control) occur in traditional system, too, but with some significant differences. Integrity and secrecy are achieved by using registered mail and locking documents up.

Before getting into the solutions themselves, it is worth spending a few moments considering where in the protocol stack network security belongs. There is probably no one single place. Every layer has something to contribute.

Computer Security Timeline

There are several key events contributed to the conception and development of computer security. The following are some of the most significant events that brought attention to computer and information security and contributed to its importance today.

1. The 1960s

Students at the Massachusetts Institute of Technology (MIT) formed Tech Model Railroad Club and began exploring and programming the school's PDP-1 mainframe computer system. The group eventually used the term hacker in the context it is known today. Although originally a complimentary word for a computer enthusiast, "hacker" now has a derogatory connotation referring to a person who breaks into computer systems. Some hackers break into a computer for the challenge. Other hackers use or steal computer resources or corrupt a computer's data.

2. The 1970s

Jim Ellis and Tom Truscott created USENET, a bulletin-board style system for electronic communication between disparate users. USENET quickly became one of the most popular forums for the exchange of ideas in computing, Networking and, of course, cracking.

3. The 1980s

The Legion of Doom and the Chaos Computer club were two pioneering cracker groups that began exploiting vulnerabilities in computers and electronic data networks.

Based on the Computer Fraud and Abuse Act, the courts were able to convict Robert Morris, a graduate student, for unleashing the Morris Worm to over 6,000 vulnerable computers connected to the Internet.

4. The 1990s

The graphical Web browser was created and sparked an exponentially higher demand for public Internet access.

Vladimir Levin and accomplices illegally transferred U.S. $10 million in funds to several accounts by cracking into the CitiBank central database. Levin was arrested by Interpol and almost all of the money was recovered. Possibly the most heralded of all crackers is Kevin Mitnick, who hacked into several corporate systems, stealing everything from personal information of celebrities to over 20,000 credit card numbers and source codes for proprietary software. He was arrested and convicted of wire fraud charges and served five years in prison.

5. Security Today

In February of 2000, a Distributed Denial of Service (DDOS) attack was unleashed on several of the most heavily-trafficked sites on the Internet. The attack rendered yahoo.com, cnn.com, amazon.com, fbi.gov, and several other sites completely unreachable by normal users, as it tied up routers for several hours with large-byte ICMP packet transfers, also called a ping flood. The attack was brought on by unknown assailants using specially created, widely available programs that scanned vulnerable network servers, installed client applications called Trojans on the servers, and timed an attack with every infected server flooding the victim sites and rendering them unavailable. Many blame the attack on fundamental flaws in the way routers and the protocols are structured to accept all incoming data, no matter where it comes from or for what its purpose.

Common Methods of Attack

Confidential information can reside in two states on a network. It can reside on physical storage media, such as a hard drive or memory, or it can reside in transit across the physical network wire in the form of packets. These two information states present multiple opportunities for attacks from users on an internal network, as well as from those users on the Internet. We are primarily concerned with the second state, which involves network security issues. An attack is an attempt to bypass security controls on a computer. The attack may alter, release, or deny data. Whether an attack will succeed depends on the vulnerability of the computer system and the effectiveness of existing countermeasures. There are five common methods of attack that present opportunities to compromise the information on your network: network packet sniffers IP spoofing, denial-of-service attacks, password attacks, distribution of sensitive internal information to external sources, and man-in-the-middle attacks.

Security and Protection Technologies

1. Traditional Cryptography

Until the advent of computers, one of the main constraints on cryptography had been the ability of the code clerk to perform the necessary transformations, often on a battlefield with little equipment. An additional constraint has been the difficulty in switching over quickly from one cryptographic method to another one, since this entails retraining a large number of people. However, the danger of a code clerk being captured by the enemy has made it essential to be able to change the cryptographic method instantly, if need be.

The messages to be encrypted, known as the plaintext, are transformed by a function that is parametrized by a key. The output of the encryption process, known as the ciphertext, is then transmitted, often by messenger or radio. We assume that the enemy, or intruder, hears and accurately copies down the complete ciphertext. However, unlike the intended recipient, he does not know what the decryption key is and so cannot decrypt the ciphertext easily. Sometimes the intruder cannot only listen to the communication channel (passive intruder) but can also record messages and play them back later, inject his own messages, or modify legitimate messages before they get to the receiver (active intruder). The art of breaking ciphers is called cryptanalysis. The art of devising ciphers (cryptography) and breaking then (cryptanalysis) is collectively known as cryptology.

The real secrecy is in the key, and its length is a major design issue. Consider a simple combination lock. The general principle is that you enter digits in sequence. Everyone knows this, but the key is secret. A key length is two digits means that there are 100 possibilities. A key length of three digits means 1000 possibilities, and a key length of six digits means a million. The longer the key, the higher the work factor the cryptanalyst has to deal with. The work factor for breaking the system by exhaustive search of the key space is exponential in the key length. Secrecy comes from having a strong (but public) algorithm and a long key. To prevent your kid brother from reading your email, 64-bit keys will do. To keep major governments at bay, keys of at least 256 bits are needed.

Although we will study many different cryptographic systems in the pages ahead, there are two principles underlying all of them that are important to understand.

The first principle is that all encrypted messages must contain some redundancy, that is, information not needed to understand the message. An exam may make it clear why this is needed. Consider a mail-order company, The Couch Potato (TCP), with 60,000 products. Thinking they are being very efficient, TCP's programmers decide that ordering messages should consist of a 16-byte customer name followed by a 3-byte data field (l byte for the quantity and 2 bytes for the product number). The last 3 bytes are to be encrypted using a very long key known only by the customer and TCP.

At first this might seem secure, and in a sense it is because passive intruders cannot decrypt the messages. Unfortunately, it also has a fatal flaw that renders it useless. Suppose that a recently fired employee wants to punish TCP for firing her. Just before leaving, she takes (part of) the customer list with her. She works through the night writing a program to generate fictitious orders using real customer names. Since she does not have the list of keys, she just puts random numbers in the last 3 bytes, and sends hundreds of orders off to TCP.

When these messages arrive, TCP's computer uses the customer's name to locate the key and decrypt the message. Unfortunately for TCP, almost every 3-byte message is valid, so the computer begins printing out shipping instructions. While it might seem odd for a customer to order 137 sets of children's swings, or 240 sandboxes, for the computer knows, the customer might be planning to open a chain of franchised playgrounds. In this way an active intruder (the ex-employee) can cause a massive amount of trouble, even though she cannot understand the messages her computer is generating.

This problem can be solved by adding redundancy to all messages. However, adding redundancy also makes it much easier for cryptanalysts to break messages.

Thus cryptographic principle number one is that all messages must contain redundancy to prevent active intruders from tricking the receiver into acting on a false message. However, this same redundancy makes it much easier for passive intruders to break the system, so there is some tension here. Furthermore, the redundancy should never be in the form of number zeros at the start or end of a message, since running such messages through some cryptographic algorithms gives more predictable results, making the cryptanalysts' job easier. A random string of English words would be a much better choice for the redundancy.

The second cryptographic principle is that some measures must be taken to prevent active intruders from playing back old messages. If no such measures were taken, our ex-employee could tap TCP's phone line and just keep repeating previously sent valid messages.

2. Secret-Key Algorithms

Modem cryptography uses the same basic ideas as traditional cryptography, transposition and substitution, but its emphasis is different. Traditionally, cryptographers have used simple algorithms and relied on very long keys for their security. Nowadays the reverse is true: the object is to make the encryption algorithm so complex and involuted that even if the cryptanalyst acquires vast mounds of enciphered text of his own choosing, he will not be able to make any sense of it at all. Transpositions and substitutions can be implemented with simple circuits.

1) DES

In January 1977, the U.S. government adopted a product cipher developed by IBM as its official standard for unclassified information. This cipher, DES (Data Encryption Standard), was widely adopted by the industry for use in security products. It is no longer secure in its original form, but in a modified form it is still useful.

2) IDEA

IDEA (International Data Encryption Algorithm) was designed by two researchers in Switzerland. It uses a 128-bit key, which will make it immune to brute force, and also to withstand differential cryptanalysis. No currently known technique or machine is thought to be able to break IDEA.

3. Public-Key Algorithms

Historically the key distribution problem has always been the weak link in most cryptosystems. No matter how strong a cryptosystem was, if an intruder could steal the key, the system was worthless. Since all cryptologists always took for granted that the encryption key and decryption key were the same (or easily derived from one another) and the key had to be distributed to all users of the system, it seemed as if there was an inherent built-in problem: keys had to protected from theft, but they also had to be distributed so they could not just be locked up in a bank vault.

Public-key cryptography requires each user to have two keys: a public key used by the entire world for encrypting messages to be sent to that user, and a private key, which the user needs for decrypting messages. We will consistently refer to these keys as the public and private keys, respectively, and distinguish them from the secret keys used for both encryption and decryption in conventional (also called symmetric key) cryptography.

Although the RSA algorithm known by the initials of the three discovers (Rivest, Shamir, Adleman) is widely used, which method is based on some principles from number theory, it is by no means the only public-key algorithm known. The first public-key algorithm was the knapsack algorithm (Merkle and Hellman, 1978). Other public-key schemes are based on the difficulty of computing discrete logarithms (Rabin. 1979). Algorithms that use this principle have been invented by El Gamal (1985) and Schnorr (1991). A few other schemes exist. Such as those based on elliptic curves (Menezes and Vanstone, 1993), but the three major categories are those based on the difficulty of factoring large numbers, computing discrete logarithms, and determining the contents of a knapsack from its weight. These problems are thought to be genuinely difficult to solve because mathematicians have been working on them for many years without any great breakthroughs.

4. Authentication Protocols

Authentication is the technique by which a process verifies that its communication partner is who it is supposed to be and not an imposter. Verifying the identity of a remote process in the face of a malicious, active intruder is surprisingly difficult and requires complex protocols based on cryptography.

5. Digital Signatures

The authenticity of many legal, financial, and other documents is determined by the presence or absence of an authorized handwritten signature. And photocopies do not count. For computerized message systems to replace the physical transport of paper and ink documents, a solution must be found to these problems.

One approach to digital signatures called secret-key signatures is to have a central authority that knows everything and whom everyone trusts, says Big Brother (BB). Each user then chooses a secret key and carries it by hand to BB's office. Thus only Alice and BB know Alice's secret, KA, and so on.

A structural problem with using secret-key cryptography for digital signatures is that everyone has to agree to trust Big Brother. Furthermore, Big Brother gets to read all signed messages. The most logical candidates for running the Big Brother server are the governments, the banks, or the lawyers. These organizations do not inspire total confidence in all citizens. Hence it would be nice if signing documents did not require a trusted authority. Fortunately, public-key cryptography can make an important contribution here. It is public-key signatures.

There are also some social issues, such as the implication of network security for individual privacy and society in general, patents. And network security is politicized to an extent few; other technical issues are, and rightly so, since it relates to the difference between a democracy and a police state in the digital era.

6. Firewalls

A firewall is a security system designed to protect an organization's network against external threats. An enterprise with an intranet that allows its workers to have access to the wider Internet installs a firewall to prevent outsiders from accessing its own private data resources and for controlling what outside resources its own users have access to. Basically, a firewall, working closely with a router program, examines each network packet to determine whether to forward it toward its destination.

There are four generally accepted types of firewalls used on Internet connections: frame-filtering, packet-filtering, circuit gateways, and proxy servers. The following is a brief description of each type.

- Frame-filtering Firewalls: A frame-filtering firewall has the ability to filter to the bit level the layout and contents of a LAN frame. By providing filtering at this level, frames that do not belong on the trusted network are rejected before they reach anything valuable, even on the firewall itself.
- Packet-filtering Firewalls: A packet-filtering firewall can look at the packet IP address (network layer) and the types of connections (transport layer). They provide filtering based on that information. A packet-filtering firewall may be a stand-alone routing device or a computer that contains two network interface cards (dual-homed system). The router connects two networks and performs packet-filtering to control traffic between the networks. Administrators program the device with a set of rules that define how packet-filtering is done. Ports can also be blocked as part of packet-filtering. If a company security policy only allows Web browsing (HTTP), but not the rather dangerous FTP, packet-filtering can achieve this by implementing appropriate rules.
- Circuit-Level Gateways: Circuit-Level gateways are a type of proxy server that provides a controlled network connection between internal and external systems. A virtual "circuit" exists between the internal client and the proxy server. Internet requests go through this circuit to the proxy server. The proxy server delivers those requests to the Internet after changing the IP address. External users only see the IP address of the proxy server. Responses are then received by the proxy server and sent back through the circuit to the client. While traffic is allowed through, external systems never see the internal

systems.

- Proxy Server: An application-level proxy server provides all the basic proxy features and extensive packet analysis. When packets from the outside arrive at the gateway, they are examined and evaluated to determine if the security policy allows the packet to enter into the internal network. Not only does the server evaluate IP addresses, it also looks at the data in the packets to stop hackers from hiding information in the packets. If any company employee wants to access a server on the Internet, a request from the computer is sent to the proxy server. The proxy server then contacts the server on the Internet using its address as the source address. The information is sent back from the Internet server through the proxy server to the actual computer that requested the data. By doing this, the IP address of the internal (company) computer is never known outside its own network. Proxy servers also log the information on who makes requests as well as the transfer details in order to make an analysis of the Internet access.

New Words and Expressions

security *n.* 安全；抵押品；证券；保证
evaluated *adj.* 估价的
ongoing *adj.* 前进的，进行的 *n.* 前进，举止，行为
breakthrough *n.* 突破
interwined *v.* 缠绕
nonrepudiation *n.* 认可
mainframe *n.* 主机；大型机
derogatory *adj.* 贬损的
connotation *n.* 内涵；含蓄；暗示，隐含意义；储蓄的东西（词、语等）
vulnerabilities *n.* 缺陷（vulnerability 的复数形式）；脆弱点
exponentially *adv.* 以指数方式
coordinating *adj.* 协调的 *v.* 协调（coordinate 的 ing 形式）
transfers *n.* 迁移；传输（transfer 的复数）
v. 转移；变换；调任（transfer 的第三人称单数）
fundamental *adj.* 基本的，根本的 *n.* 基本原理；基本原则
opportunities *n.* 机会；机遇（opportunity 的复数）
vulnerability *n.* 弱点；易损性
irreparable *adj.* 不能挽回的；不能修补的
encrypted *v.* 把……编码；把……加密（encrypt 的过去分词）
sniffer *n.* 嗅探器；嗅探犬；以鼻吸毒者
authorized *adj.* 经授权的；经认可的 *v.* 批准；授权；辩护（authorize 的过去分词）
external *adj.* 外部的；表面的；外面的；[药]外用的；外国的
cryptography *n.* 密码使用法；密码学
ciphertext *n.* [计算机]密文

cryptanalyst *n.* 密码专家；密码破译者
algorithm *n.* 算法，运算法则
customer *n.* 顾客；[口]家伙
algorithms *n.* 算法；算法式（algorithm 的复数）
substitutions *n.* 代替；代用；替换（substitution 的复数）
inherent *adj.* 固有的；内在的；与生俱来的，遗传的
encryption *n.* 加密；加密术
mathematician *n.* 数学家
categories *n.* 分类；类别（category 的复数）
authentication *n.* 证明；鉴定；证实
firewall *n.* 防火墙 *vt.* 用作防火墙
examined *adj.* 验讫；检查过的 *v.* 检查；调查（examine 的过去式）

Exercises to the Text

1. Translate the following words and phrases into English.

（1）网络安全 （2）敏感信息 （3）注册邮件 （4）公告板系统 （5）图形化网页浏览器 （6）中央数据库 （7）国际刑事警察 （8）网络数据包嗅探器 （9）口令攻击 （10）拒绝服务攻击 （11）认证协议 （12）秘密密钥

2. Translate the following paragraphs into Chinese.

(1) Confidential information can reside in two states on a network. It can reside on physical storage media, such as a hard drive or memory, or it can reside in transit across the physical network wire in the form of packets. These two information states present multiple opportunities for attacks from users on an internal network, as well as from those users on the Internet.

(2) Public-key cryptography requires each user to have two keys: a public key used by the entire world for encrypting messages to be sent to that user, and a private key, which the user needs for decrypting messages. We will consistently refer to these keys as the public and private keys, respectively, and distinguish them from the secret keys used for both encryption and decryption in conventional (also called symmetric key) cryptography.

Text 4: Introduction to HTML5 Canvas

HTML5 is the current iteration of HTML, the HyperText Markup Language. HTML was first standardized in 1993, and it was the fuel that ignited the World Wide Web. HTML is a way to define the contents of a web page using tags that appear within pointy brackets (< >).

HTML5 Canvas is an immediate mode bitmapped area of the screen that can be manipulated with JavaScript. Immediate mode refers to the way the canvas renders pixels on the screen. HTML5 Canvas completely redraws the bitmapped screen on every frame by using Canvas API calls from JavaScript. As a programmer, your job is to set up the screen display before each frame is rendered so that the correct pixels will be shown.

This makes HTML5 Canvas very different from Flash, Silverlight, or SVG, which operate inretained *mode*. In this mode, a display list of objects is kept by the graphics renderer, and objects are displayed on the screen according to attributes set in code (that is, the *x* position, *y* position, and alpha transparency of an object). This keeps the programmer away from low-level operations but gives her less control over the final rendering of the bitmapped screen.

The basic HTML5 Canvas API includes a 2D context that allows a programmer to draw various shapes, render text, and display images directly onto a defined area of the browser window. You can apply colors; rotations; gradient fills; alpha transparencies; pixel manipulations; and various types of lines, curves, boxes, and fills to augment the shapes, text, and images you place onto the canvas.

In itself, the HTML5 Canvas 2D context is a display API used to render graphics on a bitmapped area, but there is very little in that context to create applications using the technology. By adding cross-browser-compatible JavaScript functionality for keyboard and mouse inputs, timer intervals, events, objects, classes, sound, math functions, and so on, you can learn to take HTML5 Canvas and create stunning animations, applications, and games.

Here's where this book comes in. We are going to break down the Canvas API into digestible parts and then put it back together, demonstrating how to use it to create applications. Many of the techniques you will learn in this book have been tried and used successfully on other platforms, and now we are applying them to this exciting new technology.

Browser Support for HTML5 Canvas

With the exception of Internet Explorer 8, HTML5 Canvas is supported in some way by most modern web browsers, with specific feature support growing on an almost daily basis. The best support seems to be from Google Chrome, followed closely by Safari, Internet Explorer 10, Firefox, and Opera. We will utilize a JavaScript library named modernizr.js that will help us figure out which browsers support which Canvas features.

What Is HTML5?

Recently the definition of HTML5 has undergone a transition. When we wrote the first edition of this book in 2010, the W3C HTML5 specification was a distinct unit that covered a finite set of functionality. This included things like new HTML mark-up, <video>, <audio>, and <canvas>tags. However, in the past year, that definition has changed.

So, what *is* HTML5 now? The W3C HTML5 FAQ says this about HTML5:

HTML5 is an open platform developed under royalty free licensing terms. People use the term HTML5 in two ways:

- to refer to a set of technologies that together form the future Open Web Platform. These technologies include HTML5 specification, CSS3, SVG, MathML, Geolocation, XmlHttpRequest, Context 2D, Web Fonts (WOFF) and others. The boundary of this set of technologies is informal and changes over time.
- to refer to the HTML5 specification, which is, of course, also part of the Open Web

Platform.

What we have learned through conversations and project work in the past few months is that, to the common person who does not follow this closely (or more likely, the common customer who needs something done right away), *it's all HTML5*, and therefore when someone says "HTML5", they are actually referring to the "Open Web Platform."

The one thing we are certain about regarding this "Open Web Platform" is that the one technology that was definitely left off the invite list was Adobe Flash.

So what is HTML5? In a nutshell, it is "not Flash" (and other like technologies), and HTML5 Canvas is the technology that has the best capability of replacing Flash functionality on the web and mobile web. This book will teach you how to get started.

The Basic HTML5 Page

Before we get to Canvas, we need to talk a bit about the HTML5 standards that we will be using to create our web pages.

HTML is the standard language used to construct pages on the World Wide Web. We will not spend much time on HTML, but it does form the basis of <canvas>, so we cannot skip it entirely.

A basic HTML page is divided into sections, commonly <head> and <body>. The new HTML5 specification adds a few new sections, such as <nav>, <article>, <header>, and <footer>.

The <head> tag usually contains information that will be used by the HTML <body> tags to create the HTML page. It is a standard convention to put JavaScript functions in the <head>, as you will see later when we discuss the <canvas> tag. There might be reasons to put some JavaScript in the <body>, but we will make every attempt to keep things simple by having all JavaScript in the<head>.

Basic HTML for a page might look like Example 1.

Example 1 A basic HTML page

```
<!doctype html>
<html lang="en">
<head>
<meta charset="UTF-8">
<title>CH1EX1: Basic Hello World HTML Page</title>
</head>
<body>
Hello World!
</body>
</html>
```

1. <!doctype html>

This tag informs the web browser to render the page in standards mode. According to the HTML5 spec from W3C, this is required for HTML5 documents. This tag simplified a long history of oddities when it came to rendering HTML in different browsers. This should always be the first line of HTML in a document.

2. <html lang="en">

This is the <html> tag with the language referenced: for example, "en" = English. Some of the more common language values are:

```
Chinese: lang = "zh"
French: lang = "fr"
German: lang = "de"
Italian: lang = "it"
Japanese: lang = "ja"
Korean: lang = "ko"
Polish: lang = "pl"
Russian: lang = "ru"
Spanish (Castilian): lang = "es"
```

3. <meta charset="UTF-8">

This tag tells the web browser which character-encoding method to use for the page. Unless you know what you're doing, there is no need to change it. This is a required element for HTML5 pages.

4. <title>…</title>

This is the title that will be displayed in the browser window for the HTML page. This is a very important tag, because it is one of the main pieces of information a search engine uses to catalog the content on the HTML page.

5. A Simple HTML5 Page

Now let's look at this page in a web browser. (This would be a great time to get your tools together to start developing code.) Open your chosen text editor, and get ready to use your preferred web browser: Safari, Firefox, Opera, Chrome, or IE.

(1) In your text editor, type in the code from Example 1.

(2) Save the code as CH1EX1.html in a directory of your choosing.

(3) Under the File menu in Chrome, Safari, or Firefox, you should find the option Open File. Click that selection. You should then see a box to open a file. (On Windows using Chrome, you might need to press Ctrl+O to open a file.)

(4) Locate the CH1EX1.html that you just created.

(5) Click Open.

You should see something similar to Figure 47.

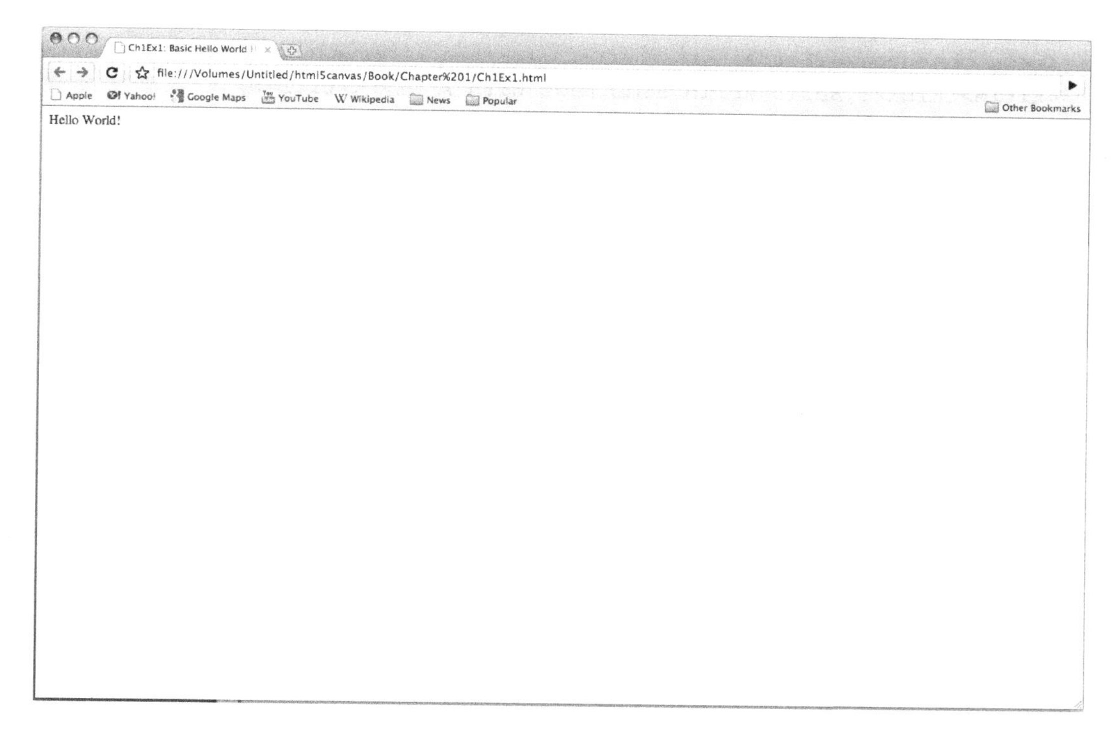

Figure 47 HTML Hello World!

Basic HTML We Will Use in This Book

Many HTML tags can be used to create an HTML page. In past versions of HTML, tags that specifically instructed the web browser on how to render the HTML page (for example, <font>and <center>) were very popular. However, as browser standards have become more restrictive in the past decade, those types of tags have been pushed aside, and the use of CSS (Cascading Style Sheets) has been adopted as the primary way to style HTML content. Because this book is not about creating HTML pages (that is, pages that don't have Canvas in them), we are not going to discuss the inner workings of CSS.

We will focus on only two of the most basic HTML tags: <div> and <canvas>.

1. <div>

This is the main HTML tag that we will use in this book. We will use it to position <canvas> on the HTML page.

Example 2 uses a <div> tag to position the words "Hello World!" on the screen, as shown in Figure 48.

Example 2 HTML5 Hello World!

```
<!doctype html>
<html lang="en">
<head>
```

```
<meta charset="UTF-8">
<title>CH1EX2: Hello World HTML Page With A DIV </title>
</head>
<body>
<div style="position: absolute; top: 50px; left: 50px;">
Hello World!
</div>
</body>
</html>
```

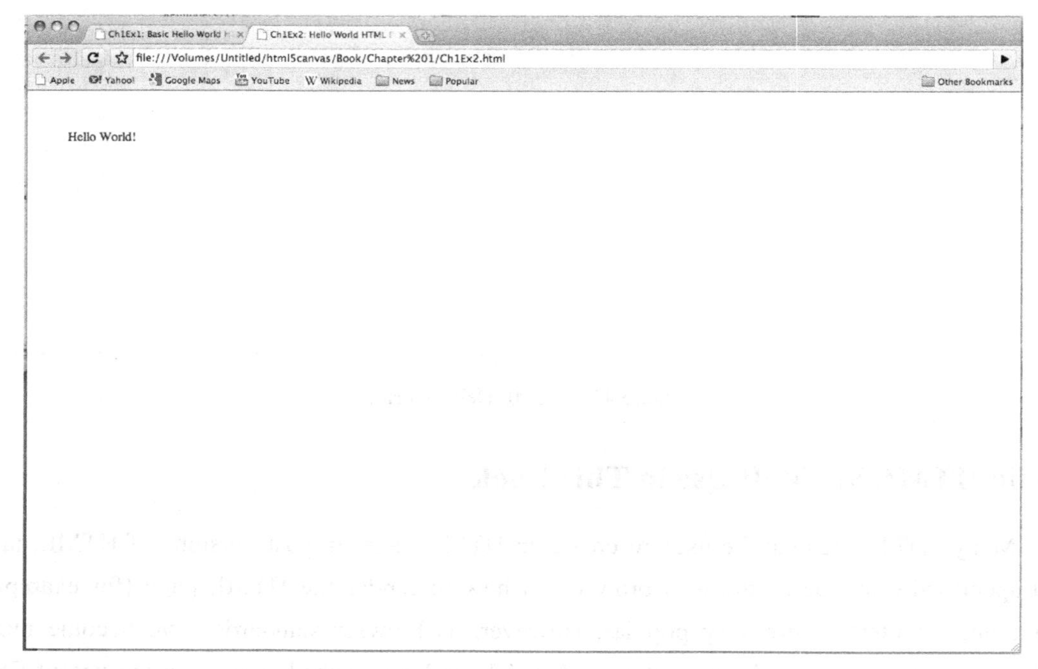

Figure 48　HTML5 Hello World! with a <div>

The style="position: absolute; top: 50px; left: 50px;" code is an example of inline CSS in an HTML page. It tells the browser to render the content at the absolute position of 50 pixels from the top of the page and 50 pixels from the left of the page.

This <div> might position the Canvas in the web browser, but it will not help us when we try to capture mouse clicks on the Canvas. In Chapter 5, we will discuss a way to both position the Canvas and capture mouse clicks in the correct locations.

2. <canvas>

Our work with <canvas> will benefit from using the absolute positioning method with <div>. We will place our <canvas> inside the <div> tag, and it will help us retrieve information, such as the relative position of the mouse pointer when it appears over a canvas.

The Document Object Model (DOM) and Canvas

The Document Object Model represents all the objects on an HTML page. It is language-neutral and platform-neutral, allowing the content and style of the page to be updated after it is rendered in the web browser. The DOM is accessible through JavaScript and has been a staple of JavaScript, DHTML, and CSS development since the late 1990s.

The canvas element itself is accessible through the DOM in a web browser via the Canvas 2D context, but the individual graphical elements created on Canvas are not accessible to the DOM. As we stated earlier, this is because Canvas works in immediate mode and does not have its own objects, only instructions on what to draw on any single frame.

Our first example will use the DOM to locate the <canvas> tag on the HTML5 page so that we can manipulate it with JavaScript. There are two specific DOM objects we will need to understand when we start using <canvas>: window and document.

The window object is the top level of the DOM. We will need to test this object to make sure all the assets and code have loaded before we can start our Canvas applications.

The document object contains all the HTML tags that are on the HTML page. We will need to look at this object to find the instance of <canvas> that manipulates with JavaScript.

JavaScript and Canvas

JavaScript, the programming language we will use to create Canvas applications, can be run inside nearly any web browser in existence. If you need a refresher on the topic, read Douglas Crockford's *JavaScript: The Good Parts* (O'Reilly), which is a very popular and well-written reference on the subject.

Where Does JavaScript Go and Why?

Because we will create the programming logic for the Canvas in JavaScript, a question arises: where does that JavaScript go in the pages we have already created?

It's a good idea to place your JavaScript in the <head> of your HTML page because it makes it easy to find. However, placing JavaScript there means that the entire HTML page needs to load before your JavaScript can work with the HTML. This also means that the JavaScript code will start to execute before the entire page loads. As a result, you will need to test to see whether the HTML page has loaded before you run your JavaScript program.

There has been a recent move to put JavaScript right before the </body> at the end of an HTML document to make sure that the whole page loads before the JavaScript runs. However, because we are going to test to see whether the page has loaded in JavaScript before we run our <canvas>program, we will put our JavaScript in the traditional <head> location. If you are not comfortable with this, you can adapt the style of the code to your liking.

No matter where you put the code, you can place it inline in the HTML page or load an *external*.js file. The code for loading an external JavaScript file might look like this:

```
<script type="text/javascript" src="canvasapp.js"></script>
```

To make things simple, we will code our JavaScript inline in the HTML page. However, if you know what you are doing, saving an external file and loading it will work just as well.

In HTML5, you no longer have to specify the script type.

New Words and Expressions

iteration *n.* [数] 迭代；反复；重复
undergo *v.* 经历
oddity *n.* 奇异；古怪；怪癖
language-neutral 语言中立的

Exercises to the Text

1. Translate the following words and phrases into English.

（1）超文本标记语言 （2）尖括号 （3）位图屏幕 （4）渐变色填充 （5）alpha 透明度 （6）像素处理 （7）开放式网络平台 （8）字符编码

2. Translate the following paragraphs into Chinese.

(1) HTML5 Canvas is an immediate mode bitmapped area of the screen that can be manipulated with JavaScript. Immediate mode refers to the way the canvas renders pixels on the screen. HTML5 Canvas completely redraws the bitmapped screen on every frame by using Canvas API calls from JavaScript. As a programmer, your job is to set up the screen display before each frame is rendered so that the correct pixels will be shown.

(2) So what is HTML5? In a nutshell, it is “not Flash” (and other like technologies), and HTML5 Canvas is the technology that has the best capability of replacing Flash functionality on the web and mobile web.

附录A 参考译文*

第一部分 数 字 媒 体

课文 1：多媒体概述

分布式多媒体系统的发展已经开始大大地影响了多媒体点播服务的发展。研究人员正在对已有的计算机技术进行移植，并开发新技术。多媒体的巨大发展前景表明多媒体将是计算机产业、通信产业和广播产业三者结合而形成的一个新的产业。

多媒体系统综合了种类繁多的信息源，如语音、图形、动画、图像、声音和全动态视频，涵盖了三大产业：计算机产业、通信产业和广播产业。

多媒体计算技术的研发主要分为两个方向：一个方向主要是独立多媒体工作站及其相关的软件系统和工具，如音乐作曲、计算机辅助学习和交互视频。另一个方向是把多媒体计算处理和分布式系统结合起来，这具有更广阔的发展前景。基于分布式的多媒体系统的潜在的新应用包括多媒体信息系统、协作系统和会议系统、多媒体点播服务和远程学习。

多媒体系统的典型特征是它综合应用连续媒体，包括语音、视频和动画。分布式多媒体系统要求相对长时间的连续数据传输（如从远程摄像机播放录像）、媒体同步、大容量存储器和适合多媒体数据类型的特殊索引和检索技术。

1. 技术要求

多媒体系统能够存储音频和视频信息，供以后在应用中使用，如用于训练，或者以实时方式进行“现场”传输。“现场”声音和视频可以是交互的，如多媒体会议系统；也可以是非交互的，如电视播放。相似地，存储的静态图像也可以以交互方式（如浏览和检索）或者以非交互方式（如幻灯片）进行使用。

多媒体应用的复杂性对计算机系统的所有组成部分提出了高要求，多媒体要求计算机具备强大的处理能力来实现编码译码、执行多媒体文件系统和相应的文件格式，计算机系统的体系结构必须提供高总线带宽和高效率的I/O处理能力。

多媒体操作系统应该支持新的数据类型、实时调度和快速中断处理。外存储器和内存储器要求具有很大的容量、快速存取时间和高传输率。新型网络及其协议必须支持多媒体所需要的高带宽、低延时和低抖动的要求。我们需要面向对象的和对用户友好的软件开发工具以及对不同协议的大型联网分布式多媒体系统很重要的检索和数据管理工具。

研究人员正在现有的计算机领域内进行工作，把现有的技术移植到多媒体中去，或者开发新的多媒体技术。这些研究包括快速处理器、高速网络、大容量存储器、新的算法和数据结构、视频和音频压缩算法、图形系统、人机界面、实时操作系统、面向对象程序设

---

* 涉及图表的内容在译文中进行了删改。

计、信息存储和检索、超文本和超媒体、脚本语言、并行处理方法和分布式系统的复杂体系结构。

2．多媒体压缩

音频、图像和视频信号产生大量的数据。显然，压缩技术在数字多媒体应用系统中扮演了关键的角色。当前的多媒体系统需要数据压缩有三个方面的原因：多媒体数据的大容量存储要求，存储器存取速度较低不能实时播放多媒体数据（尤其是视频数据）和网络带宽不能满足实时视频数据传输。

数字数据压缩采用各种算法，由软件或硬件来实现。压缩技术可分为无损压缩和有损压缩两大类。无损压缩技术可以完全恢复源数据，有损压缩技术恢复时则损失一定的精确度，但后者具有更高的压缩率，因此，在图像和视频压缩中有损压缩比无损压缩更常用。

有损压缩技术可以进一步分为预测压缩、频率压缩和基于重要性的压缩。预测压缩法（如 ADPCM）可以通过观察前面的值来预测后面的值。面向频率的压缩方法使用与快速傅里叶变换相关的离散余弦变换（DCT）。基于重要性的压缩方法利用图像的其他特征作为压缩的基础，如 DVI 技术利用了颜色查找表和数据过滤。

混合压缩技术综合了几种方法，如离散余弦变换和向量量化或者差分脉码调制。各种组织已经在 JPEG、MPEG 和 px64 标准的基础上建立了数字多媒体压缩标准。

实现压缩/解压缩算法时，关键问题是如何在硬件与软件之间取得平衡，以取得最好的性价比。大多数压缩算法使用专用视频处理器和可编程数字信号处理器（DSP）来实现。然而，功能强大的 RISC 处理器正使纯软件的解决方案成为可能。我们可以把压缩算法的实现方法分为三类：(1)取得最高性能的硬连线方案（如 C cube）；(2)强调通用处理器的灵活性的软件实现方案；(3)使用专用视频处理器的软硬件共同处理方法。

3．多媒体联网

许多应用，如视频邮件、视频会议和合作工作系统，要求网络化的多媒体。在这些应用中，多媒体对象存储在服务器中，并回放到客户站点。这些应用可能要求向远处的各个站点播放多媒体数据，或者访问大型的多媒体资源库。

在传统的局域网环境中，数据资源只在局部可用，由于许多原因，不支持访问远程的多媒体数据资源。

传统的网络不适合多媒体。以太网带宽只有 10Mbps，其访问时间没有限制，且延时和抖动都无法预测。令牌环网具有 16Mbps 带宽，也具有稳定性；从这一点来看，令牌环网可用于处理多媒体。然而，可预测的最坏情况的访问等待时间可能很长。

FDDI 网络可提供 100Mbps 的带宽，足够用于多媒体。在同步模式中，FDDI 网络具有低访问反应时间及低抖动的特点，并保证有限的访问延迟和对同步通信的可预测的平均带宽。然而，由于成本高，FDDI 网络主要用作骨干网而不是作为连接工作站的网络。

较便宜的一些选择包括增强型传统网络，如具有 100Mbps 的带宽的快速以太网和优先令牌环网。

目前的光网络技术能支持宽带综合业务数字网（B-ISDN）标准，它有望成为多媒体应用系统关键网络。B-ISDN 访问可以是基本的，也可以是主要的。基本 ISDN 访问支持 2B+D

通道，一个 B 通道的传输率为 64kbps，D 通道的传输率为 16kbps。主 ISDN 访问在美国支持 23B+D，在欧洲支持 30B+D。

推荐的 B-ISDN 网络可以在同步传输模式（STM）或异步传输模式（ATM）中处理恒定和可变比特率的通信流应用。STM 提供固定带宽的通道，因此，STM 对于处理多媒体应用中不同类型的通信流显得不够灵活。ATM 却适用于多媒体通信，它通过指定称为单元（cell）的固定长度的包，为虚拟连接在带宽分配方面提供了很大的灵活性。ATM 通过利用缓冲存储器和按照统计规律多路复用突发流量，虽然有单元的延迟和丢失，但能增加带宽效率。

4．多媒体系统

几类技术的进展使多媒体系统在技术上和经济上可行。这些进展包括：功能强大的工作站，海量存储器，高速网络，图像和视频处理（如动画和图形），音频处理（如音乐合成与音效），语音处理（语音识别和文本/语音转换）以及先进的静态、视频、音频和语音压缩算法。

多媒体系统由三个关键部分组成：多媒体硬件、操作系统和图形用户界面、多媒体软件开发工具和编辑工具。自 1989 年第一代多媒体系统成功开发以来，多媒体系统可以分为三代。

第一代多媒体系统基于 Intel 80386 和 Motorola 68030 处理器，具有位图和动画，JPEG 图像压缩技术，基于以太网和令牌环网的局域网及超媒体编辑工具的特点。第二代多媒体系统使用 i80486 和 MC68040 处理器，具有动态和静态图像、16 位音频、JPEG 和 MPEG-1 视频压缩、FDDI 网络，集成文本、图形、动画和声音的面向对象的多媒体编辑工具。

我们目前处于由第二代向第三代多媒体系统过渡的阶段，基于功能更强大的处理器，如 Pentium 处理器和 Power PC 处理器。第三代将使用全动态、具有 VCR 质量的影像，最终达到 NTSC/PAL 和 HDTV 质量。压缩算法将包括 MPEG-2、MPEG-3 和 MPEG-4，也许包括现在还处于研究阶段的小波方法。多媒体系统将使用扩展以太网、令牌环网和 FDDI 网以及新型的同步网络和 ATM 网络。编辑工具将把面向对象的多媒体集成在操作系统中。

5．应用

多媒体系统具有广泛的应用前景。已在使用的三个重要应用是多媒体邮件系统、协同工作系统和多媒体会议系统。

多媒体邮件系统比通常的电子邮件系统更复杂，实现了多种应用，如多媒体编辑和语音邮件，比仅传输纯文本的系统要求更高的传输率。

协同工作系统允许组员讨论问题，共同完成某一工作。在开会期间，用户可以查看、讨论和修改多媒体文档。

多媒体会议系统使许多参与者能够通过语音和数据网络交换各种多媒体信息。每一参与者拥有一台多媒体工作站，用高速网络与其他工作站连接起来。每一参与者能够发出和接收图像、声音和数据，参与一定的合作活动。多媒体会议应用了共享虚拟工作空间的概念，用来描述在每一工作站重复的显示。

多媒体会议系统必须提供许多功能，如多方呼叫设置、会议状态传输、音频和视频的实时控制、网络资源的动态分配、多端口数据传输、共享工作区的同步以及出错情况下的妥善处理。

6．研究方向

高速网络的研究和发展将为分布式多媒体应用提供所需的带宽。因此，我预感分布式多媒体及其应用将有巨大的发展。

分布式多媒体系统的进展已经对点播多媒体业务产生了重大的影响，如交互娱乐、视频新闻发布、视频租借服务和数字多媒体图书馆。已有众多的公司意识到，随着计算机技术和压缩技术的进步，光纤网络不久将能够用于传输数字电影。在过去一年，在娱乐、有线电视、电话和计算机公司之间形成了一些联盟，其主要着眼点就是视频点播的应用。

阻碍多媒体进一步发展的许多棘手的问题还有待研究和解决。多媒体应用系统对计算机硬件和软件资源提出了很高的要求。因此，开发功能更强的工作站将是长期要求。多媒体工作站还需要多媒体操作系统（MMOS）和先进的多媒体用户界面。多媒体操作系统应该提供抢先多任务处理、易扩展性、独立于格式的数据存取来处理连续媒体，支持实时调度。它应当是面向对象的，具有对即时得到的数据进行同步处理的能力，因此用户界面必须完善而且直观。

把用户界面集成到操作系统中能够为应用软件开发人员省去许多麻烦。其他研究方向上的挑战包括开发新的实时压缩算法（也许基于小波）、大容量存储设备和多媒体数据管理系统。人们将进一步优化具有低反应时间、低抖动特点的高速、稳定的网络以及继续研究新的多媒体同步算法。

课文2：数字图像处理简介

数字图像处理方法的研究源于两个主要的应用领域：一是为了便于人们分析并改进图像信息；二是为使机器自动理解而对图像数据进行处理、存储、传输及显示。

1．什么是数字图像处理？

一幅图像可定义为一个二维函数$f(x, y)$，这里x和y是空间坐标，而在任何一时空间坐标(x, y)上的幅值f称为该点图像的强度或灰度。当x、y和幅值f为有限的、离散的数值时，我们称该图像为数字图像。数字图像处理是指用数字计算机处理数字图像，值得注意的是，数字图像是由有限的元素组成的，每一个元素都有一个特定的位置和幅值，该元素称为图像像素、画面元素或像素。像素是广泛用于表示数字图像元素的术语。

视觉是人类最高级的感知器官。所以，毫无疑问，图像在人类感知能力中扮演着最重要的角色。然而，人类感知只限于电磁波谱的可见波段，成像机器则可覆盖几乎整个电磁波谱，从伽马射线到无线电波。它们可以对人类不习惯的图像源产生的图像进行加工，这些图像源包括超声波、电子显微镜和计算机生成的图像。因此，数字图像处理涉及各种各样的应用领域。

有时人们将图像处理界定为输入和输出内容都是图像的一门学科。我们认为这一定义只是人为的界定和限制，例如，如果按照这个定义，甚至最普通的计算一幅图像灰度平均值的工作都不能算作图像处理。另一方面，有些领域（如计算机视觉）研究的最终目标是用计算机去模拟人类视觉，包括理解和推理并根据视觉输入采取行动等。这一领域本身是人工智能的分支，其目的是模仿人类智能。人工智能领域处在其发展过程中的初期阶段。

它的发展比预期的要慢得多，图像分析（也称为图像理解）领域则处在图像处理和计算机视觉两个学科之间。

从图像处理到计算机视觉这个连续的统一体内并没有明确的界线。然而，在这个连续的统一体中可以考虑三种典型的计算处理（即低级、中级和高级）来进行区分。低级处理涉及初级操作，如降低噪声的图像预处理、增强对比度和图像锐化。低级处理的特点是输入、输出都是图像。中级处理涉及分割（把图像分为不同区域或目标）、对目标进行描述、减小图像以使其更适合计算机处理，并且对单独目标进行分类（识别）。中级图像处理的特点是输入为图像，但输出是从这些图像中提取的属性（如边缘、轮廓及不同物体的标识等）。最后，高级处理涉及在图像分析中被识别目标的总体理解和在连续统一体边缘执行与视觉相关的识别函数。

根据上述讨论，我们看到图像处理和图像分析两个领域存在逻辑重叠区域，即对图像中特定区域或目标的识别。这样，我们界定数字图像处理包括输入和输出均是图像的处理，同时也包括从图像中提取特征及识别特定目标的处理。以简单的文本自动识别为例来具体说明这一概念。首先获取一幅包含文本的图像，对该图像进行预处理，提取（分割）字符；然后以适合计算机处理的形式描述这些字符；最后识别这些字符。所有这些操作都属于我们界定的数字图像处理的范围内。

2．数字图像处理的基本步骤

把后续各章涉及的材料划分为两个主要类别是有帮助的。一类是其输入和输出都是图像；另一类是输入可能是图像，但是输出是从图像中提取的特征属性。

图像获取是第一步处理。注意，获取与给出一幅已经是数字形式的图像一样简单。通常，图像获取包括如缩放比例等预处理。

图像增强是数字图像处理最简单和最有吸引力的领域。基本上，增强技术后面的思路是显现那些被模糊了的细节，或简单地突出一幅图像中的感兴趣的特征。一个图像增强的例子是增强图像的对比度，使其看起来好一些。增强是图像处理中非常主观的领域，记住这一点很重要。图像增强处理并不比其他技术更重要，而是说我们要用增强处理引导读者更好地运用其他技术。另外，我们没有用一章专门讲述数学预备知识，而是通过说明在图像增强中用到的数学知识来介绍一些所需要的数学概念。这样处理使读者能在运用图像处理技术的过程中熟悉这些概念。一个较好的例子就是傅里叶变换。

图像复原也是改进图像外貌的一个处理领域。然而，不像图像增强是主观的，图像复原是客观的。在某种意义上说，复原技术倾向于以图像退化的数学或概率模型为基础。另一方面，图像增强则是以怎样构成好的增强效果这种人的主观偏爱为基础的。

彩色图像处理已经成为一个越来越重要的领域，因为在互联网上，数字图像的应用在不断增长。

小波是在各种分辨率下描述图像的基础。这些理论被用于图像数据压缩及金字塔描述方法。在这里，图像被成功地细分为较小的区域。

顾名思义，压缩所涉及的技术是减小保存图像所需的空间，或者减小传输时的带宽。虽然存储技术在过去的 10 年内有了很大改进，但对于传输容量我们还不能这样说，尤其对于在互联网上的应用更是如此，因为互联网是以大量的图片内容为特征的。大多数计算机用户对于与图像压缩技术对应的图像文件扩展名是很熟悉的（也许是无意识的），如 JPG 文件扩展名对应的 JPEG（联合图像专家组）图像压缩标准。

形态学处理涉及提取图像成分的工具，这些图像成分在表现和描述形状方面非常有用。

分割过程将一幅图像划分为组成部分或目标。通常，自主分割是数字图像处理中最为困难的任务之一。复杂的分割过程导致要成功解决成像问题需要大量的处理工作，因为成像要求分别识别出各个目标。另一方面，不权威不规律的分割算法几乎总是会导致最终失败。总之，分割越精确，识别越成功。

表示与描述几乎总是跟随在分割步骤的输出之后，通常这一输出是未加工像素数据，其构成不是区域的边缘（也就是区分一个图像区域和另一个区域的像素集）就是其区域内部的所有点。无论哪种情况，必须把数据转换成适合计算机处理的形式。首先，必须确定数据是应该被表示为边界还是整个区域。当注意的焦点是外部形状特性（如拐角和曲线）时，则表示为边界是合适的。当注意的焦点是内部特性（如纹理或骨骼形状）时，则表示为区域是合适的。在某些应用中，这两种表示方法是互为补充。选择一种表现方式仅是解决把原始数据转换为适合计算机后续处理的形式的一部分。还必须确定一种方法来描述数据，以突出我们感兴趣的特征。描述也叫特征选择，涉及提取特征，该特征是某些感兴趣的定量信息或是区分一组目标与其他目标的基础。

识别是基于目标描述符给目标赋以符号的过程。

关于问题域的知识以知识库的形式被编码装入一个图像处理系统。这一知识可能像详细设计一幅图像的有趣信息所在区域那样简单，这样，限制了为找寻那些信息所需要进行的搜索。知识库也可能相当复杂，如材料检测问题中所有主要缺陷的相关列表或者图像数据库，该库包含与变化侦查应用相关的地区的高分辨率卫星图像。除了引导每一个处理模块的操作，知识库还要控制模块间的相互作用。

3．图像处理系统的部件

20 世纪 80 年代中期，世界各地出售的各种型号的图像处理系统基本都是由主机和与主机相配合的外设装置构成。20 世纪 80 年代末和 90 年代初，市场已转为将图像处理硬件设计为与工业标准总线兼容并能配合工程工作站机箱和个人计算机的单板形式。除了降低价格外，这一市场转变还如催化剂一样催生了大量新公司，这些公司的任务是开发用于图像处理的软件。

虽然针对大规模图像应用（如处理卫星图像）的大规模图像处理系统一直在出售，但趋于小型化，并趋向于将通用的小型计算机和专用的图像处理硬件的混合。在下面的段落中，我们会阐述每一部分的功能，从图像感知开始。

关于感知，需要两个部件以获取数字图像：第一个是物理设备，该设备对我们希望成像的物体辐射的能量很敏感；第二个称为数字化装置，数字化装置是一种把物理感知装置的输出转换为数字形式的设备。例如，在数字视频摄像机中，传感器产生一个与光强成比例的输出，数字化装置把该输出转化为数字数据。

专业的图像处理硬件通常由刚刚谈到的数字化装置与执行其他初期操作[如算术逻辑单元（ALU）]的硬件组成，算术逻辑单元以并行的方式对整个图像执行算数和逻辑的操作。举例说明算术逻辑单元的使用，在对图像求平均值时，要求与数字化一样快的速度，目的是降低噪声。有时这种类型的硬件叫作前端子系统，它的显著特点是速度快。换句话说，这个单元可以执行要求数据快速吞吐的功能（例如，在30帧/秒的速度下对视频图像进行数字化求和的平均值），通常的主机是不能胜任这项工作的。

在图像处理系统中的计算机是通用的计算机，其范围从PC到超级计算机。在专门应用中，有时也采用特殊设计的计算机以达到所要求的性能水平。但是，我们感兴趣的还是通用图像处理系统。在这些系统中，几乎任何配置较好的PC都适用于离线图像处理任务。

图像处理软件由执行特定任务的专业模块组成。一个设计优良的软件包还包括为用户写代码的能力，如最小化就使用专用模块完成。完善的软件包允许那些模块和至少用一种计算机语言编写的通用软件命令集成。

大容量存储能力在图像处理中是必需的。一幅图像的尺寸是1024×1024px，每像素的亮度采用8比特量化，如果图像不压缩，需要1MB的存储空间。当处理数千甚至数百万幅图像时，在图像处理系统中很难提供足够的存储空间。图像处理应用的数字存储分成三个主要类别：(1)处理过程中的短期存储；(2)关系到快速调用的在线存储；(3)档案存储，其特点是不经常地存取。存储是以字节（8bit），千字节（1KB），兆字节（1MB），吉字节（1GB），太字节（1TB）为单位的。

提供短期存储的一种方法是使用计算机内存。另一种方法采用专门的板子，该存储板叫作帧缓存，它可以存储一帧或多帧图像并可以以视频帧速率（30帧/秒）快速存取。后一种方法允许实际意义上的快速图像变焦、卷动（垂直移动）和摇动（水平移动）。帧缓存通常放在专业的图像处理硬件单元中。在线存储一般采取磁盘或光介质存储。在线存储的关键特性参数是对存储数据的存取频率高。最后，档案存储是以海量存储要求为特点的，但无须频繁存取。放在类似于投币电唱机的盒子内的磁带和光盘通常使用档案存储。

现在使用的图像显示器主要是彩色（更好的是纯平屏幕的）电视监视器。监视器由图像和图形显示卡的输出驱动，这些显示卡是计算机系统的主要组成部分。对于图像显示应用不可能没有这样的要求，即作为计算机系统的一部分，其显示卡应满足商用性要求。在有些场合还要求立体显示，立体显示是采用戴在用户头上的帽子（目镜上嵌有两个小的显示屏）来实现的。

用于记录图像的硬拷贝装置包括激光打印机、胶片照相机、热敏装置、喷墨单元和数字单元（如光盘、CD-ROM光盘）等。胶片的分辨率最高，而纸张显然是书面材料的首选介质。为了描述图像，图像可显示在胶片幻灯片上或者使用图像投影设备显示在数字介质上。后面的方法作为图像描述标准正得到越来越多的认同。

网络在今天的任何计算机系统中几乎是默认使用的功能。因为图像处理应用中存在大量固有数据，所以在图像传输中主要考虑的问题是带宽。在专用网络中，这不是一个典型问题。但对于经由互联网的远程通信就不一样了。幸运的是，由于光纤和其他宽带技术的发展，这个状况正得到迅速改善。

课文3：数字音频

在早期的计算机上，我们能够听到计算机发出的声音就是伴随着出错信息的嘟嘟声，而现在各种类型的声音计算机都能够播放，包括音乐、旁白、音效和事件的原始声音（比如演讲或音乐会）。计算机中的声音元素通常称为数字音频，并且它是多媒体的基本组成部分。

我们对听到的声音有了一个大体认识，这样就可以了解音频专家所指的“波形”图的含义。

波形并不是它看上去的那样紧凑，本质上，一个波形是在声波传播过程中空气压力随时间变化的曲线图。y 轴表示空气压力，x 轴表示时间。

一个声波到达，空气压力上升，波形线也上升，而当声波经过空气压力下降则波形也下降。这些变化发生得相当快——一秒钟几千次。因此，即使是代表很短时间的声音的波形图也是相当大的。这就是为什么在计算机上看到声音的波形相当复杂并且是弯弯曲曲的。

应该知道描述波形的几个术语，当讨论数字音频时它们很有用。

零线：曲线图中间的水平线被称为零线，它表示初始状态，也就是空气没有被压缩或稀释时的状态。

振幅：振幅用来描述波形上任一点的被压缩量或被稀释。它表示相对于零线上或下的距离。通常，振幅越大，声音的音量越大。

周期：一个周期表示波形的振幅回到相同幅度所用的时间。

频率：声音的频率表示为每秒钟的周期数。频率越高，声音越尖锐。人耳能听到的声音频率是 20Hz~20kHz，通常随着年龄的增长上限会降低。其他物种有不同的听力范围，狗的平均听力范围可以高到 45 000Hz，猫能听到 63 000Hz 的声音，白鲸听见的频率可以到 123 000Hz。

数字音频的出现取决于它在记录、处理、大批量生产和声音的传播方面的有效性。现代音乐的传播通过因特网的在线商店，这依赖于数字记录与数字压缩算法。以数字文件传播的音频优于以物理介质传播的方式，它大大降低了传播费用。

数字音频利用数字信号对声音进行再现，这包括模数转换与数模转换、存储和传输。

数字音频信号始于模数转换器，它能将模拟信号转换为数字信号。模数转换器在特定的采样频率与量化精度下进行转换。例如，CD 音频的采样速率为 44.1kHz，每个声道的量化精度为 16 位。如果模拟信号没有进行限波，那么在转换之前先要通过抗混叠滤波器，以防止在数字信号中产生混叠。

声波能经过话筒，引起内部振动膜振动，这个振动引起从话筒到计算机电线中的电压的变化，这个变动的电压就是声音的模拟表示，因为这个振动平滑地从一个振幅到下一个，之间包括了所有值。在计算机内部模数转换器（通常是声卡的一部分）中，在规定的时间间隔测定话筒的模拟信号，并且在某个精确瞬间输出信号振幅的一系列数据，这个过程就叫作“采样”。然后以时间顺序排列采样数据，形成一个原始波形的“点图”。

采样速率也称作采样频率，它是从连续信号变为离散信号过程中每秒钟的采样数。对于时域信号来说，以赫兹进行衡量。常用的三个采样速率为 11.025kHz、22.05kHz 和 44.1kHz。采样速率越高，采样个数越多，所得数字音频质量越高。

数字音频的采样所包含的数据是转换成模拟信号时，提供所需信息进行声波的重现。采样之后要进行量化，量化一般采用量化位数。在数字音频中，量化位数描述了用于记录每个采样点所包含信息的位数。量化位数直接与数字音频数据的每个采样点相对应。常用的两个量化位数为 8 位和 16 位。常用的量化位数的实例是记录 16 位的 CD 音频和支持 24 位的 DVD 音频。标准的 CD 音频数据率为 44.1kHz/16，表示音频数据每秒采样 44100 次，量化位数为 16 位。CD 音轨通常为立体声，使用左右音轨，因此每秒的音频数据就是单声道的两倍。位速率为 44 100×16×2=1 411 200b/s 或 1.4Mb/s。

数字音频信号可以被存储或传输，它能够存储在 CD 光盘、MP3 播放器、硬盘、U 盘、压缩闪存中或其他数字存储设备中。音频数据压缩技术——比如 MP3，高级音频编码或 Flac——能够减少文件存储容量。数字音频可以以媒体的形式传送到其他设备上。

数字音频的最后一步是利用数模转换器转换回模拟信号。如同模数转换器一样，数模转换器以特定的采样速率和量化精度进行转换，但是这个过程可能为过采样、高采样、低采样，此时的采样速率可以不同于初始采样值。

采样与编码完成之后，音频信号就能存储或传输，音频数据可以以未压缩形式或减少文件大小的压缩形式进行存储。音频文件格式是在计算机系统上存储音频数据的格式。

有必要区分音频文件格式与编码解码器。编码解码器完成原始音频数据的编码与解码，而这些数据是以特定音频文件格式存储的一个文件。尽管多数音频文件格式只能支持一个音频编码解码器，但是一个文件格式能够像 AVI 一样支持多种编码解码器。

音频文件的三种主要格式如下：

- 未压缩音频格式，比如 WAV、AIFF 和 AU。
- 无损压缩格式，比如 FLAC、Monkey's Audio（文件扩展名为 APE）、WavPack（文件扩展名为 WV）、Shorten、Tom's lossless Audio Kompressor（TAK）、TTA、Apple Lossless and lossless Windows Media Audio（WMA）。
- 有损压缩格式，比如 MP3、Vorbis、Musepack、lossy Windows Media Audio（WMA）和 AAC。

最主要的一种未压缩音频格式是 PCM，它通常存储为 Windows 的.wav 文件或苹果机操作系统上的.aiff 文件。WAV 是一种灵活性很强的文件格式，被设计用于存储或多或少与采样速率或量化速率相结合的数据，这是一种存储和获取原始记录最有效的文件格式。无损压缩格式对于相同的记录时间将需要更长的处理过程，但在空间使用上更加有效。WAV 文件如同其他未压缩格式一样，对所有的声音编码，不管它们是复杂的声音或者只是静音，每个时间单位都有相同的字节数。例如，如果都是以 WAV 进行存储，一个存储一分钟交响乐演奏的文件与一个一分钟只有静音的文件大小是一样的。如果文件以无损压缩的形式进行编码，第一个文件将会小一些，而第二个文件将不会占用任何存储空间。然而，将一个文件编码成无损格式所用的时间将远远超过编码成 WAV 文件所用的时间。最新的无损格式已经被开发出来（比如 TAK 格式），它的目的就是要达到较快的编码速度和较好的压缩比率。

下面列出了常用的音频文件格式类型。

- WAV——主要用于 Windows 操作系统中的标准音频文件格式。通常用于存储未压缩的文件（比如 PCM 文件）、CD 音质的声音文件，这就意味着大容量存储——大约每分钟 10MB，波形文件也利用多种编码解码进行编码以减少文件大小。WAV 文件使用 RIFF 文件结构形式。
- FLAC——一种无损压缩编码解码形式。这种格式就像 zip 压缩文件格式，只不过只用于音频。如果将PCM文件压缩成FLAC 文件，那么它就能完美地对原始声音进行回放（其他的编码解码器都是有损的，即质量则有些降低）。这种无损压缩的代价就是压缩比率较低。Flac 文件被推荐用于高质量要求的 PCM 文件（例如用于广播和音乐）。

- AIFF（音频交换文件格式）——苹果机上标准的音频文件格式。就像苹果机上使用的 WAV 文件一样。
- RAW——一个 RAW 文件能够包含任何编码形式的音频，但是它通常使用 PCM 音频数据。除了技术测试之外，很少使用它。
- AU——用在 Sun、UNIX 和 Java 上的标准音频文件格式。
- VOX——VOX 格式通常使用对话式自适应脉冲编码调制形式进行编码解码。如同其他自适应脉冲编码调制格式，它压缩到4位。VOX 格式的文件与 WAV 文件相类似，除了 VOX 文件不包含文件自身的信息，因此在播放 VOX 文件之前必须指定采样速率和通道数。
- AAC——AAC（高级音频编码）格式是基于 MPEG2 和 MPEG4 标准生成的。AAC 文件通常存储 ADTS 或 ADIF。
- MP3——MPEG-1 音频标准第3层压缩格式，它广泛用于下载和存储音乐。通过减少声音的冗余部分，MP3 文件可以压缩到相当 PCM 文件的 1/10，并且保持较高的音频质量。
- WMA——微软公司的受欢迎的 Windows Media Audio 格式。
- RA——设计用于在因特网上音频流媒体传输的 Real Audio 格式。RA 格式允许文件在计算机中独立存储，而所有的音频数据都包含在文件中。

现今，数字音频作为一个存储方式趋向于比模拟音频使用更广泛。唱片和录音带仍然在使用，但是相对而言只占很小的市场份额。为什么数字音频如此流行？虽然仍然有一些关于数字音频声音质量是否比模拟音频高的争论，但是数字音频确实比较容易复制，并且以较少的质量损失处理音频文件。如今正是由于数字音频，使得不管是业余爱好者还是专业音乐家都能很轻松地制作出音乐棚中的高质量的音乐。

课文4：资产管理

影视内容创作正变得越来越依赖文件。广播公司必须以更少的投资传播更多的内容。无线广播得到卫星作为补充，有线电视提供视频点播（VOD）和实现互动，现在都与通过电信公司网络传播的交互式网络电视（IPTV）竞争。手持设备亦可接收广播电视和视频点播。

多种格式的内容的创建要针对不同传输渠道。现在的“空中下载”节目，在将来可以下载到移动设备上。而且现在的节目内容几乎同时出现在几种媒体上。以不同媒体传播的几个月甚至几年前的旧节目已经被盗版。出版商已学会如何让节目内容以公众愿意采取的消费方式，在新鲜出炉时就能赚到钱。

节目内容创作者和出版商都希望使用数字资产管理（DAM）来提高生产力，在一个以文件为基础的生产环境下提供合理的管理。

本文介绍资产管理的概念。首先，它回答了这个问题“什么是资产？”，并进行了相关方案的一些介绍——媒体管理、文档管理和内容管理。最后讨论了为什么数字资产管理对任何企业都是有用的，而不仅仅对于媒体和出版业。

DAM 证明了我们面临着相当大的技术挑战。直到 20 世纪末才出现了不太昂贵的可用系统。除了本地和远距离的计算机网络不断增加的带宽，处理成本和磁盘存储成本的下降已经成为主要因素。许多设计用来管理大型资料库的产品，可以对劳动密集型的、重复的任务实现自动化。这包括对内容进行编目和索引以及我们用来搜索内容强大的搜索引擎。

要对数字资产索引和编目，产品开发者已经投入对语音理解、符号识别和概念分析的研究。这些想法的广泛应用需要相当强的处理能力来提供内容的实时获取，在许多音视频应用方面起到了至关重要的作用。我们现在认为理所当然地引入低成本的强大的计算机工作站，使得这样的技术具备商业上的可行性。

现代商业创建了许多形式的媒体内容。传统上，各部门管理自己的媒体内容，通常有一些正式的文件结构。如果某个公司没有陷入混乱、使用错误的标志、采用过时的公司形象，并在网站上难以找到的话，全球品牌管理都要求在企业范围内实现内容共享。

内容现在已有了许多不同的传播渠道。目前的电子渠道主要有电视、iTV（交互电视）、互联网和网播、移动电话以及无线个人数码助理（PDAs，俗称掌上电脑）。除此之外还有传统的以印刷为基础的媒体，如商品目录、小册子、邮件以及展示广告。重新进行内容企划，让这些不同的渠道具备引人注目的力量，对于内容出版商和聚合器是一个巨大的挑战。宣传册设计者或视频制作者熟悉的以项目为基础的方法已经被一个人才合作网所取代。由于文件数量多，格式也各不相同，也会有诸多出错的可能。

内容和媒体资产管理（MAM）已经成为出版业核心的业务部门的应用。这些系统非常适合对借助多种新技术出版的大型内容库进行控制和管理。

1．数字资产管理（DAM）

文件管理已经在传统的出版业证明了它的价值。内容管理使大型网站能得到有效的部署。随着发行平台数量的增加，品牌管理的要求和控制成本的愿望，导致了创作和制作过程的自然结合。大型企业可以不再让出版、网络和多媒体制作孤立地运作。“富媒体”这个术语已经应用到这个融合的领域中。

对于一些企业来说，这个结合可能只适用于内容创作，它也可以延伸到市场营销的担保和企业通信管理。直接与消费者交流的公司可能需要一个互动电视广告来链接到公司网站，客户可以查看网页或者要求更多的信息。所有这些不同的媒体必须展示出一个完美准确的品牌形象。

同样运用于内容管理的原则，可作为管理富媒体资产的基础体系。富媒体的复杂性意味着 DAM 现在对提高运营效率和控制成本越来越重要。

富媒体制作可能会从传统的音频和视频材料开始，后期加上同步的图形和文本而得到提升。使用不同的创造性人才的条件下，制作流程比传统的电视生产的线性过程具备更多的并行处理。降低生产成本的压力是极为主要的，市场仍然要求以种类繁多的形式出版内容。

2．什么是资产

在快速浏览一下字典后我们会知道，这个词通常和财产相关。数字媒体内容与财产也有相同的联系。如果您对内容拥有知识产权，那么这些内容可以代表一种资产。因此，知识产权管理和保护（IPMP）、数字版权管理（DRM）应成为资产管理框架的必要的组成部分。

没有使用权的内容不是资产。

DAM 使用元数据将内容与权限信息联系起来。保护产权的相同的系统也能用来保护机密信息，因此 DAM 也可用于商业敏感信息的内部分配使用。

资产管理不仅仅是权限管理——它构成一个小的，但至关重要的整体解决方案的一部分。也许最重要的特点是，DAM 为成功的媒体资产货币化提供了一个框架。

它也存在潜在的缺点。如果系统即将用于一个成熟的公司，将有大量的已有内容。那么索引和编目存档的成本是多大？这些费用可能超过在线访问旧材料的任何潜在优势。像任何其他主要的业务决策那样，对 DAM 部署应包括成本效益分析。

有一个折中的方案能实现部分数字化，而不是全面合并档案材料的每一段内容。比如说，可以将最后访问日期作为一个标准。仅仅因为你有访问内容的权限，容易忽略它的价值，因此它完全可能存储在库。如果存储成本高，你甚至可以处理掉它。

是什么赋予资产价值呢？如果它可以转售，那么它的价值是显而易见的。但是如果它能够在成本方面重新进行有效策划，然后用在新的材料中，那么它也可以代表金融资产。举一个例子，通过对媒体档案的有权访问，一个新的广告活动能够借鉴以前的教训和经验。在研究这些项目方面，可以为公司节省时间和资源。

3．什么是资产管理

具体来说，DAM 提供对一个大型资产信息库的合理开发和管理。一个资产管理系统为作者、出版者以及媒体的终端用户提供了一个完整的工具箱，从而有效地利用资产。

媒体资产可以以不同的格式出现：音频和视频剪辑文件、照片、图形图像和文本文件。在资产之间也可能有联系和关联。这些文件可以在整个企业范围内共享。媒体可能被买卖，它可以被合并、出租或出售。

为满足所有这些要求，该系统架构结构必须是灵活的。下面列举数字资产管理系统可能具备的功能和组成部分：

- 合著
- 流程
- 存储管理
- 搜索工具
- 档案文件
- 出版工具
- 多种格式
- 广域分布
- 版本控制

资产管理可以延伸至基于 Web 的访问。这就有了在远离内容的桌面上预览和创作出版的机会。这种访问可以在企业内部的任何拨号位置或在互联网上进行。

4. 为什么要使用资产管理

即使是最精心设计的文件归档方案都有其缺点。当资料以多种格式在一个大的企业中共享时，这一点尤其明显。对一个用户而言最适合的方案往往不适合另一个用户。在不同的平台上支持多种文件格式会使得方案复杂化；文件可能位于 Windows 或 UNIX 服务器上，或专业的视频服务器、录像带和光盘上。

用户可能不知道文件名字。为了帮助找出所需的资料，文件归档方案需要一个目录。设想一个类似“帮我找到一个在拥挤的海滩上穿着红裙子看起来快乐的孩子”的要求。资产管理对类似这样的问题做出回应。

课文 5：虚拟现实与应用

1. 虚拟现实

虚拟现实（VR）是一种计算机技术，在虚拟境界创造了三维的真实的感觉。虚拟现实应用在现实生活中的许多方面，从商业规划到制造业和娱乐业。

1）虚拟现实的历史

虚拟现实拥有既有趣又复杂的富有意义的历史。大概 30 多年前，一位名为莫顿·海利格的年轻的电影摄影师想利用 72%的观众视野创造一个终极的全景体验。

由于无法获取任何财政支持，海利格无法实现自己的梦想，然而，他创造了一个叫作“Sensorama 模拟器”的装置，在 20 世纪 60 年代初期面世。这种虚拟工作站利用三维视频，由摄影师身上的三个 35mm 的相机实现。这个装置融合了全三维相机取景和立体声音响。观众可以骑摩托车，同时能感受到风（由风扇模拟）甚至道路的起伏。这个机器是简单粗糙的，但它开始引发了许多新的想法。

1966 年，伊凡·萨瑟兰，一个美国犹他州大学的研究生，继续了海利格的研究。他开创了图形加速器的理念，这是现代虚拟仿真的一个不可或缺的组成部分。军方很快认识到这个理念在飞行模拟中的潜力，并于 20 世纪 70 年代花了大部分时间设计头盔，这是一种可以模拟夜景的头盔。此外，美国国家航空和宇宙航行局（NASA）开始在太空飞行中使用这项技术的研究，后来用在了登月行动中。海利格的发明改变了计算机世界以及计算机本身的发展。

使虚拟现实成为可能的关键技术的融合，是在过去的 10 年产生的。在过去 10 年，我们可以看到对虚拟现实世界的实现中一些至关重要的领域的发展。这些发展包括液晶和 CRT 显示设备，高性能图像生成系统，追踪系统。因为由集成电路发展到 MIPS（每秒能处理百万条指令）时代，人们渐渐能生产出高速、高性能系统。因此，进行非军事研究是可能的，而且研究也由美国发展到其他国家，如日本、法国，特别是德国。具体技术的发展包括显示技术（如上所述），人机界面和成像。显示技术的发展对虚拟现实规范的进步起到了重要作用。

虚拟现实技术可分为：

- 真实环境的模拟，如通常以培训或教育为目的的建筑物和飞船的内部环境。
- 一个设想的环境的开发，通常用于游戏或教育的探索。

在个人计算机上能够实现虚拟现实效果的受欢迎的产品包括 Bryce、Extreme 3D、Ray Dream Studio、TrueSpace、3D Studio MAX 和 Visual Reality。虚拟现实建模语言（VRML）允许创作者使用文本语言语句为他们的显示和交互指定图像和规则。

2）VRML

VRML（虚拟现实建模语言）是一个在因特网上开放的、可扩展的、工业标准的三维场景（或境界）描述语言。使用 VRML 和 Netscape 公司的 Live3D，用户可以自己创作和查看分布式的互动三维世界，这里有着丰富的文本、图像、动画、声音、音乐，甚至视频。

- VRML 嵌入 HTML 文档

VRML 境界一旦被创建，它可以嵌入 HTML 文件并用<EMBED>标记。使用< EMBED>标记放置在 HTML 文档中的 VRML 境界，和使用<IMG>标记在 HTML 文档中放置二维图像类似。在下面的示例中演示了如何把一个 VRML 文件称为 example.wrl，嵌入到一个 HTML 文档中：

```
<EMBED SRC="example.wrl" WIDTH=128 HEIGHT=128 BORDED=0 ALIGN=middle>
```

- 性能

即使是最有热情的用户对于慢如蜗牛的网页都不会太有耐心。对于 VRML 的作者们，这是一个关键问题，因为 VRML 是基于计算密集型的三维图形，可能包含其他资源密集型的媒体。如同 HTML 文件，下载时间是在虚拟境界创作的一个重要因素。一旦下载完成，虚拟境界创作可能需要更多的客户端系统资源。一种快速的浏览器将在一定程度上抵消这种劣势，但重要的是要建立有效的虚拟境界。如何使用下列内容会影响网页的性能。

- 多边形

VRML 境界的图形是由多边形构成的。形状越复杂，需要的多边形就越大。例如，一个立方体通常只包括 12 个多边形，因为每一面是由两个三角形构成的。相反，一个看似简单的球体区域，可能需要 200 多个三角形。随着越来越多的对象添加到一个虚拟境界中，多边形的数量也增加了。每当用户对 VRML 虚拟境界的看法发生变化，浏览器都得重新绘制这个场景。这个境界包含的多边形越多，则重绘过程需要的时间越长。因此，尽可能减少多边形的数量是提高用户浏览速度的一种方式。

- 纹理

VRML 允许将纹理映射到形状上。纹理用在 VRML 虚拟境界可能会极大程度地增加它的大小。这将影响下载和重绘时间。因此，如果使用纹理，小纹理是能够保证较短的下载时间和较高浏览速度的一条可取的途径。此外，如果用于 VRML 虚拟境界的纹理使用的颜色较少，需要的客户资源也较少。

- 实例

如果用户定义对象，对象可以被重新使用在 VRML 虚拟境界中。这项技术可以帮助虚拟境界文件保持小的尺寸，这种技术被称为实例。虽然也有一些实例局限性，它的使用可以使 VRML 代码更容易被编写和维护，VRML 虚拟境界更容易被下载。

- 多细节层次

在现实世界中，当用户越接近一个对象，就可以看见更多的细节。多细节层次（LOD）使上述事实可在 VRML 虚拟境界中成为可能。该 LOD 节点决定在 VRML 场景中坐标定义的范围内将看到哪些对象，允许特效和逼真的模拟。

- 内联

其他虚拟境界文件可以被“拉入”到某一个境界来帮助建立一个 VRML 场景。当使用这一技术时，这些文件被称为内联。该 WWWInline 节点被用来指一个被列入的境界文件，并有选择地显示一个边框来提示用户该对象或对象组在渲染前将放置在哪里。

- 压缩

VRML 境界文件越大，下载时间越长。通过使用如 GZIP 的工具可以对虚拟境界文件进行压缩。如果一个 VRML 浏览器识别出这个文件类型，它可以自动解析该压缩文件来显示 VRML 虚拟境界。

2．虚拟现实技术的应用

虚拟现实可以说是在人类和电脑之间一个非常特别的互动。在虚拟现实中，人戴着一个“头盔显示器（HMD）”——眼镜、护目镜，或在每只眼睛前放置一个小屏幕的头盔。装上追踪系统使 HMD 和电脑以及其他的一些导航工具，如三维鼠标、魔杖或高科技手套相连，那么这个人就可以开始一场真正引人入胜的计算机之旅。

一旦身处在虚拟境界中，这个人可以看到前面、上面、旁边、下面和后面的计算机图形。该旅客会有一种强烈的视觉感受，感觉到身临计算机图形环境中。

现在常见的系统在 20 世纪 60 年代开始形成。虚拟现实的第一个实际的用途是用于开发军方飞行员使用的平视显示器。

虽然研究人员认为在完全发掘它的潜力之前，虚拟现实还有很长的路要走，但是在实际使用中，虚拟现实被应用在诸多行业，如军事、娱乐、教育和商业。在教育方面，预计在虚拟现实方面的投入会有巨大的增长。如美国华盛顿大学和北卡罗来纳大学就投入许多资源进行虚拟现实的研究，研究其在物理学上的影响及其在现实世界的应用。

在教育方面的下一步就是将它应用于改善学习进程，并创造新的学习系统。虚拟现实点燃了那些有远见的教育工作者的激情和想象力，他们认为虚拟现实是以计算机为基础的教育的下一个合理步骤。

在商业应用中，由辛格、瑞迪富森和其他人建成的使用高速专用电脑，价值数百万美元的飞行模拟器，引领了虚拟现实在商业方面的应用。F-16 战斗机和波音 777，罗克韦尔航天飞机的飞行员在真实飞行之前已经模拟了很多次。在加州海洋学院和其他商船军官培训学校，电脑控制的模拟器用来讲授油轮，集装箱船复杂的装货和卸货。在华盛顿大学一直赞助的虚拟境界的商业联合会中，虚拟境界和商业交叉发展。如果目前证明虚拟现实的确是一个强大的学习工具，今后在企业和商业的培训将对这项基本的虚拟学习方法加以利用。

第二部分　数字媒体应用软件

课文1：Photoshop简介

在打开Photoshop时，会看到以下四个项目：菜单栏、工具选项栏、工具箱、图层面板。这些被用来作为许多操作的捷径。

菜单栏——选择某一类别的菜单项——“文件”和“编辑”菜单比较常用，用来保存和打开文件以及复制和粘贴图像。

工具箱——工具箱中包含输入、选择、填充、绘制、编辑、移动和浏览图片的工具。

下面是经常使用的一些工具。

矩形选择工具——用于选择部分图形。

移动工具——用来微调选中物。

裁剪工具——用于缩减图片的尺寸。

画笔工具——用于绘制和润色图像。

油漆桶工具——用于给所选对象进行填充。

文本工具——用来添加文本。

吸管工具——用于在一张照片上选择特定的颜色。

放大镜工具——用来放大图像。

工具选项栏——大多数工具在工具选项栏里显示相应的选项。当选择不同的工具时选项是不同的。

图层控制面板——图层为组织和处理图像的各种各样的组成部分提供了一个强有力的方式。通过在一个单独的图层上放置一个单独的元素，能容易地编辑和安排那个元素，不会干涉图像的其他部分。

1．创建一个新的图像

选择“文件”|“新建”命令

以英寸、像素为单位（网页图片的尺寸标准）来设置宽度和高度。

对于这个图像，分别设置宽度和高度为600和100，单击“确定”按钮。

2．创建文本图形

单击图层面板的“新建图层”按钮。

单击“文本工具”按钮。

单击“设置前景颜色”按钮打开“颜色拾取”对话框。

选择只适用于网页的颜色。

选择喜爱的颜色并单击“确定”按钮。

在选项栏，设置字体为Arial Black并且改变字体大小为48pt。

输入Photoshop。

注意：设置前景色和背景色为黑白的快捷方式是按D键。

注意：当为网站制作图像时，使用仅适用于 Web 的颜色是一个明智的做法。这些颜色在各种显示器上看起来是相同的。

3．保存图像

以下是 Photoshop 中保存图像的几个选项。

“文件”|“保存”——保存文件，重复使用“保存”命令将覆盖原始文件。

“文件”|“另存为”——用一个新的名字，在一个新的地点保存一个文件，然后在新命名的文件上工作。

“文件”|“另存为 Web”——当使用这个选项时，将优化文件为 Web 使用。

在这里选择“文件”和“另存为”命令，现在需要选择文件格式。以下是保存图像最常用的格式。

.psd——Photoshop 文件格式，保存你的图像文件及所有图层效果，以后可以重新编辑。

.gif——图形交换格式，使用这个格式保存网页图片，可以压缩文件加快在网页上下载图片的速度。

.jpg——联合图像专家组。适合用于照片保存，网页浏览器只使用 216 种颜色。

将文件保存为.psd 格式，以便以后编辑和修改它。

课文 2：Flash 动画制作

本文将介绍一些 Flash 动画制作中最基础的知识，包括动画理论、Flash 的工作环境以及简单的动画设计。

1．动画的基本理论

在学习动画设计之前，首先要了解一个问题，也就是动画到底是什么？实际上，一系列静态的画面连续快速播放就构成了动画。动画与电影一样都是利用人眼的视觉暂留来产生动态的感觉。在 Flash 中通常把动画称为电影（movies），两者的概念是相似的，这里也就不做区分了。

动画中每一张静态的动画被称为“帧”。动画播放的速率正是由每秒播放多少帧来决定的，也就是 fps（Frame Per Second，帧/秒）。在计算机上常见的动画速率一般是 8 帧/秒到 12 帧/秒，尤其在网络上播放的动画更不要把帧速设定得太高。如果设的太高，网络频率来不及配合输送足够的影音数据，会导致播放时断时续。Flash 动画是许多帧按顺序排列而形成的。说到这里，以前没有接触过计算机动画设计的朋友一定会有疑问，如果要创作一个有 100 帧的动画，是不是要先做出 100 张图呢，这样不是太麻烦了吗？Flash 解决了这个问题，它利用了计算机强大的计算机功能，采用了一种叫作关键帧的技术，大大减少了动画设计的工作量。

下面就来谈谈这个重要的概念：关键帧。所谓关键帧，也就是该帧中的内容与先前帧中的内容有很大的区别，因而它呈现出关键性的动作及内容的变化。

2．Flash 的工作环境

为了便于学习，不至于造成名词上的误解，下面简单地介绍一下 Flash 的工作环境。Flash 的标准工作环境包括菜单栏、工具栏、舞台、时间轴窗口、工具面板。除了这几个主要的部分，通过 Windows 菜单，还可以调出素材窗口等小窗口。

1）工作区

工作区就是 Flash 的工作平台，它是一个比较大的区域，实际上涵盖了下面要说的舞台以及画图的编辑动画的工作对象。

2）舞台

舞台是 Flash 动画中各个元素的表演平台，舞台将显示当前选择的帧的内容。与工作区不同的是：动画发布后只有在舞台上的内容才能被看到，而舞台之外的工作区中的内容就如同在后台的演员和工作人员一样，不会被观众看见。

3）场景

就像戏剧可以有几幕一样，舞台上可以放下几个场景。注意舞台右上角，有两个小按钮，其中的第一个就是场景切换按钮。可以通过不同的场景之间的交互性，来创作出非常复杂的作品。

4）时间轴窗口

时间轴窗口用于对 Flash 的两个基本元素：图层和帧进行操作。在系统默认设置下，时间轴窗口以编辑栏的形式出现在舞台的上面，紧靠上框。使用者可以根据需要或个人喜好，用鼠标拖动该编辑栏到 Flash 界面的其他位置，甚至将其拖出边框，使之成为可自由移动的浮动窗口。时间轴控制窗口分为左右两个区域，它们分别是层控制区和时间轴控制区。

Flash 的层与 Photoshop 中的层的概念是差不多的，它们都是透明的，只不过在 Photoshop 中是指图层，而在 Flash 中是指动画层。使用层可以设定动画中各元素的上下叠放次序。层控制区位于时间轴窗口的左边，是进行图层显示和操作的主要区域，由几个图层功能操作示意栏和按钮组成。当前舞台中正在编辑作品的所有层的名称、类型、状态都会按照图层的放置顺序排列在图层示意栏中。在层控制区中，不但可以显示当前作品的层及所属信息，还可以对某一个或部分层进行操作，其中包括添加层、删除层和重新安排层的顺序。这里要提醒大家的是，层的使用不会增加文件的规模，相反，合理地使用层可以在设计时使各个层层次清晰且易于编辑，所以最好在初学的时候就养成使用层的习惯。

时间轴窗口的下半部分是时间轴控制区域，该区域主要由若干与左边层示意栏对应的帧序栏、示意栏、信息提示栏以及一些用于控制动画显示和操作的工具按钮组成。它用于对各帧的播放和放置进行有效控制。

利用时间轴窗口安排好素材的空间位置和时间位置能够制作出比较好的动画效果。

5）工具面板

如果没有看见工具面板，可选择 Window→Toolbar 命令。Flash 的工具面板包括 Flash 所有选择工具和绘图工具。工具栏包括两个区域：一个是选择区，它在工具面板的上半部，用于选择所需要的工具；另一个是属性区，用于对所选工具进行一些属性的设定。

6）素材库窗口

在 Flash 中，素材库窗口起着组织、管理动画内的全部基本元素的作用，每个 Flash 文件都包含各种元素，而这些基本元素全部存放在素材库内，使用户容易对这些元素进行查找、编辑和设定。

素材库窗口可以变换大小以适应用户当前的需要。如果按下“窄库视图”按钮，则素材库窗口以窄库状态出现，仅显示素材名。如果按下“宽库视图”按钮，则素材库窗口就以宽库状态出现，显示素材名、类型、使用次数等几组信息。在显示状态按钮上方有一个排序键，可以使各列按照升序或降序来排列。

在编辑 Flash 作品时，还可以调用其他 Flash 作品的素材元件库，但前提条件是作品还未转化为动画播放文件。要调用其他作品元件库，必须先打开这些文件。

7）菜单栏和工具栏

除了绘图命令以外的绝大多数命令都可以在菜单栏中实现。这与 Windows 下的程序菜单栏的使用基本相同，只要知道名称就可以了。工具栏上放置了一些常用工具按钮，同样也很易于操作。

有几个部件对于理解 Flash 界面是非常重要的。

- 标准工具栏：用于打开、保存、新建和打印的图标以及几个 Flash 专用图标。
- 绘图工具栏：类似于许多其他绘图软件包的绘图工具。
- 状态栏：将鼠标光标停留在一个绘图工具上时，显示该工具用途的简短解释。
- 活动场所：绘制电影一帧的实现工作区，可以直接在活动场所中绘制，也可以安排从外部引入的工艺图。
- 层：把层想象成一叠醋酸盐幻灯片（透明板），可以把不同元素放在不同的层上，以便使它们更易于协调。层数不影响电影文件的大小。
- 帧：一个动画由一系列帧构成。每个影像的每一帧都是不同的。因此，当这些帧快速地经过眼前的时候，观看者就看到了一个动作的形体。如果每一帧的改变都只有一点点并让这些帧快速移动的话，那么，电影中的动作看起来就显得很平滑了。一个电影就是一个帧序列。

3．绘图工具

为了能够绘制、编辑、选择和在活动场所中安排对象，可以使用绘图工具栏，绘图工具有：

选择工具——在活动场所中选择和移动对象，也可用于对象变形。

线条工具——画直线，可以调整其颜色、粗细和样式。

文本工具——在某个帧中加入文本。

椭圆工具——用所选的线条和填充色绘制椭圆。

矩形工具——用所选的线条和填充色绘制矩形。

铅笔工具——画线，光标就像是铅笔的笔尖。

画笔工具——用法和铅笔一样，可以用粗线条斜向填充。

墨水瓶工具——改变所选线条或外形的颜色。

颜料桶工具——改变所选形体的填充颜色。

滴管工具——选取一个对象的填充颜色，并使它变成当前的填充颜色。

橡皮擦工具——擦除线条和填充色。

手形工具——用于移动整个活动场所。

放大工具——用于在场所中拉近镜头。

4．元件和实例

对于大多数刚刚接触Flash的人来说，元件（Symbol）和实例（Instance）是一对比较特别的概念，但是它们不难理解。简言之，元件是可以重复使用的图形（Graph）、按钮（Button）或是影片剪辑（Movie Clip）；实例则是元件在舞台上的具体体现。如果先创建了一个动画元件，把它放在舞台上，那么舞台上的这个动画就叫作实例。

引入元件的概念对于动画的设计人员是大有好处的。首先，在处理大量元素时，不会手忙脚乱。如果电影中有大量的重复元素，而要对它们进行修改时，就不再需要对每一个实例进行修改，只要改变元件就可以了，因此编辑工作大大简化了。此外，元件的使用可以显著减小文件的大小，使用元件明显地减少文件的大小以及储存数据所需要的空间。比如，多个场景中需要使用同一个背景元件，就不需要保存全部场景的背景图像，而只要保存几个引用信息就可以了。使用元件对在网络上播放的动画来说还有一个好处，就是在用户观看时，元件只需要下载到用户端一次，而不用重复下载多个同样的元素，这样可以大大加快动画的播放速度。

5．多媒体动画的设计过程

在学习动画设计之前有必要对多媒体动画设计的步骤进行简单介绍。可以毫不夸张地说，多媒体动画设计是个相当复杂的过程，尤其在作品比较大、比较重要的时候。设计人员必须胜任整个过程，先要制定一个总体计划，否则设计工作必然是又费力气又费时间。在学习动画设计的时候必须要先制定全盘计划再动手设计。

制作多媒体动画时，内容质量的要求不同，所付出的精力和时间也不同。根据经验，制作多媒体动画的过程大体可分为三个阶段：

第一阶段，是整个制作过程中最为重要的一个部分，设计人员应首先考虑动画产品的目的、投入、内容、制作时间等，同时也要考虑动画的使用方式、用后效果、交互方式等，做好这些前期工作，对以后的制作将十分有利。

第二阶段，就是按照第一阶段的设计规划收集所需的素材并进行加工，例如编写文档、录音、制作动画等。在收集素材时应注意严格要求以保证质量。

第三阶段，用Flash将收集的素材集成并测试；在这个过程中应不断地进行测试，对发现的问题及时加以修改以保证精心制作的多媒体动画问世。

制作多媒体动画时，如果有多人分工合作，要注意相互协调每个人的工作。亲身体验制作过程，每一位参与者一定能够积累一些经验，对多媒体的制作会有更深刻的体会。

课文3：Dreamweaver简介

Adobe Dreamweaver是一款专业的Web站点开发程序，用于创建标准Web页面和动态应用。在其最新版本中，Dreamweaver已经使之进一步符合Web标准并加强了它的功能。除了用增强的层叠样式表（Cascading Style Sheet，CSS）创建基于标准的HTML页面之外，它还适用于编写各种Web格式代码，包括JavaScript、XML和ActionScript，甚至那些结合Web 2.0的方法，如Ajax。它具有众多的优点，是第一个具有选择多种服务器模型能力的Web开发工具。因此，ASP、ColdFusion或者PHP的开发人员能够轻松操作这个软件。

事实上，Dreamweaver 是由 Web 开发人员为自己量身定做的设计工具。它从头到尾都按照专业网页设计师的工作方式进行设计，从而加速了网站的构建，并简化了网站的维护。因为网页设计师很少仅使用一种软件，Dreamweaver 流畅地整合了主流的媒体软件：Adobe Photoshop、Adobe Fireworks 和 Adobe Flash。本文将介绍这个程序的基本原理，简要阐述 Dreamweaver 如何使用尖端的服务器端技术和 CSS 设计标准，将传统 HTML 和其他 Web 语言有效结合。此外，我们还将学习一些有助于用户管理网站的高级特性。

Dreamweaver 是源于现实世界的程序。人们为各种服务器模型开发了 Web 应用，Dreamweaver 为最广泛使用的模型编写代码。而现实世界也是一个变化的世界，所以，程序的可扩展结构为自定义或者第三方服务器模型打开了方便之门，例如，Dreamweaver 的最新版本提供了新的工具用于开发丰富的、交互式的 Web 2.0 页面。

此外，Dreamweaver 认识到不兼容浏览器指令的现实问题，通过产生跨浏览器兼容的代码加以解决。它包含浏览器专用的 HTML 验证，用于查看现有的或者新的代码在特定浏览器中的工作情况。

Dreamweaver 把现实世界的概念延伸到工作区中。Dreamweaver 的 CSS 呈现无所不能，它允许用户使用 Web 标准进行设计，如同不存在其他程序一样。在创建阶段，高级的“设计”视图能够快速构建整个网页，当网页发布后，能维持对浏览器的向后兼容性。比如“资源”面板这样的特性，能简化大型网站的创建和维护过程。另外，网页设计师还可以使用 Dreamweaver 的“命令”，自动创建大部分有难度的网页，并使用“服务器行为创建器”轻松地插入常用的自定义代码。

1. 连通性

连通性在 Dreamweaver 里不仅仅是一个术语，它还是一个基本概念。Dreamweaver 能够与任何受到大多数应用服务器（ASP、ASP.NET、ColdFusion、PHP、JSP，甚至 XML）支持的数据源进行连接。而且，实际连接类型相当灵活；开发人员可以选择更加易于实现但稳定性有所欠缺的连接，或者选择需要对服务器端有所了解的，同时能提供更大的可伸缩性的连接。Dreamweaver 提供了一组专用的特性，使用 XSLT （Extensible Stylesheet Language Transfomation，可扩展样式表语言转换）技术将 XML 数据转换为可供浏览器阅读的格式。Dreamweaver 还为大量应用服务器提供语言选择，并包含一个即用型的 CSS 标准设计集。

Dreamweaver 访问标准记录集（数据库的子集）以及更复杂的数据源，比如会话或者应用程序变量和存储过程。通过 cookies 和服务器端代码的执行，在 Dreamweaver 中设计的 Web 应用程序可以跟踪访问者或者拒绝他们登录。

Dreamweaver 还支持高端技术，比如 Web 服务、JavaBeans 和 ColdFusion 组件。Dreamweaver 允许查看所有技术的元素，使得编码人员快速掌握所需的语法、方法和函数。

2. 真实的数据表示

Dreamweaver 真正的创新特性之一是将需要的实际数据与处于设计阶段的网页集成。“动态数据”视图把处理中的页面发送到应用服务器，从而说明来自网页内部数据源的记录。

页面上的所有元素都是可编辑的；设计师甚至还可以修改动态数据的格式，并能立刻看到应用之后的变化。“动态数据”视图向设计师真实地显示了浏览者将看到的效果，从而缩短了工作周期。另外，通过“设置动态数据”特性，还可以在不同的条件下查看网页。

3．完整的可视化编辑器和文本编辑器

在万维网出现的早期，大多数开发人员使用简单的文本编辑器（比如 Notepad 和 SimpleText）手动编写网页。第二代 Web 编写工具将可视化设计或者 WYSIWYG（what you see is what you get，所见即所得）编辑器引入市场。这些产品使得页面布局工作更轻松，但缺少代码的完整性。即便有了最顶尖的所见即所得编辑器，专业网页开发人员仍需要手动编写网页。

考虑到这个现实，Dreamweaver 把一个优秀的可视化编辑器集成到类似于浏览器的“文档”视图中。设计师可以在“设计”视图中以图形方式工作，或者在“代码”视图中进行编程。甚至还可以选择能够同时显示“设计”视图和“代码”视图的分屏视图。任何在“设计”视图中进行的改变，都会反映到“代码”视图中，反之亦然。Dreamweaver 还允许使用任何文本编辑器，使设计师能用熟悉的代码编辑器工作。无论选择哪种方式，在可视化编辑器和代码编辑器之间，Dreamweaver 都提供了简单自然的动态流程。

Dreamweaver 用“快速标签编辑器（Quick Tag Editor）”使得可视化设计和底层代码进一步紧密融合。网页设计师经常需要不断地精细调整 HTML 代码，比如修改这里的属性或者在别处添加一个标签。而在“设计”视图中，“快速标签编辑器”显示为一个小弹出窗口，它能快速轻松地调整这些代码。

4．世界级的代码编辑

代码编写与网页的开发过程密不可分，而 Dreamweaver 的编码环境是最好的。如果你正在手动编写代码，那么你会很欣赏 Dreamweaver 提供的代码提示、代码折叠和代码完成特性。在“代码”视图一侧的“编码”工具栏中封装了很多这样的元素。这些特性不仅加快了 HTML 页的开发，而且 Dreamweaver 的基本“标签库”还将它们的使用范围全面扩展到其他代码格式，比如 JavaScript、ActionScript 和 XML。

Dreamweaver 的“代码”视图相当悦目，而且能够随意打开或者关闭句法的色彩。要想快速浏览网页，可以使用标准的行编号工具或者高级的代码导航特性；“代码导航”列出一个页面中的所有函数，选择一个函数后会立即跳转到对应的代码处。

资深开发人员与新手都会发现，Dreamweaver 的“标签选择器”和“标签检查器”是工作中必不可缺的工具。顾名思义，利用“标签选择器”，编码人员可以从全面的标签列表中选择标签，这个标签列表涵盖了多种 Web 标记语言，包括 HTML、CFML、PHP、ASP、ASP.NET 等。

通过“标签检查器”，可以综览选定标签各个方面的信息。从中不仅能看到所有相关属性的完整显示，比在属性检查器中能看到的还要多得多，另外，还可以适当修改属性值。任何应用的 JavaScript 行为都可以显示在“标签检查器”中。也许这个检查器最具创新的特性就在于与 CSS 相关，它显示任何对标签有影响的样式及完全可修改的属性和值。选择一个 CSS 样式，“标签检查器”就会变为能够方便快速地编辑 CSS 的“规则检查器”。

当然，代码远不止是一系列独立的标签，Dreamweaver 的“代码片段”面板存储了最常用的代码片段，可以直接对它们进行拖放操作。Dreamweaver 自带许多可以即用的代码片段，并提供了随时添加自定义代码的方法。

5．往返于 HTML

大多数 Web 创作程序可以修改系统中传递的任何代码，包括插入回车、删除缩进、添加〈meta〉标签、设置大小写命令等。Dreamweaver 的程序员理解和尊重这一事实，即 Web 开发人员拥有他们自己特定的编码风格。“往返于 HTML 技术”这样一个基本概念，可以确保代码不被改写的情况下，在可视化编辑器和任何 HTML 文本编辑器之间来回切换。

6．网站维护工具

Dreamweaver 的开发者意识到，创建网站只是网管工作的一部分。维护站点才是一项持续、费时的日常工作。为了简化这项烦琐的工作，Dreamweaver 提供了一组站点管理工具，包括一个重复元素库和便于团队更新文件的锁定能力。

Dreamweaver 内置的 FTP 传送引擎相当强大，它凭借在后台的出色工作能力，能更好地适应设计者的工作流程。设计师现在可以随意进行大型的发布操作，还可以在 FTP 传送过程中返回到 Dreamweaver 中继续设计页面，或者随时调出日志查看详细情况。

速度是网站维护的另一个基本方面。有了 Dreamweaver 的无站点编辑模式，即可与连接到服务器时一样快速地进行修改。如果只是想修改几个页面，则无须定义整个站点，只需设置一个服务器连接。Dreamweaver 可以通过平滑的工作流程访问、编辑和发布页面。

7．面向团队的站点构建

以前，如果要在团队开发的环境中使用 Dreamweaver，则每个网页设计师都会处于困难境地。文件难以锁定，完成的修改会被不小心覆盖；站点报告限制于一个范围，只能输出为 HTML；而且，最糟糕的是，那时候没有版本控制。Dreamweaver CS3 为长远的连通性奠定了基础，将这些问题一一解决。

Dreamweaver CS3 支持两个行业标准的源控制系统：Visual SourceSafe （VSS）和 WebDAV。Dreamweaver 很好地集成了与 Visual SourceSafe 服务器连接的功能，只需把 VSS 服务器定义为远程站点，并添加必要的连接信息；WebDAV 的名气可能要比 VSS 小一些，但它提供了相当强大而且更有用的内容管理解决方案。更为重要的是，Adobe 开发了源控制解决方案作为系统架构，允许其他第三方内容管理或者版本控制的开发人员使用 Dreamweaver 作为其前端。

ColdFusion 开发人员一直享受着远程开发服务（Remote Development Services，RDS）所带来的好处。现在，在 Dreamweaver 中也加入了 RDS 连接。通过 RDS，开发者团队可以致力于保存在远程服务器上的同一个站点。而且，无须创建站点就能直接与 RDS 服务器进行连接。

Dreamweaver 的站点报告工具也同样是可扩展的。Dreamweaver 自身包含了生成报告的能力，报告内容涉及可用性问题（比如缺少替换文本）或者工作流程问题（比如谁检查了什么文件）。用户还可以基于具体的项目开发自定义报告。

课文 4：3ds Max 的建模类型

建模是纯粹的创建过程。无论是雕刻品、砖瓦建筑物、施工工程、木雕、建筑物还是高级注塑物，创建对象的方式多种多样。3ds Max包括许多不同的模型类型，而使用这些建模类型的方法就更多了。

爬山有许多条路，建模也有许多方法。使用砖块、立方体和球体等基本体对象可以创建山脉模型，也可以将其创建为多边形网格。随着经验的逐渐增加，会发现使用某种方法建模某些对象更容易，而对于另一些对象，使用其他方法建模才更容易。3ds Max提供了多种不同的建模类型，可以处理各种建模情况。

1．参数化对象与可编辑对象

3ds Max中的所有几何对象都可以分成两大类：参数化对象和可编辑对象。参数化意味着对象的几何体是由称为参数的变量所控制的，修改这些参数就可以修改对象的几何形状。这种强大的概念为参数化对象赋予了很大的灵活性。例如，球体对象有一个参数称为半径，改变这个参数就可以改变球体的大小。3ds Max中的参数对象包括“创建”菜单中的所有对象。

可编辑对象没有这种参数灵活性，但是它们可以处理子对象并编辑函数。可编辑对象包括“可编辑样条线”“网格”“多边形”“面片”和NURBS（非均匀有理B样条）。可编辑对象列在“修改器堆栈”中，在其基础对象前有“可编辑”这个单词（NURBS对象例外，只称它为“NURBS曲面”）。例如，可编辑网格对象在“修改器堆栈”中显示为“可编辑网格”。

实际上NURBS对象完全是个另类。使用“创建”菜单进行创建时，它们是参数化对象，选定“修改”面板后，它们就是可编辑对象，拥有许多子对象模式和编辑函数。

可编辑对象不是创建的，而是由另一个对象转换或修改而来的。将基本体对象转换成不同的对象类型时（例如，“可编辑网格”或NURBS对象），它就失去了其参数化本质，不能通过修改其基本参数来改变。但是可编辑对象也有其自身的优点，可以对子对象（例如，网格的顶点、边和面）以及任何参数化对象不能编辑的部分进行编辑。每种可编辑对象类型都有大量专用于该类型的功能。

有几个修改器允许维持对象的参数化本质并编辑子对象。这些修改器包括“编辑面片”“编辑网格”“编辑多边形”和“编辑样条线”。

3ds Max包括以下几种建模类型。

- 基本体：诸如立方体、球体和棱锥等基本参数化对象。基本体分为两组：“标准基本体”和“扩展基本体”。“AEC 对象”也被认为是基本体对象。
- 图形和样条线：诸如圆形、星形、弧形和文字之类的简单矢量图形及类似“螺旋线”的样条线。这些对象都是完全可渲染的。“创建”菜单包括许多参数化图形和样条线，可以将这些参数化对象转换成可编辑样条线对象，对其执行更多的编辑操作。
- 网格：渲染对象时，复杂模型由多个多边形面平滑连接而创建。这些对象只可以作为“可编辑网格”对象使用。
- 多边形：由多边形面构成的对象，与网格对象类似，但它具有其唯一的特性，这些对象只能作为“可编辑多边形”对象使用。
- 面片：以样条线为基础；使用控制点可以对面片进行修改。“创建”菜单包括两个参数化“面片”对象，但是大多数对象也可以转换成“可编辑面片”对象。
- NURBS：代表“非均匀有理 B 样条线”。NURBS 和面片类似，因为它们也有控制点。这些控制点定义了曲面如何在曲线上延展。
- 复合对象：各种建模类型混杂成组，包括“布尔”“放样”和“散射”对象等。其他复合对象擅长对某种特殊类型的对象进行建模，例如“地形”或“水滴网格”对象。

- 粒子系统：一些小对象构成的系统，这些小对象将作为一个组共同工作。该系统对于创建诸如雨、雪和火花等内容十分有用。
- 毛发：建模成千上万的圆柱体对象来创作可信的头发可能很快使系统陷入困境，因此头发要使用把每根头发表示成样条线的单独系统来建模。
- 织物系统：织物（具有摆动、自由流动的性质）在某些情况下类似于水，在另外一些情况下又类似于固体。3ds Max 包括一组用于处理织物系统的专用修改器。

毛发和织物通常被看作是一种效果或动态模拟而不是建模结构，因此将它们列在此列表中，应该是对建模类型的一种扩展。

尽管有了这些选项，在3ds Max中建模仍使人有些惧怕，但随着对3ds Max使用的增多，你会了解如何使用这些类型的对象。对于初学者，首先可以从基本体或导入的对象开始，然后再慢慢扩展，将其转换成可编辑对象。一个3ds Max场景可以包括多种对象类型。

2．转换成可编辑对象

在“创建”菜单和“创建”面板的所有命令中，没有用于创建可编辑对象的菜单或子类别。

要想创建可编辑对象，需要导入它或从另一种对象类型进行转换。在视口中右键单击该对象并从弹出的方形菜单中选择“转换为”子菜单即可转换该对象，也可以在“修改器堆栈”中右击基础对象并从弹出菜单中选择所要转换为的对象类型。

转换之后，所选类型对象的所有编辑特性都可在“修改”面板上访问，但此对象将不再是参数化的，其“半径”和“段数”等常规参数将无法再访问。但是，3ds Max还提供一些专门化的修改器，如“编辑多边形”修改器，此修改器允许在访问可编辑对象的编辑特性的同时保持基本体对象的参数特性。

如果已对某个对象应用了修改器，那么使用“塌陷”命令之前，“修改器堆栈”弹出菜单中的“转换为”菜单选项不可用。

弹出菜单中包括了“转换成可编辑网格”“可编辑多边形”“可编辑面片”和NURBS的选项。如果选定了一个图形或样条线对象，则该对象也会被转换成可编辑样条线。使用任一“转换为”菜单选项都可以塌陷“修改器堆栈”。

可以在不同类型之间多次转换对象，但每次转换都可能细分对象，因此不推荐进行多次转换。

在3ds Max中，两种类型对象之间的转化是自动完成的，但是如果把一个“转化”修改器应用给一个对象，会出现一些参数用于定义该对象将如何转化。例如“转化为网格”修改器包含一个选项“使用不可见边”，该选项决定使用不可见边拆分多边形。如果禁用该选项，整个对象将以三角形拆分。“转化为面片”修改器包含一个使方形转化为方形面片的选项。如果禁用该选项，所有方形将以三角形拆分。

“转化为多边形”修改器包含的选项有“保持多边形为凸面体”“限制多边形大小”“需要平面多边形”和“移除中间边顶点”。启用“保持多边形为凸面体”可以拆分任何凹面体多边形。“限制多边形大小”选项用于确定最大允许的多边形大小，这可用于去除网格上的五边形和六边形。“需要平面多边形”选项用于当相邻多边形之间的角度超过指定的阈值时，多边形以三角形拆分。“移除中间边顶点”选项用于移除不可见边的交叉点。

所有“转化”修改器还包括保持当前子对象选择（包括任何软选择）和指定“选择级别”的选项。给对象应用了“转化”修改器之后，必须塌陷“修改器堆栈”才能完成转化。

课文5：Premiere Pro的窗口

Adobe Premiere Pro的用户界面是视频编辑工作室和电子图像编辑工作室的结合体。如果你对电影、视频编辑、音频编辑非常熟悉，那么使用Premiere Pro中的项目、监视器和音频窗口工作时就会感到非常得心应手。如果使用过Adobe After Effects、Adobe Live Motion、macromedia Flash或者Macromedia Director这些程序的话，会对Premiere Pro 时间线、数字工具和palette（面板）感到很熟悉。如果对视频编辑和计算机一无所知，请不要担心，Premiere Pro中有效设计的调色板、窗口和菜单，会让你迅速进入状态。

为了帮助你尽快上手，本文将对Premiere Pro窗口、菜单和面板加以概述。以之对程序工作区进行详细介绍，并将此作为你自己规划和制作数字视频的便捷的参考资料。

当你第一次启动Premiere Pro，屏幕上会自动显示几个窗口，每个窗口都很引人注目。为什么需要一次打开多个窗口？制作视频是一件多方面的事。每次制作，你可能需要获取视频，编辑视频，并创建字幕、转换和特效。Premiere Pro的窗口使这些工作相互独立，组织有序。

虽然Premiere Pro程序的主窗口在屏幕上不时地自动打开，但是有时你可能想关闭其中的一个窗口。要关闭一个窗口，只需单击关闭窗口的×图标。如果尝试关闭项目窗口，Premiere Pro会认为你要关闭整个项目，并提示在关闭前存盘。如果想打开时间线、监视器、调音台、历史、信息，或工具窗口，选择窗口菜单，然后单击所需的窗口名称。如果在屏幕上有一个以上的时间线，你会看到它在“窗口”|“时间线子菜单”中列出。

如果为窗口和面板设置了特定大小和具体位置，可以选择保存这个设置，通过选择“窗口”|“工作区”|“保存工作区”。对工作区命名，并将其保存，工作区的名称出现在窗口➪工作区子菜单中。任何时候你想使用该工作区，只需单击一下其名称即可。

1. 项目窗口

如果你曾经对包含很多视频和音频剪辑以及其他作品元素的项目进行过处理，那么你很快就会喜欢上Premiere Pro程序的项目窗口。项目窗口为你的作品元素提供了一个快捷的鸟瞰式的图景，并且能够使你从项目窗口中预览作品片段。

在工作的时候，Premiere Pro会自动加载项目到项目窗口。当加载一个文件时，视频和音频片段会自动加载到项目窗口bin中（bin是在项目窗口中的一个文件夹）。如果导入一个包含多个剪辑的文件夹，Premiere Pro会创建一个新的bin，以文件夹的名称命名。当要获取声音或视频时，你可以在关闭剪辑片段之前快速添加媒体到一个项目窗口。稍后，你可以通过单击bin按钮创建自己的bin，在这点上你可以把制作元素从一个bin拖动到另一个bin中。新建分类按钮使你能够快速创建新的字幕和其他的制作元素，如颜色底纹或彩条和声调（用来校正色彩和声音）。项目窗口还包括一个新建分类按钮，允许添加新的序列、脱机文件、字幕、彩条和音调、黑屏、颜色底纹，或倒计时向导（universal counting leader）。如果您单击图标按钮，所有的制作要素在屏幕上显示为图标，而不是以列表格式出现。单击列表按钮返回到项目窗口列表视图显示。如果想快速添加项目窗口的元素到时间线，可以先选择它们，然后单击序列自动化按钮。

如果通过单击并拖动窗口边框来扩大工作窗口，你会看到Premiere Pro列出了每段剪辑的起止时间、入出点以及时长。在项目窗口，制作要素根据当前的排序归类，可以改变制作要素的排列顺序，使它们按照任意列标题进行顺序安排。要按列的种类进行排序的话，只要单击列的类目即可。第一次单击的时候，制作条目以升序排列。要按降序排列，再次点击列标题即可。排列顺序通过一个小三角形来表示。当箭头向上时，排序顺序为升序；当它指向下，排序顺序为降序。

为了将制作材料有序组织起来，可以创建bin来保存同类的元素。例如，可以创建bin用以存放所有的声音要素，或创建一个bin用以存放所有采访剪辑文件。如果bin被填满了，可以通过从默认的缩略图显示切换到列表显示，从而一次看到更多的要素，列表视图列出了每个项目，但不显示缩略图。

如果想在项目窗口中的小监视器里播放一段剪辑的文件，单击文件，然后再单击小监视器旁边的向右指示的迷你箭头即可。如果想保留空间和隐藏项目窗口的小监视器，在项目窗口菜单中选择“视图”|“预览区”。这样就执行了监视器的播放和关闭。

可以通过右击项目窗口中的剪辑文件并选择“速度/持续时间”命令来改变文件的速度和持续时间，也可以通过右击剪辑文件并选择“在源窗口中打开”命令，从而迅速地将剪辑文件放置在监视器源窗口中。

2．时间线窗口

时间线窗口是时间线视频制作的基础。它为整个项目提供了形象的时间综览。幸运的是，时间线不仅只供查看，它也是互动的。可以使用鼠标，通过把从项目窗口中的视频和音频剪辑文件、图形、字幕拖动到时间线来实现自己的制作。使用时间线工具，可以安排、剪切和扩展剪辑文件。通过单击和拖动工作区域条两端的工作区域标记——时间线顶部的浅灰色条的边缘，可以指定Premiere Pro预览或输出的时间线部分。在下方的工作区有一个很细的彩色条表明项目预览文件是否存在。红条表示没有预览，绿条表示视频预览已创建。如果存在音频预览，会出现一个更细的浅绿色的条。（要创建预览文件，选择“时间轴”|“渲染工作区域”，或按回车键渲染工作区域。）

“预览工作区域”有助于确保项目以项目帧频进行播放。此外，如果创建了视频和音频效果，预览文件存储渲染后的效果。因此，下一次播放时，Premiere Pro就不用再重复处理该效果。

毫无疑问，在时间线窗口中使用的最贴切的隐喻是将它的视频和音频轨道比作平行的双杠。Premiere Pro中提供了多个并行的轨道，使你既可以预览，又可以实时地对作品进行概念化操作。例如，并行视频和音频轨道使你可以把视频当作音频播放。平行轨道还使你能够创建透明效果，透过视频轨可以看到另一个视频轨的一部分。时间线中还包括隐藏或查看轨道的图标。当预览作品的时候，单击“切换轨道输出”按钮（眼睛图标）隐藏轨道；再次单击它可以使得轨道可见。

单击音频轨道的“切换轨道输出”按钮（扬声器图标）可以打开和关闭音频轨。“眼睛”图标下方是另一个图标，设置在轨道中剪辑文件的显示模式。单击“设置显示”图标，可以选择是要看到时间线上的实际剪辑的帧还是只显示剪辑文件的名称。在窗口的左下方有时间缩放滑块，可以更改时间线的时间间隔。例如，缩小时会将项目显示在较少的时间线空间，放大时会将作品显示在一个面积较大的时间线空间。因此，如果你正在查看时间线上的帧，“放大”的操作会显示更多的帧。还可以点击在时间线顶部灰色条块的边缘来放大或缩小。

其他按钮——轨道选项对话框、对齐到边缘开关按钮、查看边缘开关按钮、切换轨道移动选项开关、切换同步模式按钮——在窗口左下方，能够让你为同步轨道和边缘对齐改变选择。

3．监视器窗口

监视器窗口主要用于在创作过程中预览作品。当预览您的作品时，单击“播放”按钮来播放在时间线上的剪辑文件，然后按一下“循环播放”按钮从第1帧开始。可以单击并拖动穿梭滑块跳转到一个特定的剪辑区域。当单击时，监视器窗口的时间读出器显示出你在剪辑文件中的位置。监视器窗口也可以用来设置剪辑点的入出点。入出点决定了剪辑文件的哪一个部分会出现在你的项目中。

在监视器窗口提供了三种视图模式。要切换模式，选择在监视器窗口菜单中的双视图、单视图或修剪（Trim）视图（还可以通过单击在监视器窗口底部的Trim图标来打开Trim视图）。

- 双视图。监视器窗口的设置类似于一个传统的录像带编辑工作室。该源剪辑文件（影片片段）出现在监视器窗口的一边，节目（需编辑的视频）出现在窗口的另一边。这种模式主要是用于创建三点编辑和四点编辑。
- 单视图。当预览作品时，窗口显示一个监视器。使用这种模式类似于在电视屏幕上观看你的作品。
- 修剪（Trim）视图。这种模式允许精确编辑。一个剪辑出现在一个监视器窗口，其他剪辑出现在第二个监视器窗口。

所有的监视器窗口模式都提供图标，可以快速编辑入出点，并对视频进行逐帧查看。

4．音频混合器窗口

音频混合器窗口使你能够混合不同的音轨，并创建交叉淡入淡出和摇摆（摇摆使你能够平衡立体声声道，或从左声道切换到右声道，反之亦然）。Premiere Pro的早期版本可以使用户感受音频混合器的实时处理能力，这意味着你可以一边查看视频轨道，一边混合音频轨。

使用面板调节装置，可以通过单击并拖动音量控制器来提高和降低三个轨道的音量。还可以输入数字到分贝音量指示框（在音量控制器区域的底部）来设置音量。圆形旋钮式的调节装置使你可以摇摆或平衡音频。可以通过单击和拖动旋钮图标来改变设置。在音频混合器顶部的按钮可以播放所有音轨，可以选择要监听的音轨，或选择要静音的音轨。

在播放音频时，在音频混合器窗口底部熟悉的控制器用来启动和停止录制。

5．效果窗口

效果窗口可以快速将效果和过渡应用于音频和视频。效果窗口提供了一系列有用的效果制作和过渡。例如，视频效果折叠夹包括对图像改变对比度、扭曲和模糊的效果。举例来说，在效果窗口中有众多的文件夹，其中扭曲文件夹的功能效果，是通过扭曲和收聚来对剪辑文件进行扭曲效果处理。

应用一个效果很简单，只需从面板中直接单击该效果并拖曳到时间线中的剪辑文件。通常情况下，这样做将打开一个对话框，您可以指定效果选项。

效果窗口允许你创建自己的折叠夹，并将效果移至其中，这样可以在每个项目中快速得到想要的效果。

Premiere Pro的“过渡”折叠夹，它也出现在效果窗口，有70多种过渡效果。一些效果，如溶解组，可以实现从一个视频剪辑文件到另一个的平稳过渡。其他转换效果，如卷页特效，可实现一个特殊的效果从一个场景跳转到另一个。

如果要在整个制作过程中使用相同的效果转换，可以创建一个折叠夹，命名并将该过渡效果保存在自定义折叠夹中，以便快速取得。

6. 效果控制窗口

效果控制窗口允许你快速创建和控制音频和视频效果以及过渡效果。例如，只需在效果窗口中选中某种效果并直接拖动到时间线的剪辑文件上，或者直接添加到效果控制面板，就可以对剪辑文件添加效果。通过单击和拖动时间线以及改变效果设置，可以随时间改变效果。当改变设置时，可以沿时间线创建关键帧。

如果要为剪辑文件制作特效，选择剪辑文件和打开效果控制窗口可以看到不同的效果。

课文 6：After Effects 的工作区和面板

我们无从知道，这些年来，After Effects 艺术家为了查看他们所需要的内容，浪费了多少时间和精力来移动调板，但升级到 7.0 的所有用户都将告别这种四处移动窗口的日子。

在介绍这个界面的使用之前，我们先来定义界面中的元素，这对我们是有帮助的。该界面包括一个主应用程序窗口，在 Mac 系统上，这个窗口包含打开的项目名称，而在 Windows 中，它还包含一个菜单栏。除此之外，还可以创建其他浮动窗口。

这两种窗口都包含框架，它们由分隔条进行分隔。每个框架包含一个或多个面板。如果一个框架内有多个面板，其顶部将显示各个面板的选项卡，但只能看到前面选项卡的内容。单击选项卡，可以将对应面板移到最前端。

有些面板是视图，这些面板的选项卡中包含一个下拉菜单，允许选择其中显示的内容。这些面板还包含锁定选项（通过一个小小的锁定图标进行设定，该图标也在选项卡上），可以防止面板自动切换到不同的显示。

Standard （标准） 工作空间包含整个项目的主要元素。

- Project （项目）面板：包含合成图像中使用的所有资源（源素材、静态图像、创建的纯色图层、音频，甚至可以是合成图像自身）。
- Composition （合成图像）面板：这个视图主要用于对视频进行可视化编辑工作。
- Timeline （时间线）面板：用于组织各个合成图像（或称场景）中的元素。
- Info（信息）、Audio（音频）、Time Controls（时间控制）和 Effects & Presets（特效

和预设）面板：用于帮助对合成图像进行处理。

After Effects 中的所有可用面板都列出在 Window 菜单下，有些甚至还列出了预设的键盘快捷键，供我们快速调用。在旧版本中，这些面板实际上是浮动窗口，但 Adobe 决定不改变菜单的名称，因为这些名称在许多应用程序中已成为标准。

所以，After Effects 的基本工作流是创建新的合成图像，它通常包含 Project 面板中的项目，然后在 Composition 面板和 Timeline 视图中操作该合成图像。

然而，为了全面了解 After Effects 工作流，可以使用 Workspace（工作空间）下拉菜单切换到 All Panels（所有面板）版面。这将显示出各种工具的折叠面板，如画笔工具（绘图和画笔笔尖形状工具）、文字工具（字符和段落）、运动跟踪工具（跟踪控制）等。

仔细查看工作空间，可发现 After Effects 工作流中其他一些关键内容：用于对图层添加特定效果的 Effect Controls（特效控制），用于处理合成图像内各个图层或查看源素材的 Layer（图层）和 Footage（素材）调板，提供基于节点的项目浏览的 Flowchart（流程图）窗口，输出作品的 Render Queue（渲染队列）。

上面这些讨论基于还未对 After Effects 工作空间做任何调整的假设。如果选择 Workspaces 时所看到的与上述不同，则可以使用 Window | Workspace | Reset（复位）命令，将工作空间恢复到以前的设置（即使已修改和保存过默认设置，也会看到保存过的版本）。

All Panels（所有面板）布局用处不大，其中塞满许多你可能从不会用到的面板，但它可以让你从总体上了解在 After Effects 用户界面中可以使用哪些面板。Adobe 根据用户对 After Effects 的通常使用方式定义了其他默认工作空间：Animation（动画）、Effects（特效）、Motion Tracking（运动跟踪）、Paint（绘图）和 Text（文字）。Minimal（最小）工作空间仅包含 Timeline 和 Composition 面板（这是工作中最常使用的面板）。

工作空间过去常常是 After Effects 未充分利用的一项功能，但现在它们已成为用户关注的焦点。定制工作空间最好的方法就是仔细查看自己的用户界面，下面介绍定制工作空间时应注意的几点事项。

如果您还认不出面板，After Effects 中的每个面板都通过左上角的选项卡进行标注，这个选项卡包含面板标题和一个用来关闭面板的×图标（标题的右侧）以及标题左侧的抓手区域。右上方是一个较小的选项卡，其中包含另一个抓手区域以及一个用于打开面板菜单的三角形图标。

在把面板拖到用户界面（UI）中的其他位置时，我们可以拖动这些抓手区域。请在任意一个工作空间中尝试拖动面板。

如果将鼠标指针放置于两个面板之间的边界上，它将变成分隔条拖动光标。它允许调整相邻面板的尺寸。但如果你想暂时扩大当前工作面板所占用的空间，则还有另一种更好的选择。

按下~键将使当前处于激活状态的面板（面板周围有黄色线条标识）占据 After Effects 的整个窗口。无论显示器尺寸有多大，在处理合成图像时会多次使用这个快捷键。它最常用于在 Composition 或 Layer（图层）视图中显示更多的图像内容，能自由使用界面的其他区域。

如果将太多面板分组到一起，就无法看到它们的整个标题，这时面板上方将显示一个滚动条，沿着它们可以前后滚动面板。滚动条很小，极易被忽略。

当重新排列面板后将无法撤销所做的修改。但选择 Workspace 菜单底部的 Reset（复位）菜单项可以复位任何指定的工作空间。

如果在定制的工作空间之间来回切换过，就可能发现对工作空间所做的修改都被保持。如果将 Standard 工作空间调整得一团糟，之后切换到 Effects 工作空间，然后再切换回 Standard，将发现 Standard 工作空间仍保留先前的杂乱状态。如果习惯使用 6.5 版本及其以前版本中的旧工作空间，那么在看到 Reset 选项之前你可能会感到很沮丧。

如果从没有选择过 Reset 菜单项，定制的工作空间会一直保持，至少保持到首选项被复位或不经意间对它们修改之前。因此，如果遇到自己喜欢的定制工作空间，则请保存它（使用 Workspace 菜单下的新建工作空间菜单项）。我们可能至少保存一个比例和方向最适合显示器配置的工作空间，如果今后改变主意，随时都可以覆盖或删除它。

第三部分　通信和电信技术

课文 1：信号与系统

本书介绍了理解和分析连续时间系统和离散时间系统所必需的工具和数学方法。我们尽力深入探讨了这些工具和方法在解决实际工程问题中的应用。我们的原则是全书采用系统方法来介绍连续时间信号和离散时间信号，而不用传统电路理论的框架。我们认为系统观点是介绍这些知识和扩大学生视野的更自然的方法。而且，以系统观点介绍离散时间信号和系统的分析是最自然的，它使全书内容的展开具有一致性。当然，我们还得紧密依靠学生的电路理论基础来选例题。

本书内容组织直截了当，前六章讲述时域和频域连续时间线性系统。用于时域分析的主要工具是卷积积分。频域方法包括傅氏变换和拉氏变换。该书还介绍了状态变量法。其余各章研究离散时间系统，包括 Z 变换分析法、数字滤波器分析与综合法以及离散傅氏变换和快速傅氏变换（FFT）算法。

该内容组织方式使本书适于用两门三学分的课程授完。第一门课讲授连续时间信号和系统，第二门课讲授离散时间信号和系统。或者该书作为三季度长课程的教材。这样，第一门课讲解连续时间系统的时域和频域的分析。第二门课讲解状态变量、抽样及 Z 变换和离散时间系统的入门知识。第三门课讲解数字滤波器的分析和综合，并介绍离散傅氏变换及其应用。

学生应具备的基础知识是学到微分方程的数学知识和电路理论的一般入门知识。掌握有关矩阵代数的基本概念会有助于学习，但并非必须具备。附录 A 总结了第 5、6 章所用的有关矩阵关系式。我们觉得在多数电气工程课程的安排中，该书内容最好放在大学三年级讲授。

本书首先介绍了信号与系统模型的基本概念及系统的分类。第 1 章就首先介绍了周期信号的频谱表达这一概念，这是因为我们觉得让学生一开始就从时域和频域两个角度思考是重要的。

第2章讲解了卷积积分及其利用叠加原理在固定线性系统分析中的应用。卷积积分的求值是通过详细的例题进行讲述以便巩固概念。本章还讨论了冲击响应的计算及其与系统的阶跃响应和斜升响应的关系。第2章还包括可任选的几章内容及例题，其内容是如何写出模拟固定线性系统的控制方程以及线性固定系数微分方程的解法，供复习之用，若舍去不讲也不失内容的连续性。

第3章介绍了傅氏级数和傅氏变换。我们强调的内容是用三角级数逼近一个周期函数及用正弦和余弦的正交性得到展开系数这一基本方法。其原因是大部分同学是首次接触傅氏级数。另一种广义的正交函数法放在本章末尾一节讲解，不作必读内容，可供感兴趣的读者阅读。我们在考虑信号失真的情况下讨论了系统正弦稳态响应下的传输函数这一概念。然后介绍了傅氏变换及其在频域频谱分析和系统分析中的应用。这里还介绍了由无失真传输所引出的理想滤波器的概念。在最后任选几节中介绍了吉布斯现象、窗函数和傅氏级数的收敛性。

第4章介绍了拉氏变换及其性质。尽管附录B总结了复变量理论以供教师增加内容时选用，但考虑到大部分学生是首次接触该内容，所以我们力求使讲解尽可能简单。由基本变换对推出拉氏变换式的方法是通过例题讲解的，就像用部分分式展开得到拉氏逆变换的方法一样。最后供选择的几节介绍了用复数反向积分对拉氏逆变换求值的方法，并介绍了双边拉氏变换。

第5章详细论述了拉氏变换在网络分析中的应用，并介绍了如何通过观察写出拉氏变换网络方程的方法，以供学生复习阻抗和导纳矩阵概念时使用。这些知识是学生在电阻网络的电路课程中已学过的。本章还详细介绍了传输函数和用来确定稳定性的罗斯检测法。本章最后讲述了固定线性系统的波特图和框图代数。

第6章阐明了状态变量的概念及系统分析的状态变量法，还用时域法和拉氏变换法对状态方程求解，并讨论了其解的重要性质。最后，用一个例子说明了如何把状态变量法用于电路分析。

最后三章介绍了离散时间信号和系统分析的概念。

一本内容全面的解题手册现已出版，它包括了所有习题的解答，可作为教师的辅助材料。附录E给出了部分习题的答案以供学生使用。

很多人为本书的编写出版做出了贡献，有些为人所知，有些则是为人所不知的，在此作者对他们表示感谢。首先要感谢长期刻苦学习的学生们。他们曾不得不以不同编写阶段的手稿作为课本。他们的许多批评是无价的、值得感谢的。很多在密苏里——罗拉大学电气工程系工作的同事使用该书的草稿授课，并提出了很多改进建议。为此，我们感谢卡尔逊教授、卡本特教授等。卡尔逊教授校阅了很多草稿并提出了宝贵的改进建议。另外，我们还要感谢其他学校参与校阅工作的教授，他们的批评意见很有价值。但是，最终结果的任何缺点都只是作者的责任。还要衷心感谢我们的秘书，他们以认真的态度、熟练的打字技术使最终的手稿问题最少。还要感谢国家工程协会，因为第7、8章的大部分内容是在该协会的一系列研习班最先讲授的。

最后，并且也很重要的是要感谢我们的妻子们和家人对这项有时似乎没有尽头的工程所付出的耐心。

课文 2：数据通信

通信是人类固有需求的一部分，人类一开始就使用不同的技术和方式进行交流，而环境和可利用的技术也限定了人类通信的方法和方式。数据通信涉及在通信双方间传递信息。让我们共同来学习关于数据通信的基本知识。

1．信号

1 信号和 0 信号不能通过网络链路直接传送。它们必须进一步被转换成传输介质能够按受的形式。传输介质通过沿着一条物理线路传输能量来工作。所以，一个 1 信号和 0 信号的数据流必须被转换为电磁信号形式的能量。

2．模拟和数字

数据和代表它们的信号都能够采用模拟的或者数字的形式。模拟涉及连续的某些事物——组特殊点的数据以及各点之间所有可能的点。数字涉及离散的某些事物。

信息可以是模拟的或数字的。模拟信息是连续的。数字信息是离散的。

信号可以是模拟的或数字的。模拟信号在一个范围内可以有任何值；而数字信号仅有有限数量的值。

3．信道的某些特性

首先是传输率。一个信道的传输率是由它的带宽和速度决定的。带宽是一个通道能支持的频率范围。因为传输的数据能被分配到不同频率，所以带宽越宽频率越高，同一时间能被传输的数据就越多。

传输数据的速度通常以每秒多少比特或以波特率来表示。比特每秒（bps）是在一秒被传输的比特数。波特率是被传输的信号每秒变化的次数。一般仅当 1 比特以一个信号变化被传输时，每秒比特和波特率才是相同的。

第二是传输方向。数据传输方向被分为单工、半双工和全双工。在单工传输中，数据仅沿着一个方向流动。单工方式仅被用于像无线电通信一类的从不需要计算机响应的发送设备。在半双工传输中，数据能够沿着两个方向流动，但在一段时间内仅沿一个方向流动。半双工通常用于终端和中央计算机之间的通信，例如对讲机。在全双工传输中，数据能够同时沿着两个方向传送。普通的电话线路就是全双工传输的一个例子，通话双方可以同时讲话。全双工传输用于大部分交互式计算机应用和计算机到计算机的数据通信。

第三是传输模式。传输模式包括异步和同步。在异步传输模式中，单独的字符（由比特组成）以不规则的时间间隔传输，例如当用户输入数据时，为了辨别一个字符在什么地方停止而另一个字符在什么地方开始，异步通信模式使用一个起始位和一个停止位。一个称为奇偶位的附加位有时包含在每个字符的末端。奇偶位用于错误校验。异步传输模式用于低速数据传输以及用在设计适合个人计算机的大多数的通信设备中。

4．串行和并行传输

数据以两种方式传输：串行和并行。在串行数据传输中，比特以串行或连续的一串流动，就像汽车通过一个单车道的桥。串行传输是大多数的数据通过电话线传输的方法。为此，外接调制解调器典型地通过一个串行端口连接到一台微型计算机。串行端口著名的技术是 RS-232C 连接器与异步通信接口。

使用并行数据传输时，各个位同时通过各自的线路流动。换句话说，它们像多辆汽车在一条多车道的高速公路上以同样的速度一起运动。并行传输通常只适用于短距离通信，特别是不使用通过电话线的通信。然而，它是一个将数据从系统单元传送到打印机的标准方法。

课文 3：数据传输媒介

数据必须通过某种东西，才能由此及彼。一条电话线，一根电缆，或者空气都可称为传输媒介或信道。但是，在数据传输之前，必须转化为适合通信的形式。可以转化为适合通信的三种基本数据形式是：

- 电脉冲和电荷（用于通过电话线传送声音和数据）；
- 电磁波（类似于无线电波）；
- 光波。

通信的形式或方法决定着数据可以通过信道传输的最大速度以及会出现的噪声等级——例如，光波传输速度比电磁波快，某些类型的卫星传输系统比通过电话线传输的噪音要小。显然，有时要求数据传输得越快越好，有时则不然。像电报线路传输数据较慢的信道，它是窄波段信道。大多数电话线路是声波信道，它们的波段范围比窄波段信道要宽的多。宽波段信道（如同轴电缆、光纤、微波电路和卫星系统）高速传输大量数据。

用于支持数据传输的传输媒介有电话线、同轴电缆、微波系统、卫星系统和光缆。理解这些媒介的性能将帮助你区别这些媒介的不同的速率和使用费用，进而决定在一个特定的情况下使用哪一种媒介是最适合的。

1．电话线路

最早类型的电话线也叫作“裸线”，为挂在电线杆上的无护套的铜导线，用玻璃绝缘体加以保护。因为它是非绝缘体，所以这种电话线极易受电磁波干扰；这些导线不得不分成以 12 英寸长为一段，以最大限度地减小这个问题。虽然裸线在一些地方仍能见到，它几乎完全被电缆和其他类型的通信媒介所代替。电缆是加绝缘层的导线。一对加绝缘层的导线互相缠绕在一起——称为双绞线或电缆——可以 1000 对或更多对的做成一束。这些直径很大的电缆如今被广泛用作电话线，并且经常能在大的建筑物里和城市道路下面见到。尽管这种缆线比起裸线是一个很大的进步，但它仍有很大局限性。双绞线对某种电磁干扰（噪声）很敏感，这限制了数据不失真传输的实际距离（想要完整地接收数字信号，必须每 1~2 英里用放大器及相关电路重新放大，这种放大装置称为增音器。虽然增音器确实增大了信号的强度，弥补信号因长距离的衰减，但它们的价格可能很高）。多年来，双绞线用于传输声音和数据；但是，更新的、更先进的传输媒介正在代替它。

2．同轴电缆

同轴电缆（也称屏蔽电缆）比双绞线更贵，是一种有厚绝缘层的铜导线，它能传输大量数据——每秒可达 100 000 000 比特或同时包括大约 1800～3600 次的电话呼叫。绝缘层是由一层金属筛网和较厚的橡胶或塑料的绝缘材料组成。在地下或水底的同轴电缆，类似于用在连接电视机和电视台之间的电缆。同轴电缆也可束成一个大电缆，因其性能优良，且无须每 2～4 英里刷新或加强信号，这种通信线路已非常普及。同轴电缆经常用作局域网络连接的主要通信媒介，比如在同一幢建筑物里。同轴电缆也用作海底电话线路。

3. 微波系统

和用导线及电缆不同，微波系统用大气作为媒介来传输信号。这些系统主要用于大批量及长距离的数据和语音通信，类似于无线电波，但频率范围更高。微波信号通常是直线信号，因为它们不能随地球的弯曲而弯曲；相反，它们必须由微波塔或每 20～30 英里设置的中继站实现点到点的转发。两座塔之间的距离取决于附近地表的弯曲程度。地球表面一般说来每英里弯曲 8 英寸。微波塔有一个抛物面天线或一个圆锥形天线。天线的长度随信号需要覆盖的范围而变化。一个长距离天线可达 10 英尺或更长；直径在 2～4 英尺的抛物面（dish）天线对于短距离的转发也就足够了，这种天线在城市建筑物顶上能经常看到。每个塔上的设备把接收过来的信号加以放大，并把信号发送到下一个基站。

应用微波进行声音和数据通信的主要优点是不需要直接的电缆连接（显然，电话线和同轴电缆必须实际接到通信系统的每一点上）。现在，有一半以上的电信系统通过微波传输信号。但是，当传输微波的空气波达到饱和状态时，将由其他方式来满足将来的需要，例如光纤和卫星系统。

4. 卫星系统

卫星通信系统采用千兆赫频率传输信号——周期为每秒 10 亿次。卫星通常必须处于地球同步轨道，距离地表 22 300 英里，所以它同地球一起每天旋转一圈。对一个地面观测者来说，它似乎总是固定在一个地方。卫星利用太阳能，里面有多达 100 个转发机（转发机是一个小的，特制的无线电设备），用于接收、转发信号。卫星作为地面卫星接收站（称作地面站）之间的中继站。

虽然建立卫星系统的价格非常高（由于卫星的价格和把卫星送入地表轨道及发射失败保险等有关问题的费用），但是卫星通信系统已经成为远距离传输大量数据的最流行和效益最好的方式。卫星通信的最大优势是单个卫星信号的覆盖范围。在特定的轨道上放置 3 颗卫星就可以覆盖整个地球表面，还会有些重叠。

然而，卫星传输的确存在一些问题：

- 信号经长距离传输会衰减，并且天气条件和太阳的活动都会产生噪声干扰。
- 一颗卫星只能用 7～10 年，然后它将偏离轨道。
- 每一个人都能接收到卫星信号，所以重要的数据必须用保密或译成密码的形式进行传输。
- 依据卫星传输的频率，地面微波站能以同样频率的操作来“堵塞”，或阻止传送。

当然，还有一种非常重要的传输媒介——光纤，我们将在下面做详细探讨。

课文 4：交换技术

不管通信网络提供的是一台计算机与另一台计算机之间的连接还是终端与计算机之间的连接，通信网络可以分成两种基本类型：电路交换（有时叫作面向连接的）和分组交换（报文交换的变形，有时叫作无连接的）。电路交换网络运行时在两点之间形成一条专用连线（线路）。美国电话系统使用电路交换技术，即一个电话呼叫建立一条线路，从发起呼叫的电话机通过本地交换局、穿过中继线到一个远程交换局，最后到达目的电话机。在线路存在时，电话设备对话筒的输出重复采样，把采样进行数字编码，并通过线路把它们传送到接收方。发话方确信采样一定会被传输和重新生成，因为线路提供了一条被保证的64kbps（千比特每秒）数据路径，这个速率是发送数字化的语音所必需的。电路交换的好处在于它的容量有保证：一旦建立一条线路，没有其他网络活动会减少这条线路的容量。电路交换的缺点是代价大：线路的费用是固定的，与通信量无关，例如，一个人要为一个电话交付固定的费用，即使两边没有交谈。

在报文交换中，传输单元是一个被精心定义的数据块，该数据块称为报文。除了要发送的内容外，报文还包括报头和校验项。报头含有源地址和目的地址的信息以及其他的控制信息，而校验项用于误码控制。交换单元是一台被称为报文处理器的计算机，它具有处理和存储的能力。报文独立并异步地传输，在源点与终点间选择自己的传送路由。首先报文由主机送往与之相连的报文处理机。一旦报文被完全收到，报文处理机就检查其报头，并相应地决定该报文传送的下一个输出信道。如果这个所选信道忙，则该报文排队等待，直到此信道空闲时开始发送。在下一个报文处理器中，报文被再次收到、存储、检查，并在某个输出信道上再发送出去。相同的过程继续进行着，直到报文交到目的地为止。这种传送技术亦被称为存储转发传输技术。

分组交换是报文交换的一种变形。在分组交换中，报文被以指定的最大长度分成若干个被称为分组的段节。与报文交换一样，每个分组都含有一个报头和校验项。分组以存储转发方式独立传送。

分组交换类型的网络通常用于连接计算机，它采取完全不同的方法。在一个分组交换网络中，网络上传输的数据被分成一个个小的片，叫作分组，分组被多路复用在大容量的机器间的连接上。一个分组，通常含有几百个字节的数据，载有使网络硬件知道怎样把它发送到指定目的地的标志信息。例如，一个要在两台机器间传送的大文件，必须被分成许多分组，在网络上一个一个地传送。网络硬件把分组传送到指定目的地，在那里，软件把它们重新组装成一个文件。分组交换的主要优点是计算机之间的多路通信可以并行进行，机器间连接被正在通信的各对机器所共享。

在电路交换的情况下，建立电路总要对开始的连接付出代价。只有在这种情况下，即一旦电路建立后，信息的传送确保持续、源源不断，以便分摊初始花费，才能提高价格效率比。传统方式的语音通信就属于这种情况，因此电路交换技术才为电话系统所采用。但是，计算机的通信具有突发特性。突发性是报文产生过程和报文高度的随机性所造成的，也是用户对时延要求很短造成的结果。用户及设备不怎么经常用到通信资源，但是当用到时，就要求相当迅速地反应。如果要建立一个固定的专用的端到端电路以连接两个用户，则必须对该电路分配足够的传输带宽以合乎对时延的要求，其结果是电路的利用率很低。如果对每个报文传输要求都要建立和释放大带宽的电路，则与报文传输的时间相比，电路建立的时值将很大，造成很低的电路利用率。因此，对突发性用户（峰值速率与平均速率的比值很高为该用户的特征），存储转发传输技术提供了一个更低价高效的解决办法，因为只有在报文传送的时间里，报文才占据一条特定的链路。在其他时间，报文是被存储在某个中间交换机中，因而此时的链路可用于其他传输。这样，与电路交换相比，存储转发方式的主要优点是通信带宽的动态分配，而且这种分配是以网络中的特定链路和特定报文（对一对特定的源点-终点来说）为基础的。

分组交换除具有以上讨论的优点外，还具有一些缺点，它的缺点是随着网络活动的增加，一对通信的计算机所获得的网络容量就会减少，也就是说，每当一个分组交换网络超载，那么，使用这个网络的计算机在可以继续发送分组之前，必须等待。

尽管分组交换网存在不能保证网络容量的潜在缺点，但分组交换网络已变得非常流行。采用分组交换的动机是从成本和性能方面考虑的。因为多机可以共享网络硬件，所以只要求较少的几个连接，费用低；而且工程师们已经能制造高速网络硬件，容量通常不成为问题。因此，许多计算机互连都使用分组交换，所以本书后面，术语“网络（network）”将仅指分组交换网络。

课文 5：ATM

ATM 是基于 ITU-T 的宽带综合业务数字网（B-ISDN）标准的成果。它最初设想作为在公共网络上用于声音、视频和数据的高速传输技术。

ATM 是一种综合了电路交换（保证容量和稳定的传输延迟）和分组交换（灵活性和间歇传输的效率）的优点的信元交换和多路技术。它提供从每秒几兆位（Mbps）到每秒几吉字节（Gbps）的可升级的带宽。ATM 依靠其异步特性而比诸如时分多路复用（TDM）等同步技术的效率更高。

ATM（异步转移模式）既是复用技术，又是交换技术。最初，人们是想用 ATM 来处理高比特率的数字信号，事实却证明它是一种通用技术，可以用来传输和交换任何类型并具有各种比特率的数字化信息。

无论传输的信息是什么，ATM 都以称作“信元”的短的分组来传送信息。信元是由固定的 48B 加上 5B 的信头组成的。信元寻找路由是基于带有双重识别的逻辑信道原则：信元头包含了信元所属的基本连接识别符，这种基本连接称作虚电路（VC），另一种是连接所属的 VC 组识别符，称作虚路径（VP）。

ATM 既与电路方式有关又与分组方式有关。由于使用简单的协议，信元至网络节点的转移可完全由硬件处理完成，这就缩短了转送时间，提高了传输路径的速率，使比特速率甚至可以达到每秒几百兆比特。另一方面，ATM 保留了分组方式所有的灵活性：只传送所需要的信息，提供简单、独特的复用方法而不管不同信息流的比特率，并且允许比特率变化。

ATM 网络可以近似地看作是由三个覆盖功能层组成：业务和应用层、ATM 网络层和传输层。应用层提供端到端的业务。应用层使用 ATM 网络层的逻辑连接，当信元通过由逻辑连接（称作虚连接）共享的传输链路时，ATM 网络层依次对信息流复用并寻找信息流的逻辑路由。传输层提供物理链路并处理信元的实际物理传输。

ATM 网络能够传输和交换话音、数据和视频服务，从接入的角度看，这些业务使用传统的数字接口并具有同样的服务质量。这就意味着任何两个终端间的物理连接都可由等效的逻辑连接代替，逻辑连接可在共用的传输链路中与其他的逻辑连接复用。资源可在所有连接中动态共享。

与同步时分复用技术相比，同步复用技术僵硬地将业务与传输资源相连，而异步技术的优势是根据其确切的需要来占用传输链路。

ATM技术将网络传输的应用和业务与所使用的传输资源完全分开。构成虚网络的能力意味着物理网络可以由许多用户动态实时地共享，因而使网络结构得到低价高效地使用，对高比特率业务也一样。对所有网络层的投资都是适应未来需要的，因为不同的应用在出现新的需求时可及时在同一网络结构中进行重新分配。ATM提供一种独特的方式将传输不同业务的网络协调成单一的物理网络。

随着数字化和图像编码技术的进步，交互视频业务和更通常的多媒体业务开始出现。这些业务将会对网络产生很大的影响。今天，ATM是唯一能够提供这些业务所需的高比特率和灵活性的传输技术。

ATM远比任何其他电信技术更能满足运营公司和用户对当前和未来业务的需求。与其他有可能在某些应用领域与ATM竞争的技术相比，ATM主要由于其通用性，无论是比特率还是传输的信息类型都具有特殊的优点。ATM对所有比特率的信号都可提供交换功能，这一点特别适合于高比特率和可变比特率信号。

ATM的独特性将使它成为卓越的多媒体业务的自然载体，特别是对于可变比特率的视频，使它成为能够提供如视频点播新业务的未来信息高速公路必不可少的一部分。在很短的时间内，运营者对ATM产生了很大的兴趣，主要是由于将连接的概念与实际资源分开所引入网络的灵活性和虚拟性，这就简化了网络的管理功能并能最佳地使用网络资源，特别是通过统计复用和建立虚拟专用网络。

当然，在ATM技术普遍使用之前会有很长的路要走，但是这场正在进行的技术革命将会深刻地影响数据处理和视频处理，影响电信世界。这一革命所产生的影响无疑会比在模拟网络中出现数字技术的影响要大得多。

总之，ATM远比任何其他电信技术更能满足运营公司和用户对当前和未来业务的需求。

课文6：光纤

虽然卫星系统被认为是现阶段长距离通信的主要媒介，但人们把光纤技术看作通信工业的革命，因为它具有低廉的价格、较高的传输容量、较低的误差率以及信息的保密性等优点。光缆正在取代铜导线成为建筑物和城市的主要通信媒介；大通信公司正投入大量资金用于能够传输数字信号的光纤通信网，以便增大信息和容量。

在光纤通信中，信号被转化为光的形式并由激光器通过绝缘的、非常细的玻璃（1英寸的1/2000）或塑料纤维发出。光脉冲再现电子数据代表中的“开”状态并且每秒钟出现近10亿次——一根光纤每秒可传输近10亿比特。同样重要的是，光缆没有尺寸上的不方便：一根光缆（由绝缘纤维束在一起）只有1/2英寸厚，能够支持同时有近250 000次的对话（不久将加倍到500 000次）。但是，因为数据是由光脉冲的形式进行通信的，因此要使用特殊的通信设备。

光缆对电子噪声不敏感，所以比通常的电话导线和电缆具有较低的误差率。另外，它们实际的数据通信速度比微波和卫星系统高 10 000 倍。光纤通信对数据盗窃也能有防止作用，因为在光纤里的窃听器监听或改变正在传输的数据可很容易被检测到。事实上，美国中央情报局现阶段就采用这种方法。

光纤看上去像一个普通玻璃圆柱体，由纤芯和包层组成。目前，常用的光缆有三种类型：单模、多模和塑料光纤（POF）。

单模光缆由直径为 8.3～l0μm 的一条或若干条光纤组成，并且只以一种模式进行传输。单模光纤的直径相对狭小，典型的尺寸为 1310nm 和 1550nm。传输带宽要大于多模光纤，但其光源要求具有较为狭窄的光谱宽度。

单模光纤具有较高的传输速率，并且比多模光纤的传输中继距离长 50 多倍，然而它的费用仍然很高。单模光纤的纤芯比多模光纤细得多。其细小的纤芯和单个模式的光波实际上消除了由重叠光脉冲产生的任何信号失真，还能使信号的衰减降至最小，并且是所有类型光缆中传输速率最高的。

与单模光纤相反，多模光纤的纤芯直径远大于传输光波的波长（常用的芯径为 62.5μm）。光波在光缆纤芯中传输时的典型波长为 850nm 或 1300nm，并被分散成众多的路径和模式。然而，在长距离光缆（超过 3000m）传输时，多个路径的光波会在接收端造成信号扭曲、失真，并导致传输数据不清晰和不完整。

多模光纤有两种类型：阶跃型多模光纤和渐变型多模光纤。阶跃型多模光纤是首代光纤产品，但由于存在多个不同路径长度引起的模式色散从而使得它在大多数的使用中传输速率太低。阶跃型多模光纤目前除了在塑料光纤以外已很少使用。渐变型多模光纤正如其名，它的折射率从纤芯到包层逐渐减小以弥补不同路径长度造成的模式色散。它能提供比阶跃型多模光纤多数百倍的带宽——最高达到 2GHz。

塑料光缆是一种新型的以塑料为基础的光缆，在短距离通信时其性能与石英玻璃光缆相似，但费用要低得多。

课文 7：无源光纤网络

本文阐述了使用无源光纤网络（PON）的宽带接入网结构最近的发展。人们已经广泛认识到利用无源光纤网络进行高带宽用户互联网传输的潜力以及其接入技术上的优点。在过去的几年中，PON 在标准化过程和应用方面已经取得了长足的进步。

接入网，也被形象地称为“第一里（first-mile）”网络，它连接了从网络服务提供商的中心局到企业以及住宅用户。这种网络，在文献中也被称作用户接入网或者本地环路网。住宅用户要求“第一里”接入方案具备高带宽、多媒体互联网服务以及与当前网络相比较为优惠的价格。类似的，企业用户则要求宽带基础设施能够提供从工作域网络到互联网主干路线的连接。

1. 接入网带来的挑战

近年来，人们一直在关注着高容量骨干网络的发展，到目前为止，骨干网网络运营商能提供高容量的OC-192连接（10Gbps）。然而，目前新一代的网络接入技术却不尽如人意，比如数字用户环路（DSL）在最好的情况下仅仅能够提供1.5Mbps的下行带宽和128kbps的上行带宽。由此可见，要向终端客户提供宽带服务（比如网络视频传输，网络互动游戏以及网络视频电话等），接入网是技术瓶颈。

此外，数字用户环路（DSL）还存在一个限制因素，由于信号的衰减，任何DSL用户与中心机房的距离不得大于18 000英尺，这也是为什么通常情况下DSL不提供超过12 000英尺的服务。因此，据估计只有不到60%的住宅用户可以享用DSL技术。尽管DSL技术在不断进步，例如高比特率DSL（VDSL）技术，它能够支持高达50Mbps的下行带宽，但是却有着更加严格的距离限制，比如VDSL所能支持的最大距离仅为1500英尺。

另一个针对终端用户的有效的宽带接入方法是通过有线电视网络（CATV）。有线电视网络可以在同轴电缆上牺牲部分射频信道用于数据传输，从而提供因特网服务。但是有线电视网络主要是被设计用来传输广播服务的，因此并不适合用于分配接入带宽。在高负荷的状况下，有线电视网络的性能通常令终端用户很失望。

显而易见的，更快的接入网技术是为下一代宽带应用所设计的。下一代接入网络技术的革新将会使光纤离住户更近，比如FTTx模式——光纤到家（FTTH）、光纤到路边（FTTC）、光纤到楼（FTTB）等——有可能为终端用户提供史无前例的接入带宽。这些技术致力于将光纤直接引入住户，或者非常接近住户，从而替代VDSL。FTTx技术主要是基于无源光纤网络（PON）的。本文将回顾一下近几年来PON技术的主要发展：EPON、APON、GPON和WDM PON。最后我们将回顾一下无源光纤网络使用中所遇到的问题。

2．无源光纤网络的结构

无源光纤网络是一个点对多点的光纤网络。通过一个或多个1:*N*的光分路器，一个中心机房的光纤线路终端（OLT）连接到多个位于远程节点上的光纤网络单元（ONU）。从OLT到ONU的网络是无源的，即不需要任何电源支持。网络中无源元件的存在使得网络容错能力更强，而且一旦当基础设施架设完毕，将大幅降低运营费和维护费。

人们考虑将无源光纤网络（PON）用于接入网络已经有较长的一段时间了。一个典型的无源光纤网络使用单波段对所有下行带宽传输（从OLT到ONU），而使用另一个波段进行上行传输（从ONU到OLT），通过粗波分复用（CWDM）技术复用到单根光纤上。

3．以太无源光纤接入网络

以太无源光纤网络（EPON）是一种基于PON的网络，它使用以太网帧标准封装数据（在IEEE 802.3标准中有明确定义）。它使用一个标准的8b/10b线路编码（在编码中8位用户比特编码为10位线路比特），以标准以太网数据率工作。

4．为什么以太网越来越突出

第一代PON的标准由国际电信联盟（ITU-T）G.983制定，使用ATM作为介质访问控制（MAC）层的协议。当1995年开始进行标准化时，当时电信界普遍认为异步传输模式（ATM）将是骨干网络的主流技术。异步传输（ATM）在其高效的语音和数据服务方面拥有巨大优势，同时提供了业务与性能的保证。然而从那时起，以太网就很受欢迎。其线卡价格低廉，特别是在今天，以太网被广泛地应用在局域网中。而由于接入网重点是终端用户以及局域网，因此异步传输模式变得越来越不适用于连接基于以太网的局域网了。

此外，高速千兆以太网正在高速发展，而这使万兆以太网产品变得可行。以太网是一个非常高效的MAC协议，相比之下，ATM在不定长度的因特网协议（IP）数据封包上耗费了许多资源。最新通过的服务质量（QoS）技术使以太网能够有效地支持语音、数据以及视频的传输。这些技术包括全双工传输模式、优先级（802.1p）和虚拟局域网（VLAN）标签（802.1Q）。802.1p是一种对信息流按照优先顺序划分优先级的标准，而802.1Q则定义了VLAN的结构。尽管802.1Q并不直接定义任何服务质量（QoS）支持，但它却定义了一个帧格式扩展，允许以太网的帧携带优先信息。因此，与异步传输的无源光纤网络（ATM PON）相比，EPON更有可能成为未来接入网的主要技术。

5. EPON的运行原理

在下行方向（OLT到ONU），以太网的帧传输是从OLT通过一个一对N的无源分路器然后到达每一个ONU，一般来说，N典型的取值是在8～64之间。数据包由OLT广播，并且由基于逻辑链路标识（LLID）的目的ONU所提取，ONU在网络上成功注册后，系统就会分配一个逻辑链路标识。

在上行方向，由于无源光合路器的方向性，来自任何一个ONU的数据帧只会到达OLT而不会到达任何其他的ONU。因此，在上行方向，EPON的作用跟点对点结构十分相像。然而并非如真正的点对点传输那样，在EPON中，从不同ONU出来的数据包在同时传送中可能会发生冲突。所以在上行方向，ONU需要采用一些仲裁机制从而避免数据冲突以及公平的共享信道容量。由于合路器的方向性，ONU不能够检测合路器与OLT之间光纤段中的数据冲突，所以基于竞争的介质访问机制（如CSMA/CD）很难实施。OLT可以进行冲突检测，并且可以通过发送停发信号来通知ONU；然而在PON中的传播延迟则大大降低了该方案的效率（一般来说，从OLT到ONU的距离为20km）。为了增加上行方向帧传输的确定性，出现了不同的非竞争方案。

所有的ONU都与一个共同的时间参考系同步，每一个ONU都在该时间系所分配的时隙中传输数据。而每个时隙都能够传输多个以太网帧。ONU在时隙到来前应该对来自用户的帧进行缓冲。当时隙到来，ONU将会全速地传输帧缓冲区所存储的数据。如若缓冲区中不存在用以填满整个时隙的帧，则传送一个空闲模式的信号。

因此，时隙分配是非常重要的一步。可能的时隙分配方案包括静态分配（固定的时分多址TDMA）及动态分配。动态分配主要基于每个ONU的瞬时队列长度（一种统计学的多路技术方案）。在动态时隙分配方案中，OLT收集ONU的队列长度并动态分配时隙。尽管这种分配模式将会在OLT和ONU之间产生更多的信令开销，但是这种集中智能却能够更加充分地利用带宽。当然，更加先进的带宽分配方案也是存在的，包括利用通信优先权的概念、服务质量（QoS）、服务等级协议（SLA）以及过预订率等。

课文8：电视基本原理

1. 声音和光谱

视频是声音和光结合的产物，二者都是由振动或频率组成的。我们被周围各种形式的振动环绕：看得见的、摸得着的、听得到的以及许多我们无法感知到的其他种类。我们位于一个广阔的频谱的中间，这个振动的频谱的震动次数从每秒零次到几百万次不等。我们用来衡量每秒振动次数的单位称之为赫兹（Hz）。

声音振动发生在较低的频谱区域，而光振动则在较高的频率区域。声音频谱为 20～20 000Hz。光振动范围从 370 万亿赫兹至 750 万亿赫兹。当提到光时，我们常探讨的是波长，而不是振动。

由于光的频率很高，速度快（光的传播速度为 300 000km/s），所以光的波长极短，不足千分之一毫米。（所以）振动频率越高，波长越短。

并非所有的光束都具有相同的波长。在可见光谱的波长从 780nm 到 380nm 不等。我们看到的各种颜色都对应不同的波长。最长的波长（对应于最低频率）是我们所看到的红色，这种规律遵循于我们已知的彩虹的颜色：橙色、黄色、绿色、蓝色、靛蓝和紫色，紫色是波长最短的光（频率最高）。白色不是一种颜色的光，而是其他颜色的混合体。我们无法看到的波长的光（红色区域以下和紫色波长以上的区域）分别是红外线和紫外射线。如今，红外线被用于作为红外遥控装置等应用。

注意：可见光之所以是可见的是因为我们可以看到发光源和被照亮的物体。光束本身无法看到。比如说，在薄雾中车灯的光束之所以能被看见，是因为小水滴组成的雾气反射了光线。

2．亮度

除了具有不同的颜色（即不同的频率），光的亮度也有所不同。就亮度而言，一个台灯发出的灯光比一个卤素灯少，但即使卤素灯的灯光也无法和明亮的太阳光相比较。亮度取决于可得到的光的数量，它可以被测量并用数值记录。在过去，它被表示为赫夫纳烛光，但现在勒克司被用来表示亮度的数量。

亮度值：

蜡烛灯在 20cm 范围内	10～15Lux
路灯	10～20Lux
一般室内灯	100Lux
办公室日光灯	300～500Lux
卤素灯	750Lux
阳光，日落前 1 小时	1000Lux
日光，多云的天空	5000Lux
白天，晴空	10 000Lux
明亮的阳光	> 20 000Lux

亮度是黑白电视的基本原则。所有在黑色和白色之间的灰度，都可以通过调整亮度到特定的值来表示。

3．混色

这里有两种颜色混合方法：相加和相减混色法。就像颜料的着色混合的方法，被称为相减混色。彩色光混合的方法被称为相加混色。彩色电视的原理是基于相加混色。三基色可以用于创建色谱中所有的颜色。

4．相加混色

在视频中，彩色光谱包含三基色，即红、绿、蓝。通过这三个组合，其他所有颜色的频谱（包括白色）都可以被混合出来。

红色+蓝色=品红
红色+绿色=黄色
蓝色+绿色=青色（蓝绿色）
绿色+品红=白色
红色+青色=白色
蓝色+黄色=白色
红色+蓝色+绿色=白色

用这种方法创造出颜色是基于混色原理，或者说是色光相加，这是为什么它被称为相加混合的原因。通过不同比例的三基色混合可以创造出所有可能的颜色。

白色光是由 30%的红色、59%的绿色、11%的蓝色混合而出的。这也是彩色电视机用于呈现黑白图像时所需的设置比例。通过保持该比例，设置亮度为具体的值，可以呈现不同的灰度梯度。

30%红色+59%绿色+11%蓝色=白色

5．光线的折射

光线的折射是混合色的逆过程。研究表明，白光是由所有可见光谱的颜色混合而来。我们用一个打磨过的三角形的玻璃棱镜来说明折射原理。一束光线通过棱镜折射两次，使其路径发生改变。

波长较长的红色光束比起波长较短的紫色光束，其折射较不明显，所以使得颜色分散开来。第二次折射的分光效果比第一次更明显，结果产生了一个颜色带，由以下的颜色组成：红、橙、黄、绿、蓝、靛、紫。这里各种颜色之间没有明确的界线，但是有成千上万的颜色过渡。彩虹就是一个大自然中光的折射原理的完美例子。

6．色温

色温涉及这样的一个事实，当一个对象被加热，它会发出一种颜色，直接与该物体的温度相关。色温越高，光越“蓝”，色温越低，光越“红”。光的色温可以用绝对温度开尔文（K）来衡量。日光灯的色温为 6000～7000K，人工照明的灯的色温相对更低，约 3000K 左右。实际上，色温的范围是从 1900K（烛光）到 25 000K（湛蓝的天空）不等。电视的色温设置为 6500K，是模拟“标准日光灯”。

7．人眼特性

一幅图像消失后，人眼往往会保留该图像的影像约 80ms 的时间。在电视和摄影中利用这一优势，可以用一系列的静止图片（每秒 25 帧）创造出一个不断运动的动态影像的错觉。另一个人眼特性是：比起对于颜色色调的差异，人眼对黑白亮度的差异更敏感，而且人们的眼睛对于各种颜色的敏感程度并不一样。眼睛对黄色/绿色区域最敏感，其次是红色和蓝色。

课文9：电视接收机

看电视是北美人喜欢的消遣，所以电视接收机的制造占了当今电子工业很大的比例。由于电视机的制造已彻底标准化，所以不同厂家的电视机几乎没有什么不同。现在北美所使用的系统与所有北美国家、日本和其他几个国家相同。欧洲国家常用的制式与美国的不同点主要是使用的扫描频率不同。

现在大部分电视广播是彩色的，尽管还能见到黑白电视机，可它们通常用于特殊场合，如保安监视和微机系统。彩色系统和所有彩色接收机与黑白传输是兼容的。

1．阴极射线管

接收机的阴极射线管完成的功能与发射机的摄像管相反，它是把电信号转换成图像信号。这里暂时省略了反射和聚焦装置。代表图像的电信号被加到控制电极，控制电极的工作方式与电子管的控制栅极类似。信号使得控制电极的电动势围绕一个平均值变化。电动势的负值越大，电子束的强度就越小，相应地在屏幕上形成的光点的亮度越小。同样，电动势的负值变小时，屏上光点的亮度就越大。荧光屏是由磷构成，该材料受到电子束轰击时就会发光。

在现代显像管中，磷的背面涂有一层铝薄膜，且该涂层一直延续到管子的内壁与最终阳极相连。该涂层的目的是增加屏幕的亮度（对一定的电子束强度而言）并保证亮度均匀（即不出现暗块），该涂层的电动势与最终阳极相同，它可防止电子束的电子在屏幕上堆集，否则这些电子就会排斥入射电子束，从而降低亮度。进一步讲，若出现电子堆积，屏幕表面的电子分布就会很不均匀，就会产生暗块。铝涂层还以另一种方式改进亮度，即把那些会丢失在管内的光反射到显示端。在穿透铝涂层时，电子波束会损失一些能量，但由于涂层很薄，所以能量损失足以被亮度的增加所补偿。

在发射机摄像管和接收机显像管中，都要使电子束对屏幕进行扫描。电子束通常由屏幕的左上角开始以几乎水平的路线到达右边。横扫过每一行时都要有一个小的向下偏移，这样才能最终扫过整个平面。当光束到达右边时，它又被快速地偏转回左边（这叫行回扫），从这里又开始新的一行扫描。该回扫是由名叫水平同步脉冲的一个行同步信号启动的。这个由发射机发出的特殊脉冲信号能保证接收机的回扫出现在与发射机回扫相同的时刻。

当扫描波束到达右下角时，它还必须再回到左上角以便重复整个扫描过程。所以，这时必须发射第二个同步信号以保证接收机和发射机在同一时刻开始一个完整的扫描，这叫作垂直同步脉冲的场同步信号。

2．隔行扫描和图像扫描重复速率

对目标区域的完整扫描会在电视系统的接收机中产生一个完整的图形。在这一段时间内必须传输大量的信息。如果图像重复速率很高，则传输所需要的带宽就会过大。在电视中，如果图像扫描速率太低，移动景物会出现停停走走的抖动，好像慢动作的移动画面。而且接收机显像管所用的磷留存的时间相对较短，而使画面在两次扫描间消失，扫描将产生以图像速率的抖动。图像的速率必须足够高以使观察者眼睛的正常视觉暂留能覆盖闪烁，而使一系列的图像汇成连续的运动，这个最小图像速率已知是每秒35～50个画面。

人们还发现，在电视系统中，若扫描速率接近但不准确等于电源频率时，交流输电电路所得到的电压将会对扫描电路放大器进行调制，而产生烦人的图像失真和抖动。这种干扰可通过使图像扫描速率等于电源频率的整数倍来降低到最小。在美国系统中，图像或(帧)的扫描速率是 30Hz，为电源频率 60Hz 的约数。

30Hz 的图像重复速率太低了，且若扫描是按简单的顺序方式，那么由于扫描间图像消失所产生的闪烁将会变得很烦人。所以，每帧图像的扫描分为两个单独的场，且整个区域在一个图像的周期内扫描两次。每个图像共有 525 行，所以第一场扫 262.5 行，第二场扫剩余的 262. 5 行。

在第一个 262.5 行的末尾出现一个垂直回扫，且第二个扫描从图像的顶行中间开始。结果第二场的扫描行处于第一场各行的中间。所以两场合在一起形成完整的一帧扫描。现在的实际闪烁速率是 60Hz，远高于门限，而帧或图像的重复速率为每秒 30 个，低得足以将图像信号频率降到最小。欧洲的电源通常工作在 50Hz，且这种系统的图像速率是每秒 25 个，场速率为每秒 50 个。

在每一个完整的图像帧周期中出现两个垂直回扫。且对于每个回扫，主摄像机控制单元都要产生一个同步脉冲（垂直同步脉冲），它用来启动摄像机和接收机扫描光栅的垂直回扫，以使二者保持相互同步。垂直同步脉冲等于场频：$F_V = 2P$，其中 P 是帧或图像重复速率，单位是图像数/秒。垂直同步脉冲被叠加在一个较长的“消隐”脉冲顶部。该脉冲用来关掉接收机 CRT 的电子波束（在回扫期间）以便看不到回扫行。

3. 电视接收机的框图

先来看天线，它通常是兔耳形的，但也可是多元八木天线或是电缆系统传输线（共用电视天线），第一部分就是调谐组件。输入信号被耦合到射频放大级，这是甲类调谐放大器，然后再传到混频—振荡器电路。超高频信道使用了二极管混频器，它和工作在倍频模式下的分隔振荡器电路连在一起。二次谐波被用来给混频器提供本地振荡器信号。调谐是通过旋转开关完成，该开关可为每一信道的电路连接互不相同的一套线圈和电容，还可在内部调节每个信道的振荡器频率。另外还有微动调节器用来进行微调谐，这可在前面板控制。

所有接收机和所有信道所使用的中频已标准化，统一在 41～47MHz。由于两个边带在转换时（对高于信号频率的振荡器而言）发生翻转，图像载波出现在 45.75MHz，声音载波出现在 41.25MHz。图像放大器系列的通带被设计成能通过所有这些频率，由于有残余边带，其通带有些倾斜。通带特性及增益由中频放大器系列来实现，该系列可包括 3～5 级参差调谐放大器。在两个或多个中频级和射频调谐级加有自动增益控制信号。图像检波器由位于中频系列末尾的简单包络检波器完成。

检出的图像信号再馈入音频电路系列的输入端，在此处，调谐电路把 4.5MHz 的声音载波分离出。声音中频放大级是设计用来通过并限制一个以 4.5MHz 为中心频率的 200kHz 的频段。伴音检波电路通常是比例检波器，检波器之后是伴音放大级电路，它又与扬声器相连。

所检出的图像信号还通到图像放大器，以提高图像信号的电平，去掉伴音载波，并为显像管阴极提供直流偏置。加在显像管阴极的图像信号及屏上产生的光的强度都随波束强度变化而变化。

检出的图像信号还通到同步电路，在此图像信号部分用限幅法去掉，只剩下同步脉冲。被限幅的同步脉冲经放大再去触发水平扫描振荡器。振荡器产生的斜升电压经放大再去驱动水平输出变压器和水平偏转线圈。大的回扫脉冲经整流后为显像管的靶子提供 10～20kV

的高压，并提供选通信号来控制AGC电路。

第四部分 计算机系统

课文1：关于计算机

从第一批如房间大小的大型机，到今天功能强大的台式电脑、膝上型电脑和手提电脑，所有计算机都进行着相同的信息操作。随时间变化的是所处理的信息内容、信息的处理方式以及信息处理的速度和效率。

1. 计算机的功能

计算机是一种电子设备，在存储程序指示下可以自动、快速、精确地处理数据。它可以接收、存储、处理数据并通过输出设备如屏幕或打印机输出运算结果。

一台计算机具有四个功能：输入、处理、存储和输出。

输入：输入设备允许用户把数据输入到电脑中。主要的输入设备有键盘、鼠标和其他的一些设备，比如，扫描仪等。

处理：中央处理器（CPU）是计算机的“大脑中枢”。它包括引导计算机从ROM（只读存储器）或者随机存储器（RAM）中读取指令的电路。通过执行这些指令来处理信息。

存储：随机访问存储器是短期存储的存储器，它是易失性的存储器，因为一旦关掉电源或电源突然中断，随机访问存储器中的信息就会自动消失。内存在计算机内部的主板上。主板上有内存的存储部分、电子线路和其他部分，包括中央处理器。只读存储器（ROM）不是易失性的，这意味着即使突然断电或关闭电源，信息不会丢失。当计算机重启后，只读存储器会调用硬盘内部的存储信息。

输出：输出设备如显示器或者打印机，它使你可以查看或者使用你输入的信息。

计算机技术是电子技术和计算技术结合的产物。现如今，计算机的发展已经上升到一个全新的高度，最显著的特征就是计算机与沟通交流的融合，引导着我们进入一个美妙的因特网世界。当代的计算机拥有超越逻辑判断、自动操控和大容量存储的功能。其结果就是在某些领域中，计算机可以代替人类劳动力参与有职业危险的工作。

2. 计算机的发展

数字式电子计算机ENIAC是世界上第一部大型电子计算机，于1946年问世。它最初的设计目的是在第二次世界大战中用来实现复杂的弹道计算，但是它还没有被派上用场，战争就结束了。战后，人们仍旧使用它，为研制氢弹、天气预报、宇宙射线的分析、内燃发动机、随机数、风管道设计执行计算。ENIAC内有17 468个电子管，70 000个电阻器，10 000个电容器，1500个继电器，6000个手动的电路开关和五百万个焊接点。它占地1800平方英尺，重30吨，耗电量160 000W。所以，只要它一工作就使得整个宾夕法尼亚地区的灯都很昏暗。

第一代（1940—1956年）：电子管时代——电子管利用各种电路元件和磁鼓来记忆。由于电子管坏得很频繁，所以，这些计算机不太可靠。这种计算机的使用成本很高，另外，因为耗电量巨大而产生大量的热，更使它经常发生功能障碍。第一代计算机依靠机器语言来执行操作，而且，一次只能解决一个问题。1951—1962 年间美国统计局使用过的 ENIAC 计算机是第一代计算机的一个例子。

第二代（1956—1963 年）：晶体管时代——晶体管代替电子管，宣告了第二代计算机的到来。晶体管发明于 1947 年，但是直到 20 世纪 50 年代才被广泛应用在计算机领域。晶体管时代的第一台计算机是为了研究原子能工业而研制的。晶体管比电子管更高级，跟第一代计算机相比，第二代计算机更小、更快、成本更低、功能更强而且更可靠。尽管晶体管仍然会产生大量的热量，致使计算机受到损伤，但是相比电子管，晶体管是一个巨大的进步。第二代计算机从含义模糊的二进制机器语言转换到使用汇编语言，使程序员可以使用单词书写指令。高级程序设计语言也在这时得到开发，例如，COBOL（面向商业的通用语言）和 FORTRAN（公式翻译程序语言）的早期版本。

第三代（1964—1971 年）：集成电路时代——集成电路（IC）的发明是第三代计算机的标志。晶体管已被缩小并放置在称为半导体芯片的硅片上，这使得计算机的速度和效率得到了极大的改善。由于比以前的计算机更小更便宜，计算机的大众普及率开始增加。

第四代（1971 年至今）：微处理器——随着可以把成千上万个集成电路植入一个芯片的技术的诞生，微处理器给我们带来了第四代计算机。大规模集成（LSI）和超大规模集成（VLSI）电路可以把从几百至几百万个晶体管集成到一小片芯片上。过去要塞满一整间屋子的零配件现在可以浓缩到手掌大小。英特尔 4004 芯片问世于 1971 年，集成了所有计算机的零件——从中央处理器和内存到数据输入输出的控制，再到一个小小的薄片上。1981 年 IBM 公司推出了家庭用户使用的计算机。1984 年，苹果公司推出了 Macintosh 系统。随着越来越多的日常生活产品开始使用微处理器，对它的应用开始进入了很多生活领域。随着计算机变得越来越强大，可以将计算机连接起来构成网络，这最终促使了互联网的发展。第四代计算机还目睹了图形用户界面（GUI）、鼠标和各种手持操控设备的发展。第四代计算机的主存储器容量增加而价格更便宜，这直接促进了软件的类型和应用。软件应用如文字处理、电子表格、数据库管理程序、画图程序、桌面出版系统等开始成为可以购买的商品。

第五代（现在及未来）：人工智能——基于人工智能原理的第五代计算机设备，尽管部分已经投入实际应用，如声音识别等，但它们还在研发阶段。平行处理和超级控制机的应用加大了人工智能成为现实的可能。量子计算、分子计算和纳米技术的发展将在未来可预见的时间内将计算机发展推动到一个全新的高度。第五代计算机的发展目标是通过与其他设备建立连接，实现自然语言操控、自我学习和全自我管理。

随着科学技术的不断发展，其他的新型计算机也将在未来出现。

课文 2：计算机硬件

硬件是指在数据处理的整个环节中能起到基本处理功能的物理设备。硬件可以包括计算机本身以及诸多辅助设备包括键盘、鼠标、显示器、机箱等。硬件由软件所操控。

各类计算机系统

这部分介绍目前所使用的各类计算机系统。

1．超级计算机

超级计算机问世于 1970 年，在最高容量计算机中，是运行速度最快、存储容量最大的计算机。它们的价格非常昂贵，研制超级计算机需要花费几百甚至上千万美元。普通用户不大可能见到超级计算机。它们被安置在特殊的有空调的房间里，通常只供研究使用。其中包括世界范围的天气预报、气候现象分析、石油勘测、航天器的设计、核武器时期的预测评估和数学研究。与通常只有单个中央处理器的微型计算机不同的是，超级计算机拥有成百上千个处理器，每秒可以进行数万亿条计算。

尽管对于普通用户来说，微型计算机已经足够用了，但是在一些工作场所要组合使用几种计算机。例如，在银行，可能有一台大型机用以处理和存储复杂的数据，而其他的微型计算机处理一些专门任务。此外，即便是当今的微型计算机也不能分开使用，它们通过强大的网络连接起来，以便相互沟通。

2．大型机

这是一种大型而且价格昂贵的计算机系统，它一次可以支持数以百计，甚至数以千计的用户。因为大型机一次可以支持多个程序，所以，在某些方面它的功能比超级计算机更强大。但是，超级计算机可以以更快的速度运行单个程序。

超级计算机和大型机的主要区别是，超级计算机以最快的速度运行程序，而大型机用于在同一时刻内运行尽可能多的程序。大型机比小型计算机更大、更快，也更贵。它们通常用于银行、保险公司、航空公司、大型企业和政府机构。

3．服务器

服务器是网络上负责管理网络资源的计算机或者设备。例如，文件服务器是专门用于存储文件的计算机和存储设备，网络上的用户都可以在这台服务器上存储文件。打印机服务器是管理一台或者多台打印机的计算机，网络服务器用来管理网络的运行。数据库服务器用于处理对数据库的查询请求。

4．工作站

它是一种用于运行工程应用软件、桌面出版应用、软件开发和其他要求有一定规模的计算能力和对图像质量有较高要求的应用程序的计算机。

和个人计算机相似，大多数工作站是单用户的。虽然工作站也可以作为独立的系统，但是通常将它们连接起来搭建局域网。

在网络中，工作站指任何连接到局域网上的计算机系统。工作站可以是工作站计算机或者是个人计算机。

5．个人计算机

一种为个人用户设计的小型的、低价的计算机。从价格而言，个人计算机的价格可以从几百美元到几千美元。这些都是基于微处理器技术，使得生产商得以将整个中央处理器（CPU）放在一个芯片上。商业上将个人计算机用于文字处理、会计统计、桌面排版以及运行表格和数据库管理系统等应用程序。游戏是家庭个人计算机最普遍的应用。

6．嵌入式专用计算机

一类作为系统或者机器的组成部分的特殊的计算机系统。典型的嵌入式计算机系统构建在一块微处理器板之上，运行的程序存储于 ROM 之中。部分嵌入式系统包括了操作系统，但是许多是专用的系统，整个系统逻辑由一个程序实现。

组成计算机的各部件

一台计算机的组成包含各种硬件，它们配合软件一起工作以实现计算、整合数据和与其他计算机建立连接。计算机包含五类基本的组成部分：主机、输入/输出设备、存储设备和通信设备。

机箱，又称主机，是金属或塑料材质的盒子，用来保护计算机内电子设备免遭破坏。机箱中的线路装置通常是组成主板的一部分。主板上最重要的两个部分是中央处理单元和内存（一级存储和主存储）。

1. 中央处理单元

中央处理单元，即中央处理器，是用来阐释和实现基础计算机操作的电子设备。在微型计算机系统中，CPU 是计算机的核心。现在英特尔生产的奔腾芯片或处理器是最常见的处理器。其他常用的 CPU 是由摩托罗拉生产的、用于苹果计算机的处理器。

来自输入设备或计算机内存的信息通过总线传给中央处理器，这是计算机翻译命令和运行程序的部分。中央处理器是一个微处理器芯片，亦即一块含有数百万个微小的、精密布线的电气元件的硅。信息存储在称为寄存器的中央处理器存储单元中。寄存器可看作是中央处理器暂时存储指令或数据的微型便笺簿。在程序运行的时候，一个叫作程序计数器的特殊寄存器，通过保存下一个将要执行的程序指令的存储单元来跟踪记录接下来执行哪个程序指令。中央处理器的控制单元协调中央处理器的功能并为其计时。它利用程序计数器从内存中查找和检索下一个要执行的指令。

在一个典型的操作序列中，中央处理器在适当的存储设备中找到下一个指令。然后，该指令从计算机的内存沿着总线进入中央处理器，在这里指令存储在一个特殊的指令寄存器中。当前的指令由译码器进行分析，译码器确定该指令将做什么。该指令需要的任何数据通过总线予以检索，并放置在中央处理器的寄存器中。中央处理器执行该指令，产生的结果存储在另一个寄存器中，或者通过总线复制到特定的存储单元。这整个序列的步骤称为指令周期。经常的情况是，几个指令可能同时进行，每个处于其指令周期的不同阶段，这叫作流水线处理。

CPU 有两个重要部分：控制单元和算术/逻辑单元。两个部件通过一条公共总线连接（一条公共总线也将计算机的其他部分和这两个部分连接起来）。

2. 内存

内存是数据和操作指令暂时存储的地方。内存有两种存储方式：物理内存和逻辑内存。逻辑内存就像可以画出来的表格一样，依照内存的获得和存储方式而存储，而物理内存是真正的硬件存储。系统内存被用作操作系统的主要暂时存储器或桌面。只读存储器（ROM）提供计算机每次启动时使用的操作指令。高速缓冲存储器（Cache）给 CPU 提供快速内存，就像桌子上的一个放东西用的架板。随机存取存储器（RAM）是由 CPU 控制的大多数操作过程的主要存储处，接收用户的数据输入。

3. 输入设备

输入设备是一种计算机外围设备，它允许用户将数据，程序、指令以及用户的响应输入计算机，并将输入内容用适当的形式进行传输，以便于处理。输入设备根据用户的具体应用范围和要求有所区别。输入设备有键盘、鼠标、输入笔、触摸屏、麦克风、扫描仪、数码照相机和条码阅读器等。下面将详细介绍一些输入设备。

1）键盘

计算机键盘是计算机中最重要的基本系统组件之一。它们用来键入数据到计算机。计算机键盘最早是模仿了打字机，使打字员更容易学习使用计算机。这个设计满足了计算机科学的需求。可以在屏幕显示出来的键入，这种设计对适应打字机的打字员来说很不习惯。并且，打字机只有50个键，计算机键盘有100多个键。标准的键盘有101个按键。一些特殊设计的键盘使打字变得更容易。

计算机键盘通常分为四个部分：字母键、功能键、光标和数字键。字母键包括字母、数字和标点符号，这些按键的排列类似于打字机。功能键是指从F1、F2到F12甚至F15等。它们通常用于比较常见的指令如“打印”或“退出程序”。每一个功能键的精准的目标性适用于多个程序。光标键是用来在屏幕上移动光标指针的。光标指针是屏幕上那个实心的小标志，指示着在屏幕上下一秒会出现些什么。当你在打字时，光标通常会停留在你所输入的最后一个字符的右边。光标键包括有箭头标志的按键，用来控制上、下、左、右、向上翻页（PgUp）、向下翻页（PgDn）、回首页（Home）、结束（End）。数字键包括数字和标准计算器要用到的数学计算键。

2）鼠标

大部分现代计算机使用鼠标作为指示器。一般来说，如果鼠标有两个按键，那么左边的用来选择物体和文本；右边的用来进入菜单，有第三个键的鼠标可以运行一些特殊软件程序。鼠标通常有根接入到计算机主机的连接线。在鼠标底部有滚动球，把鼠标的滑动译为数字信号。有些牌子的鼠标如微软鼠标，在左右鼠标按钮中间有个滑轮，可以滚动查看文件的内容。

鼠标受到欢迎是因为相对纯输入控制来说它具有点击功能，因而更容易操作。而且，用箭头键在绘图方面或在屏幕上移动文件时并不很方便。鼠标由一个塑料罩和底部的滚动球以及塑料罩上一个或多个按键构成。当你在桌面上移动鼠标时，内部的传感器记录滚球的滚动，并移动屏幕上的光标与之对应。对鼠标的控制有三种最基本的方式。第一，点击按键使鼠标与相关事物产生关联——例如点击你想改变的某幅画面的一部分。第二，你可以拖动鼠标，方法是按住按键并且移动。拖动功能可以实现在屏幕的任意一部分移动某个文件。第三，你可以双击鼠标按键，双击的间隔时间不能超过0.5s。双击通常可以实现屏幕上对某文件的选取。

当然，相对所有工具而言，最简单的点击方式是手指。事实上，触摸屏早已被广泛使用——百货商场的广告宣传、信息亭、摇奖游戏机等，都是些目标使用者可能对计算机不是很了解的地方。在这些机器上你只要简单地在显示屏上点触你想要选择的区域，就像按自动售货机的按钮一样——想要哪个点哪个。触摸屏的原理是当你的手指点触到屏幕时，会打破覆盖在屏幕表面的红外线网，这样就可以告诉计算机你点击的是哪里了。

3）扫描仪

扫描仪使用激光束反射光线来将文本、图画、相片之类的图像转换为数字形式。然后，这些图像可以被计算机处理，显示在一台显示器上，存储在一个存储装置上，或传递到另外一台计算机上。互联网用户将相片扫描到计算机之后，通过网络传递给朋友或者在网页上发布。

输出中央处理单元工作产生的结果可以说是计算机存在的意义所在，也是我们所需要的信息，即原始的输入数据通过计算机的处理后变成了信息。最常见的输出格式有文本、图表、音频和视频。以文本为形式的输出方式为例，包含能创造单词、句子和段落的字母。图表可能是非文本信息，例如，图片、表格或照片的电子版等表现形式。

4．输出设备

输出设备也是人与计算机系统之间实现解释和通信的设备。输出设备从 CPU 中取出输出结果，然后将其转换成人们可读的形式。常用的输出设备可能是显示器、打印机、头戴式耳机、扬声器、数据投影仪、传真机和一些多功能设备。接下来讨论三种输出设备：显示器、打印机和声音输出设备。

1）显示器

显示器（或称显示屏）提供了一个方便但短时的可视信息的渠道。早期的显示器只是简单的改装电视机。尽管普通电视机还用来玩一些游戏，但是现在大部分的计算机程序都需要更高质量的显示器。

显示器最重要的两个特征标准是大小和清晰度。显示器的大小是根据对角线测算的，并且仍然引用英寸为测量单位。比较流行的大小为 15 英寸（38cm）和 17 英寸（43cm）。更大的显示器用户用计算机工作起来眼睛更为舒适，这对于桌面出版系统和计算机辅助设计工作尤为重要。显示器的清晰度取决于它的分辨率，分辨率则是根据像素的大小测量的。像素（图像和元素的简称）是显示在屏幕上的最小单元，它可以被任意的打开或关闭并能改变色彩。分辨率越高（像素越多），显示器也需要更高分辨率的配置。

当用户打字的时候，显示器显示输入的信息，这个叫作输出信息。当计算机需要更多的信息时，它通常通过一个对话框，在荧屏上显示一个信息。有两种计算机显示器：CRT（阴极射线管）显示器和 LCD 液晶显示器。

2）打印机

打印机接收显示器屏幕上的信息，并且把它们显示在纸上或者其他硬介质上。有各种不同类型的打印机，它们的打印质量也不同。它们可以分为三种类型：喷墨打印机、激光打印机和热敏打印机。

不同种类的打印机，它们的工作原理截然不同。有些是喷墨，有些在热敏纸上加热，有些用硒鼓，还有些用激光来显示影像。激光打印机的工作原理是，激光束从快速旋转的八角形的镜子映射到一个光敏滚轴。喷墨打印机在打印纸上直接喷射细小的液态墨滴。点阵式打印机的底层涂料上有一个可以移动的打印头，上面有一排小针头。针头把打印机的色带印到纸上，呈现出一块点的矩阵或模型。随着打印头在纸上来回移动，点的矩阵既可以是字母也可以是图像。

3）声音输出设备

声音输出设备是计算机用户用来播放音乐、进行发声翻译语言和与计算机系统进行信息连接的组件。其中运用最广的是扬声器和头戴式耳机。这些设备通过系统单元里的声卡相连接。声卡是用来捕捉和回放录制声音的。大多数扬声器都可以通过音调和音量控制器来调整设置。使用扬声器时，在能听到声音的范围内人们都可以听见输出的声音。而头戴式耳机的作用则是让其使用者独享输出声音。

存储设备，又称二级存储，可以安全地返回或储存电子数据以备将来使用。它在内存之外提供了另外一个存储空间。储存设备和内存不一样，因为数据可以永久存放。存储设备对于大体积数据和当计算机关闭时必须保存的数据来说是非常关键的。存储设备的种类包括磁盘、硬盘、光盘、磁带、大容量磁盘和个人计算机卡。

通信设备可以促进网络，也就是计算机与联网的计算机小组之间的联系。这样的联系可以分享资源，包括硬件、软件和数据。计算机体系中的交流成分可以最大效率地发挥计算机的使用范围和各方面的功效。计算机可以通过使用通信设备的一种——调制解调器，再通过电话线或电缆实现上网。调制解调器有内置和外置两种。

课文3：计算机软件

计算机软件，也称为计算机程序或程序，由一系列的指令构成，它告诉计算机硬件做什么以及怎样去做。人们称之为计算机软件是为了与计算机硬件形成鲜明的对比。计算机硬件包含了各种用来存放或运行软件的物理连接设备与装置。

在计算机中，软件被安装在随机存取存储器内并被中央处理器所执行。从最低层来讲，软件由针对特定处理器的机器语言构成。而机器语言则由成批的二进制数组成，这些二进制数是一些处理器命令（目标代码），它们用来改变计算机的初始状态。软件是按顺序排列的一组指令，以特定的序列改变计算机硬件的状态。相比于机器语言，软件通常用高级程序语言来编写，因为对于人们而言高级语言使用起来更容易、更有效（更接近自然语言）。高级语言需要通过编译或直译转变成机器语言目标代码。软件也可能用汇编语言编写，本质上，它是一种使用自然语言字母表来描述机器语言的更易于记忆的表示方法。汇编语言必须通过汇编程序转译成目标代码。

术语“软件”在这层意义上于1958年首次被约翰怀尔德•图基使用。在计算机科学和软件工程方面，所有的计算机程序都可以说是计算机软件。向存储设备读入不同的指令序列从而控制计算的这一理念是由查尔斯•巴贝奇发明的，这一理念作为他差分机理论的一部分。有两种基本的软件类型：系统软件和应用软件。

系统软件包括使计算机硬件与应用软件相结合的操作系统。操作系统的目的是提供一个应用软件可以方便而且高效运行的环境。操作系统的任务分为以下六大类：

- 处理器管理——在任务被发送到CPU之前，将它们分解为可管理的块，并对它们进行排序。
- 内存管理——整理写入或读出RAM的数据流，并决定何时需要虚拟内存。
- 设备管理——给每个连接到计算机、CPU和应用程序的设备提供接口。
- 存储管理——引导数据永久性地存储到硬盘或其他存储设备上。
- 应用接口——提供软件程序和计算机间标准的通信和数据交换。
- 用户界面——提供用户与计算机通信和相互作业的途径。

除了操作系统外，系统软件还包括实用程序、设备驱动程序和语言翻译程序。

一个计算机程序如果不是系统软件那就是应用软件了。应用软件包括中间件，它把系统软件与用户接口相结合。应用软件也包括帮助用户解决应用问题（如需要排序）的实用程序。

系统软件

系统软件是软件的一个子类。它由控制计算机及其设备操作的程序集构成，在用户和计算机硬件之间起到接口的作用。系统软件并不是一个单一的程序，它是一个程序集合或系统，可以在不需要用户干预的情况下处理数以百计的技术细节。

1. 操作系统

操作系统由一系列的程序组成，这些程序控制并执行计算机硬件的操作，也为用户提供了与计算机沟通的接口。

操作系统的简称是 OS，是协调计算机资源的主要软件，它提供了用户与计算机接触的桥梁，并且能安全有效、抽象地运行应用程序。举个例子，通过只允许单个应用程序在任何一个时刻直接向打印机传输数据确保了打印机的安全使用。操作系统也可以促使 CPU 更加高效地运作，它将等待输入输出操作的程序暂时停止，为能更高效率地利用 CPU 进程的程序提供了运行空间。操作系统还有一个优势就是创造了方便的抽象概念（例如文档，而不是磁盘位置），使得用户和程序员的区别根据基本的硬件细节明晰起来。

注意以下几点：

- 操作系统的核心在于对基础硬件的直接调控，核心使得低等级的设备、内存和处理机管理得以工作（例如，处理硬件设备的中断问题、在多程序中共享处理器、为程序分配内存等）。
- 基础的与硬件无关的内核通过一个系统调用库服务于更高等级的程序（例如，创建文件、开始程序的执行、为另一台计算机打开逻辑网络连接）。
- 应用程序（例如，文字处理软件、电子表格软件）和系统实用程序（简单但有用的与操作系统有关的应用软件，比如在一批文件中寻找文本的程序）发挥了系统调用的功能。它们被设置时利用操作系统与外部的接口（一种含有操作系统命令行界面的文本方式窗口）或者是一个能直接与用户联系的图形用户界面。

不同的操作系统之间依靠系统调用而被区分开来，操作系统为系统使用和用户界面提供了平台，也为内核的任务执行提供了策略。

2. 实用程序

实用程序，也被称作服务程序，执行与管理计算机资源相关联的特殊任务。典型的这类程序包括操作系统与外部最主要的接口（Shell）、文字编辑器，编译器，有时也包括文件系统。

Shell 是操作系统与外部和用户之间连接的最主要的程序，它是一种命令解释程序，也就是可以提示用户输入命令让计算机完成用户想要完成的任务，它能读取并编译用户输入的内容，并且直接让操作系统完成指定的任务。这些指令可能将会调用其他的实用程序，例如文字编辑器和编译器，或是其他应用程序，也可能是文件系统的运行，还有可能是某些系统的运行，如登录和退出。Shell 的种类繁多，从相对简单的命令总线解释程序的 DOS 到功能更为强大的 Bourne Shell 程序，还包括在 UNIX 环境中运行的更为复杂但在苹果和 Windows 系统中相对简单实用的图形用户界面的 C Shell。用户要学会熟悉所使用的计算机中可以利用的 Shell 程序，因为它是运用计算机装置的最主要途径。

文字编辑器是用来输入程序或数据并将其保存到计算机中的程序。这些信息以单元的形式组合起来，称为文件，就和办公室里的档案橱柜类似，而它们存储在磁盘中。现在可以使用的文字编辑器有很多种，例如在 UNIX 或 Linux 系统中使用的可视界面和 EMACS 文本编辑器。用户们应该对自己操作系统内的这些程序了如指掌。

在现今的计算机环境下，大多数的程序都是用高级语言（如 C 语言）编写的，然而计算机的硬件不会直接读取和识别它们。相反，CPU 运行的程序都是由一种叫机器语言的低级语言编写的，这就需要一种叫作编译器的程序，用来将高级语言转换成硬件可以识别的语言。

最后，操作系统还有一个很重要的实用功能，就是为用户管理文件系统。文件系统是用户的程序、数据、文本材料和图形图像等的集合地。文件系统为用户提供了组织管理文件的渠道，对文件加以命名并将它们归入目录或文件夹中加以管理。对于文件典型的操作包括创建新的文件，清除、重命名和复制文件。

大部分的操作系统都含有一些可以执行与计算机管理、设备管理和程序管理相关的特殊实用程序。以下这些是经常用到的实用程序。

- 文件管理程序使得用户能更加方便地管理文件。在 DOS 的鼎盛时期，其纯文本全自动打印功能得到了很大的发展。许多程序被编写来帮助用户搜索文件，创建和组织目录，复制移动及重命名文件。更新版本的与操作系统（例如 Windows 95）一同发展的图形界面减少了用户对于替换文件管理程序的依赖性。
- 磁盘碎片整理程序是一种能定位并消除冗余碎片的实用程序，能将文件和未使用的空间重新排列整合，从而优化计算机的运行。
- 内存管理文件管理着计算机程序在随机存取存储器存放当前数据的场所。通过移动某些常驻内存项目为其他程序提供空间。内存管理文件能将未使用的碎片加以聚集到一个集合，这在很大程度上提高了内存的使用率，因此也创建了一个可使用内存的集合。
- 备份程序是用来将所选文件或整个硬盘驱动器恢复到其他的存储设备里。这是保存任何可能意外丢失数据的较为保险的方法。备份程序会将数据进行压缩以占用尽可能少的空间。
- 数据恢复程序被用来恢复被删除或损坏（毁坏）的文件。
- 数据压缩程序能将因格式化而产生的松散空间进行压缩。
- 反病毒程序也是计算机中不可或缺的程序。它们时刻监视着计算机以防病毒的出现，病毒是一些能自我复制并将其传播到其他计算机中的危害程序，对于文件更是具有毁灭性的危害。
- 屏幕保护程序在键盘停止打字的时间区间内监视其屏幕上出现活动的图像以保护屏幕的程序。

3．设备驱动程序

设备驱动程序是用来允许特定的输入或输出设备与计算机系统的其他部件相互交流的专门程序。它是为使操作系统或设备与外接设备如打印机、视频卡、声卡调制解调器、适配器和光盘驱动器能连接而专门编写的。驱动器将操作系统或用户输入的指令转换成与计算机相连的设备所能识别的指令，而不是直接作用在驱动器上。

任何时候当新的设备与计算机系统连接时，为了使这种新的设备能正常使用，用户必须安装相应的新驱动器到计算机里。Windows 操作系统提供了上百种不同种类的设备驱动系统软件程序。

4. 语言翻译程序

语言翻译程序是用来将程序指令转换成计算机能识别并能进行处理的语言。

语言翻译程序是一种将程序员用某种语言（如 C 或 C++）编写的程序转换成机器语言的软件，而机器语言则是计算机可以识别的语言。所有的系统软件和应用程序都必须先转换成机器语言才能被计算机运行。有三种类型的语言翻译程序：汇编程序、编译器和直译器。

- 汇编程序将用汇编语言编写的源程序翻译成机器语言。
- 编译器是能在用高级语言编写而成的程序转换成机器语言之前对其进行识别的软件。高级语言编程指令被称作源代码。编译器就是将这些源代码转换成机器语言，在这种情况下生成的就是目标代码。编译器能对这些代码进行保存接着便可以运行它们，这意味着目标代码可以不必进行重编译。编译器将执行以下多项或全部操作：词法分析、预处理、语法分析、代码生成和代码优化。
- 直译器这种软件一次只能转换一个高级语言语句，但这个过程是连续进行的。与编译器不同的是，直译器在将程序转换成机器语言前必须先对其进行分析，这样它能将程序代码指令的准确与否及时地反馈给程序员。

应用软件

应用软件是软件的另一个子类别。它由能为用户完成特定任务的各种程序组成。应用软件可作为生产力/商业领域工具，以协助图形和多媒体项目；支持家庭、个人和教育活动，并方便沟通。特定的应用软件产品称为软件套件，可从软件供应商获得。虽然应用软件也可作为共享软件、免费软件和公共领域的软件，但这些软件的功能通常比零售包装软件要少。大体上，软件可以分为几类。

- 商业软件允许用户执行一些与商业有关的任务，如支付账目、跟踪记录产品，汇款和撰写报告和信函。常见的商业软件有 Microsoft Works 和 Lotus Notes。
- 教学软件用来教授或者培训用户。这些软件包括百科全书、参考书和教学计划。常见的教学类软件有“大英百科全书”和“微软魔法校车”。
- 娱乐/游戏软件是用于娱乐用户，它的目的是使用户开心。这些软件包括游戏软件，常见的有微软的“帝国时代”。
- 实用软件用于用户执行与存储和操作用户信息相关的日常工作。这些软件包括诸如调度、钟表、媒体播放器和通信工具。实用软件的例子如 McAfee 病毒扫描和阿卡达备份软件。

尽管软件应用因它们特定的命令和功能而不同，但它们大多具有一些共同特点。用户界面是用户工作的应用的一部分。用户界面控制如何输入数据或指令以及信息如何在计算机屏幕上显示出来。许多当今的软件程序都有图形用户界面。图形用户界面将文字、图形和其他视觉图像结合在一起，使得软件更加容易使用。现在，大多数应用程序都有一个在窗口中显示信息的界面。窗口是一个简单的矩形区域，可以包含一个文档、程序或者消息。在计算机屏幕上一次可以打开和显示多个窗口。

大多数软件包都有显示命令的菜单。通常，菜单显示在屏幕顶部的菜单栏里。当选中一个菜单项时，就会出现一个下拉菜单。这是一个与选中菜单有关的选项或命令列表。帮助菜单是菜单栏上的命令之一。它为各种各样的帮助功能提供评估。这些功能包括一个目录表、关键字索引和帮助定位到所需信息的搜索功能。在菜单栏下面有工具栏。工具栏包含按钮和菜单，它们提供对常用命令的快速访问。对大多数应用程序而言，标准工具栏和格式工具栏都类似。

对大多数用户来说，无论是在家里还是在企业，都喜欢面向任务的软件，有时也被称为生产软件，它可以帮助用户提高工作效率，使他们的生活变得更方便。集体的业务任务集是有限的，并且对执行这些任务的一般路径的数目也是有限的。因此，在大多数情况下任务和软件解决方案分成了几类，在大多数的商业环境中都可以发现它们。这些主要类别包括文字处理（包括桌面排版）、电子表格、数据库管理、图形和通信。对每一类的简要说明如下：

1．文字处理/桌面出版系统

应用最广的个人计算机软件是文字处理软件。这类软件允许用户创建、编辑、格式化、存储和打印一个文档中的文字和图形。在这个定义中，编辑、格式化和存储三个词说明了文字处理和简单打字之间的差别。由于备忘录或文档可以存储在磁盘上，所以，下次还可以对它进行检索、修改、重新打印或操作。因此，文字处理可以大大地节省时间。存储文档中未被修改部分不须重新输入，整个修改后的文档可以像新的一样被重新打印。流行的文字处理软件是Word、WPS等。它们具有一些共同的特性。

- 字换行和回车键：文字处理的一个基本特点是换行。当完成一行时，文字处理就会决定并且自动地将插入点移动到下一行。若要开始一个新段或空一段，可以按回车键。
- 替换和查找：查找命令允许在文档中寻找任何字符、词或短语。查找时，插入点就会移动到所查项目首次出现的位置。如果还想查，程序就会继续寻找所查项目出现的所有地方。替换命令会自动地用另外的字来替换搜寻的那个字。寻找和替换命令对于发现和修改错误是非常有用的。
- 剪切、复制和粘贴：使用文字处理，通过高亮显示选择要移动的文本部分。使用菜单或工具栏选择命令来剪切所选的文本，这样所选择的文本就会在屏幕上消失。然后移动插入点到新的位置，选择粘贴命令，把剪切的内容重新插入到文本。用类似的方法，可以复制选中的文本到另外一个地方。

随着文字处理软件包的功能数量的增加，文字处理已进入到桌面排版领域。通常，遇到高级排版需求时，桌面排版软件包要比文字处理软件包好，特别是当它涉及排版和色彩再现时。许多杂志和报纸今天依靠桌面排版软件。企业用它来制作具有专业水准的通讯、报告和小册子，以改善内部沟通，并给外界一个好印象。

2．电子表格

表格由列和行组成，几个世纪以来一直被用作为商业工具。做一个人工表格是乏味的，当有更改时，可能需要重做大量的计算。电子表格仍然是一个表格，但由计算机来实现。它是用于组织和管理数字并且显示选项以供分析的电子工作表。电子表格是由金融分析师、会计师、项目承包人以及其他和操纵数字数据有关的人员来使用的。电子表格允许用户尝试各种假设分析的可能性，这是一个很有用的特性，可以通过使用存储的公式处理数字，并且计算出不同的结果。

电子表格包括几个部分。电子表格的工作区域在顶部有列标的字母，在左边有行标的数字。行和列的交点被称为单元格。单元格存有单一的信息。单元格的位置被称为是单元格地址。例如，“A1”就是电子表格的首位置，即最顶部和最左部的位置。单元格指针——也被称为是选择器——指示在表格中数据在哪里输入及修改。单元格指针可以到处移动，其移动方式非常像在字处理程序中移动插入点的方式。Excel 是常见的电子表格软件，它具有一些电子表格程序的共同特性。

- 格式：标号通常用于标记工作表中的信息，通常是一个字或符号。单元格中的数字被称为值。标号和值可以用不同的方式显示和格式化。标号可以在单元格内居中或居左、居右。值可被显示小数点位置、美元或百分数。小数位数可以被改变，列的宽度也可以被改变。
- 公式：电子表格的优点之一是可以通过使用公式来处理数据。公式是计算的指令，它们能使特殊单元格内的数字之间建立联系。
- 函数：函数是自动完成计算的内部公式。
- 重新计算：重新计算或 what-if 分析是电子表格重要的特性之一。如果改变了表格中的一个或多个数字，所有相关的公式将会自动重新计算。这样就可以替换由公式改变的单元格内的值，并且重新计算结果。对于较复杂的问题，重新计算让用户能够存储长的、复杂的公式和许多改变的值，并且很快地产生替换。

3. 图形

当已经获得标准的计算机输出结果时，向商业领域人士展示图形可能是没意义的。然而，图形、图和图表可以使人们更容易地比较数据和最新趋势，并更迅速地做出决策。而且，视觉信息通常比一个充满数字的页面更加生动。

Adobe 的 Photoshop 和微软的 PowerPoint 应用程序软件用于制作图形。它们可用于两种方式：制作原始图和创建直观的教具以支持一项口头陈述。

4. 通信

从一个在家使用个人计算机的员工的角度来看，通信意味着用户可以将电话连接到自己家的电脑上，并且与办公室的计算机进行交流，或者获取存储在其他地点的别人的计算机上的数据。微软的 Internet Explorer 应用软件用来写电子邮件、浏览万维网并参加因特网讨论小组。

课文 4：操作系统

任何一台计算机中最重要的程序就是操作系统或 OS。操作系统是一个由很多小程序组成的大程序，它控制 CPU 与硬件间的通信，也方便不懂得编程语言的用户操作计算机。换句话说，操作系统使用户更友好地操作计算机，它是控制计算机的基本软件。

操作系统对于其生产厂家及其运行的硬件环境通常是特有的。一般说来，安装一台新计算机系统的同时也购买了与该硬件相对应的操作系统。用户需要有效地支持其处理工作的可靠的操作系统软件。尽管各厂家的操作系统软件各不相同，但特性都是相似的。对于现代硬件系统，由于其复杂性，因此要求其操作系统满足某些特定的标准。例如，考虑到该领域的现状，操作系统必须支持某种形式的联机处理。

操作系统的功能

大多数的操作系统都拥有相似的功能，而这与计算机的大小无关。下面讨论操作系统常见的一些功能。

1. 进程管理与协调

CPU 执行着大量的程序，然而 CPU 最主要的作用在用户程序的执行上，因为其他的系统活动也需要 CPU 的支持，这些活动称为进程。进程指正在被执行的程序。典型地，一次批处理作业就是一个进程；一个分时用户程序也是一个进程；一个系统任务，如信息缓存也同样是一个进程。

一般说来，一个进程为完成它的任务需要一些资源，比如，CPU、内存、文件、输入输出设备等。一旦进程建立起来，这些资源就会分配给它。除了在创建进程时所获得的各种物理和逻辑资源外，还可以获得一些初始化数据（输入数据）。比如说，一个进程的作用是展示在终端显示器上一个文件夹的状况，比如说一个进程，其功能是在终端显示屏上显示一个文件，如 F1，它将会接收到一个名字为 F1 的文件作为输入信息，并执行相应的程序，以获得想要的信息。

需要强调的是，独立的一个程序并不是一个进程；一个程序是一个静态的实体，然而一个进程则是一个动态的实体。两个进程可能与同一个程序有关联；不过，它们仍被看作两个分开执行的进程。在一个系统中，一个进程是一个工作单位。这个所谓的系统由一批进程组成，一些（进程）是操作系统相关的进程，用以执行系统代码，其他的则是用以执行用户代码的用户进程。所有的这些进程有可能同时执行。

操作系统负责以下与进程管理有关的活动：

- 用户和系统进程的创建和删除。
- 进程的终止和恢复。
- 提供进程同步机制。
- 提供死锁处理机制。

2. 内存管理

内存是现代计算机系统操作的核心部分。它是一系列字母或字节的庞大组合，每个字母或字节都有其对应的地址。CPU 从内存中获取信息同时也将信息存储在里面。为了使程序得到执行，它必须被映射到绝对地址并载入内存。通过序列读取或写入特定的内存地址，CPU 和程序的交互作用得以实现。

为了提升 CPU 的利用率和计算机响应用户的速度，一些进程必须保存在内存中。根据特定的情境，有许多不同的算法。特定的系统内存管理方案的选择取决于很多因素，但最主要取决于系统硬件的设计。每种算法需要其对应硬件的支持。

操作系统负责以下与内存管理有关的活动：

- 随时追踪内存正在被使用的部分以及其使用者。
- 当存在可用内存空间时，决定哪些进程被装入内存中。
- 根据需要分配和去分配存储空间。

3. 辅助存储器管理

计算机系统的主要目的在于执行程序。这些程序，连同它们访问的数据在执行时必须存在于主存储器中。由于主存储器太小而无法永久性地容纳所有这些数据和程序，计算机系统提供辅助存储器来支持主存储器。大多数现代计算机系统使用磁盘作为主要的在线存储信息的工具，也可以同时存储数据和程序。大多数程序，如编译器、汇编程序、排序例程、编辑器、格式程序等在载入到内存之前都是被存储在磁盘里的，随后使用磁盘作为进程的起点和终点。所以磁盘存储的妥善管理对于一个计算机系统至关重要。

操作系统负责以下与磁盘管理有关的活动：

- 剩余空间的管理。
- 存储空间的分配。
- 磁盘调度。

4．文件系统管理

文件管理是操作系统中最具可视性的服务之一。计算机能以多种不同的物理形态存储信息，磁带、磁盘和磁鼓是最常见的几种类型。这些设备都有各自的特点和物理组织。

为了方便地使用计算机系统，操作系统对于信息的存储提供了一种统一的逻辑视图。操作系统从它的存储设备的物理特性抽象到一个逻辑存储单元，这就是文件。一般说来，文件是一连串的位、字节和行，或是通过用户或是创建者定义具有一定含义的录音。文件通过操作系统被映射到物理设备上。通常文件表示程序（以源程序和目标程序形式）和数据。数据文件可以是数值的、字母的或是字母数字的。它可以是任何一种形式，例如文本文件或者是严格的格式化形式。

操作系统管理海量存储设备，如类型系统和磁盘来使文件这一抽象的概念生效。同时文件为了网站的使用一般组成目录。最后当多用户能够访问文件时，它将能够掌控访问对象以及文件以何种方式被访问。

操作系统负责以下与文件管理有关的活动：

- 文件的创建和删除。
- 目录的创建和删除。
- 支持对文件和目录的操作。
- 文件在内存和磁盘做映像。
- 稳定（非易失性）存储器中文件的备份。

5．输入和输出管理

操作系统的目的之一是为了向用户隐藏一些特定硬件设备的特性。例如，在UNIX系统中，输入/输出设备的特性被它自己向操作系统隐藏起来。I/O系统包括：

- 一个缓冲区缓存系统。
- 一个一般设备驱动器代码。
- 特定硬件设备的驱动。

只有驱动程序才知道这个特定设备的特点。

6．协同网络通信

网络通信功能过去是由一个独立的网络操作系统来处理和操纵。而现在，一些通信功能包含了正规的操作系统，用以满足个人的网络连接需求。商业性的网络管理仍需要一个独立的网络操作系统，比如Netware或者Windows NT。

7．防护和安全

在操作系统中，那些多种多样的进程必须受到保护以避免彼此活动的干扰。为了做到这一点，使用众多机制以确保文件、存储片段、CPU和其他资源只被那些从操作系统获得适当授权的进程所操纵。比如，存储器寻址硬件必须确保一个进程只能在它自己的地址空间内所执行。计时器要确保没有进程可以在它让出（CPU）的时候取得CPU的控制权。

8. 提供用户界面

用户通过用户界面和软件进行互动。一个用户界面承担着如何让你去访问数据和指令以及信息是如何展示在显示器上的任务。目前有两种类型的用户界面，称为命令行用户界面和图形用户界面。许多操作系统使用两者的结合来规定用户是如何和计算机进行互动的。

操作系统的类型

在操作系统的庞大家族里，根据它们所控制的计算机类型和它们所支持的应用程序的种类，操作系统通常分成四类。

1. 实时操作系统

实时操作系统用来控制机器、科学仪器和工业系统。实时操作系统的一个很重要的部分是管理计算机资源，这样它能根据资源的每一次发生数量，精确地产生一个相应的动作。

2. 单用户、单任务操作系统

顾名思义，这一操作系统是为单用户在某一时刻有效地完成一个任务而设计的。用于Palm手持电脑的Palm OS就是现代化的单一用户、单一任务操作系统的很好样例。

3. 单用户、多任务操作系统

当今，大多数人在其台式机和笔记本电脑上用的操作系统就是这一种。微软公司的Windows和苹果公司的Mac OS平台都是单用户、多任务操作系统的例子。例如，Windows操作系统用户就完全可能做到一边从互联网上下载文件，一边打印电子邮件的信息，一边还在用文字处理器编辑文本。

4. 多用户操作系统

该操作系统允许许多不同的用户同时使用一台计算机的资源。多用户操作系统必须保证不同用户的需求的平衡，每一个用户所使用的程序有独立的足够的资源，这样一个用户的问题不会影响到整个用户群体。UNIX、VMS和大型机的操作系统（如MVS）都是多用户操作系统的例子。

操作系统如何工作

操作系统是管理、控制计算机活动的系统软件。它监控CPU的操作、控制输入、输出及数据存储活动，并且提供各种支持服务。它是计算机系统的主要管理者。操作系统决定调用解决某个问题所需要的计算机资源及资源的使用顺序。

操作系统控制不同的计算机进程，如运行电子表格程序或从计算机内存中存取信息。一个重要的进程是解释命令，使得用户可以和计算机通信。一些命令解释器是面向文本的，需要键入命令或通过键盘上的功能键选择命令。另一些命令解释器使用图形，并允许用户通过指点图标进行通信。图标是屏幕上的图片，代表特定的命令。初学者一般会发现面向图形的解释器比较容易使用，但许多有经验的计算机用户更喜欢面向文本的命令解释器，因为它们的功能更强。

操作系统可以使用称为虚拟内存的技术，来运行所需主存空间大于实际可用主存空间的进程。为了实现这种技术，硬盘空间被用来模拟所需的额外内存。然而，访问硬盘比访问主存耗时多，因此计算机的运行速度会变慢。

计算机打开后即在其内存中寻找指令。这些指令告诉计算机如何启动。通常，在这些最先运行的指令集中，有一套称为操作系统的特殊程序，它是使计算机工作的软件。它提示用户（或其他机器）提供输入和命令，报告这些命令及其他操作的结果，存储和管理数据控制软件与硬件动作的顺序。用户需要运行一个程序时，操作系统将该程序装入计算机内存并运行该程序。流行的操作系统，如微软公司的 Windows 以及麦金托什系统（Mac OS），拥有使用微小图片或图标代表各种文件与命令的图形用户界面。用户用鼠标点击图标或按键盘上的组合键即可访问这些文件或命令。有些操作系统允许用户通过语音、触摸或其他输入方式执行这些任务。

操作系统的小型引导程序存储在只读存储器上，并在系统启动时提供必要的指令，用于将操作系统的核心装入内存。操作系统的这个核心部分称为内核，它提供最基本的操作系统服务，如内存管理和文件存取。在计算机运行的整个时间内，内核一直驻留在内存中。操作系统的其他部分，如定制实用程序，需要时才装入内存。

设计操作系统的初衷是为了促进应用程序与计算机硬件之间的相互作用。当然这个初衷现在仍然存在，只是其重要性越来越体现在操作系统对于计算机的定义方面了。不少深陷苹果机、个人计算机和 UNIX 操作系统机器斗争门的用户对操作系统存在争论，而不是自身的硬件平台。操作系统为用户和硬件计算机之间架起了一座抽象的沟通桥梁。用户和应用程序不能直接掌握硬件的情况，必须通过操作系统来获取信息。这一抽象性使得硬件的某些细节逃离了用户和应用程序的眼睛，使得用户并不能获悉硬件方面的变化情况。

常见的操作系统

现在有很多操作系统。它们包括从最新的到最旧的版本。

Windows XP 是微软公司最流行的操作系统。XP 版本的发布意味着所有微软的桌面版本 Windows 系统都是基于 Windows NT/2000 的代码库。对于正在使用 Windows 3.1、Windows 95、Windows 98、Windows ME 版本的用户，我们都建议使用这个新版本。作为 Windows 2000/NT 4 的升级产品，又因为 XP 是基于 NT/2000 的代码库，所以，它是具有吸引力的选择。

Windows CE 用于掌上电脑之类的小型设备，一些较重要的应用程序的后期版本都能够在这些设备上得到运行。你可以将自己的小型计算机与其相连接，以实现文件和数据的同步。

Windows 2000 是 Windows NT 的升级版，无论是家用还是作为商业上的工作站用都适合。Windows 2000 拥有一种能够使硬件进行自检的技术，并且在 Windows NT 的基础上进行了改进。

Windows ME 是 Windows 98 的升级版，但它历来为程序设计中的错误所困扰，这使家庭用户感到很灰心。

Windows 98 是微软的一款操作系统。它之前为人所知是因其代码名称 Memphis，曾一度被称为 Windows 97。它是为当今一款功能强大的微型计算机而设计的先进的操作系统。相比于 Windows 95，Windows 98 有一些重要的优势，包括高性能、因特网一体化、操作简便和影视声音功能。

Windows NT 是一种可以在很大范围内的、功能强大的计算机和微型计算机上运行的操作系统。它是非常复杂和功能强大的操作系统。Windows NT 由微软公司开发，并不是为了替代 Windows，确切地说，它是设计给功能非常强大的微型计算机和网络的另外一个可供选择的高级操作系统。Windows NT 和 Windows 相比较具有两个主要的优点。

① 多道处理：除了应用程序是同时独立运行之外，它类似多任务。比如，你可能正在打印文字处理文档并且同时在使用数据库管理程序。在这种多任务操作情况下，文件打印的速度会受到使用数据库管理程序这一要求的影响。使用多道处理，要求数据库管理程序并不影响打印文档。

② 网络。在许多商业环境中，工作人员经常使用计算机相互进行交流，并且通过网络共享软件。这是可以通过专门的系统软件来实现和控制的。Windows NT 具有网络功能和嵌入到操作系统内的安全检测功能，这使得网络安装和使用相对容易。

Windows 95 是较老的 Windows 3.x 版本之后的第一个 Windows 版本。它提供了一个更好的界面程序和更好的库函数。

MS-DOS（微软磁盘操作系统）是一种在 20 世纪 80 年代和 20 世纪 90 年代初非常流行的操作系统。IBM 的 DOS 版本称为 PC-DOS，而微软版本称为 MS-DOS。这两个操作系统除了细微的差别外是完全相同的。

UNIX 最先被 AT&T 所发展壮大。由于它的多用户、多任务处理环境、稳定性、可携性和强大的联网能力使它成为流行的操作系统达 20 多年。UNIX 继续在强大的工作区受大众欢迎。许多老一代的计时员喜欢 UNIX 和它的命令行界面，但是这些命令要被初学者记住实为不易。X-Windows 是 UNIX 的图形界面，有些人认为它比 Windows 98 更容易工作。

Linux 是一款和 UNIX 类似且越来越受欢迎的操作系统。它是一种开放源代码程序，是由芬兰大学的莱纳斯 •托瓦于 1991 年发明的。开放源代码意味着基本计算机代码对于任何人都是开放的。程序员能直接工作于代码并新增功能。

Mac OS 是为运行在 Macintosh 计算机上而设计的操作系统。它提供高质量的图形用户界面，并且很容易使用。虽然其市场份额远远少于 Windows 的操作系统，但这是一个非常强大且易于使用的操作系统。Apple Macintosh System 7.5 是为使用 Motorola 公司的 PowerPC 微处理器的 Apple 计算机设计的，对于 Apple 机来说是一个重要的里程碑，它像 Windows NT 和 OS/2 一样是功能强大的操作系统。System 7.5 具有网络功能并能读出 Windows 和 OS/2 文件。它具有几个优点：

① 容易使用。图像用户界面使用 Macintosh 在许多计算机新手用户中普及，这是因为它容易学习。

② 高质量的图像。Macintosh 已经建立了高标准和图像处理，这就是为什么 Macintosh 在桌面印刷系统中流行的主要原因。用户可以很容易地将图片和文字材料组织成比较专业的通信稿、广告等。

③ 一致的界面。Macintosh 应用程序具有一致的图形界面。在所有的应用程序中，提供给用户相似的屏幕显示、菜单和操作。

④ 多任务。像 Windows、Windows NT 和 OS/2 一样，Macintosh 系统使得你实现多任务，即几个程序可同时运行。

⑤ 程序之间的通信。Macintosh 系统允许应用程序之间共享数据和命令。

OS/2 是指操作系统/2,它最先是由 IBM 和微软联合开发的。像 Windows 98 和 Windows NT 一样，OS/2 是专为非常强大的微型计算机所设计并且具有一些先进的特点。

① 最小的系统配置：像 Windows NT 一样，OS/2 需要大的内存和硬盘空间，但 OS/2 需要的相对较少。

② 窗口应用：OS/2 和 Windows NT 一样，没有很多专门为它编写的应用程序。OS/2 同样可以运行 Windows 程序，但它运行这些程序的速度比 Windows NT 更快一些。

③ 共同的用户界面：专门为 Windows NT 以及 OS/2 编写的微型计算机应用程序，具有一致的图形界面。在应用程序里，提供给用户类似的显示、菜单和操作。此外，OS/2 为大型计算机、小型计算机和微型计算机提供一致的界面。

Windows Vista 是微软的最新一代操作系统。和它的前辈 Windows XP，Service Pack 2 相比，它提供了许多增强的功能。一启动计算机进入 Windows，就会发现有很大的不同。新的“开始”菜单、新的桌面背景，还有在屏幕右侧的新的工具条，所有这些都会告诉用户正在进行一场新的体验。现在移动鼠标单击“开始”菜单按钮，并且启动一个新程序，屏幕上不会再有直达屏幕边缘的多层数的菜单。相反，每次打开一个菜单，它都覆盖在前一个已打开菜单的上面，这样很容易找到想要的程序。

课文 5：用树莓派构建一个 LAMP 服务器

配置树莓派 Web 服务器类似于把 Xubuntu 配置成 LAMP Web 服务器，但增加了一些需要对树莓派做些不同处理的步骤。LAMP 服务器是最普通的 Web 服务器配置之一，它的标准如下：

- Linux——操作系统。
- Apache——web 服务器（HTTP）软件。
- MySQL——数据库服务器。
- PHP 或 Perl——编程语言。

所有这些配置是在命令行中完成的。这可能不像点击几个图标一样简单，但它有许多优点，包括远程管理和安装服务器的能力。这也意味着计算机可以花更多的时间服务于网页，并且花更少的处理器时间绘制图形用户界面，而这毕竟是 Web 服务器的重点。

1. 为什么要使用树莓派

我认为使用树莓派有如下一些理由。

学习网络编程。树莓派基金会的目的是教孩子们编程，学习基于应用程序的网络编程是一个有用的技能。是否在学习 Web 编程之前先学习编写桌面应用程序，或者反过来，这一点是有争议的。但是有一点可以肯定这是一种有用的技能。

作为一个接口。作为收集来自各种元件的信息的设备，树莓派是有用的。Web 服务器是访问信息的好方法。

专用网络设备。你可以把它用于家庭专用网络服务。

作为测试或开发服务器。当开发一个 Web 应用程序时，用专用的服务器来测试应用程序是有用的，理论上，硬件和软件应该和应用程序一致，但如果这是不可能的话，那么树莓派会是一种廉价的替代品。

2．确保树莓派的安全

最重要的事是保证树莓派的安全。这包括有一个默认的用户名和口令，一旦连接到互联网，它允许任何人登录并且有一个自由浏览的设备。

登录以后要为树莓派用户修改口令，发布passwd命令，并且跟随提示修改口令。

3．操作系统性能调优

性能调优是一件费时的事，但对树莓派而言，只有一个选项可以提高服务器性能。通过配置并重启以后，可以提高它的性能，而不用必须重启服务器。

树莓派拥有256MB（或更高版本有512MB）的RAM。但是，这个RAM是由显存和主系统内存共享的。默认情况下的64MB被分配给显存。

4．配置网络

下一步是给树莓派一个静态的IP地址。默认情况下，树莓派会请求一个动态的IP地址。但是，这个IP地址以后可能会改变，以至于很难链接到Web服务器上。因此，要为树莓派提供一个静态的IP地址。而且，这个IP地址可以在本地网络上使用，但不是在互联网上。

5．启用SSH

SSH（安全Shell）是一种网络协议，它允许你通过远程输入命令，登录并远程控制计算机。顾名思义，它是安全的，因为它对网络通信内容加密（这样别人就看不到你的密码等内容）。

6．使服务器在互联网上可用

接下来，需要配置路由器，以允许通过防火墙ssh登录到树莓派并进行网络通信。

一定要记着修改树莓派用户默认密码。如果你还没修改默认密码，那么现在就修改；否则，任何人都能登录你的树莓派。

作为家庭用户，在本地网络中的IP地址是一个在因特网上不能使用的私有的地址范围，而ISP会提供一个用于路由器的动态的IP地址。为了因特网和树莓派有通信，需要使得树莓派的IP地址看起来似乎来自于路由器，这个过程称为网络地址转换（NAT）。

允许通过的端口是80端口（http），如果你想要从因特网上登录到计算机，那么是22端口（ssh）。

在最后阶段，确保路由器IP地址上要有一个DNS入口点。

7．安装Apache网络服务器

Apache网络服务器可以从Debian库中下载，可以使用apt工具来完成。

首先，确保已经更新了软件库。如果没有，运行sudo apt-get update命令，确保新的包或版本存在。Apache可以通过输入sudo apt-get install apache2命令安装。

8．安装MySQL

MySQL数据库服务器也可以从Debian库中获得，并用sudo apt-get install mysql-server命令安装。在安装过程中，有一个密码请求提示，这个密码用于MySQL的root用户。

9．安装PHP

Perl是作为操作系统的一部分安装的，因此只增加PHP。

10．安装完毕

一旦安装完毕，就可以通过在浏览器中输入路由器IP地址或DNS入口来访问网页。

首先应该获取一个返回页，以表明网页正常工作，当然，目前这里没有加载内容。测试 Web 服务器和 PHP 是否正常工作，需删除文件/var/www/html/index.html *，然后创建带有 this page 的/var/www/html/index.php 文件。

第五部分　编 程 语 言

课文 1：关于程序设计语言

就像世界上人类通过多种不同的语言进行交流一样，人与计算机之间也存在着各种不同的语言进行交流沟通。在计算机科学中，程序设计语言是用于编写可由计算机运行的一系列指令（计算机程序）的人工语言。它们是人们向计算机传达指令的标准化通信技术。与英语等自然语言相类似，程序设计语言有词汇、语法和句法。然而，自然语言不适合于计算机编程，因为它们会引起歧义，也就是说，它们的词汇和语法结构可能以多种方式进行解释。用于计算机编程的语言必须有简单的逻辑结构，它们的语法、拼写和标点符号规则必须精确。计算机语言使程序员能精确地指示计算机执行哪些数据、如何存储和传输这些数据以及面对不同的环境计算机应该执行哪些工作。

1．程序设计语言的发展

计算机编程语言已经有多年的发展历史，并且在未来将继续发展。随着量子计算机和生物计算机的发明，未来的计算机编程语言可能会变得更加符合人们的语言习惯。下面来详细介绍这五个不同阶段的语言。

2．第一代语言：机器语言

有人指出程序指令包括一个特殊的二进制数字组合，不同型号和类型的计算机使用不同的二进数位代码来代表指令。这些代码被称为机器代码或指令代码。这个等级的计算机有一个可以运行的基本指令集，即我们所知的指令系统。典型的这类指令系统包括：

- 基本算术运算；
- 各种比较运算（例如等式运算等）；
- 字符序列处理运算；
- 输入/输出运算。

在早期的计算机编程中，所有程序都必须用机器代码编写。例如，一个简单的（包含三条指令）的程序可能是这样的：

0111 0001 0000 1111

1001 1101 1011 0001

1110 0001 0011 1110

一条机器语言指令一般会告诉计算机四件事：（1）到计算机主存（随机存储器）的什么位置去找一两个数字或简单的数据片；（2）要执行的简单运算，如将两个数字相加；（3）在主存的什么位置存放该简单运算的结果；（4）到什么位置去找下一条要执行的指令。尽管所有的可执行程序最终都是以机器语言的形式被计算机读入的，但它们并非都是用机器语言编写的。由于指令是 0 和 1 的序列，所以直接用机器语言编程是极端困难的。

机器代码因此也具有许多明显的缺点：

- 仅仅从它的编码不能直观明显地获知机器代码所要执行的内容，因此理解或编写机器代码比较困难。
- 机器代码的编写极其费时并且很容易出错。
- 存在着许多不同的机器代码（一种代码适用于一类计算机）。

这些不足在很大程度上限制了计算机在使用机器代码时的应用范围。

3．第二代语言：汇编语言

开发汇编语言的初衷是为了弥补机器代码编程的不足，它运用符号代码取代了二进制指令。由此，程序设计变得不再那么困难。以下是汇编语言的一个例子：

```
MOV AX 01
MOV BX 02
ADD AX BX
```

汇编语言中的每一行的程序都与其机器代码的指令相对应，因此为了使通过汇编语言编写的程序得到执行，必须用一种被称为汇编程序的转换工具将其转换成机器代码。

汇编语言与机器语言具有某些共同的特征。例如，对特定的位进行操控，用汇编语言和机器语言都是可行的。当尽量减少程序的运行时间很重要时，程序员就使用汇编语言，因为从汇编语言到机器语言的翻译相对简单。汇编语言也用于计算机的某个部分必须被直接控制的情况，如监视器上的单个点或者单个字符向打印机的流动。

尽管汇编语言的应用具有许多优势，但是这种语言在使用时仍存在很大的缺陷：（1）不同的计算机有各自相对应的汇编语言；（2）汇编程序设计对于细节的要求非常严格，因此编写起来也会非常费时而且枯燥；（3）编程过程中出现错误的可能性并没有得到很大程度的减少。

但是在计算机的一些应用方面，如与外围设备进行连接，汇编语言还是必需的。

4．第三代语言：高级语言

20 世纪 50 年代初，高级语言出现了。它的主要目的是为了能更有效、更少出错地来编写计算机程序。高级语言是相对复杂的一系列语句，它们的使用来自人类语言的词汇和句法。高级语言比汇编语言和机器语言更像正常的人类语言，因此用高级语言来编写复杂的程序比较容易。这些程序设计语言可以更快地开发更大和更复杂的程序。高级语言的优点如下：

- 用高级语言编写的程序比计算机的指令系统更倾向于人类的表达方式，程序用一种半英文的方式编写，同时算术计算也用一种与数学相似的方式编写。
- 程序员能更好地将重心放在需要解决的问题上，而不用机器语言或汇编语言编程时需要注意的冗长细节。
- 高级语言不再只适用于一种类型的计算机，它可以很方便地在其他计算机上使用。

由于计算机只能使用机器代码执行特定的程序，因此用高级语言编写的程序，例如 Java，不能在计算机上直接运行。要执行一个使用高级语言编写的计算机程序，必须先对它进行编译或解释。因为这个原因，与用汇编语言编写的程序相比较，用高级语言编写的程序可能运行时间长，占用内存多。

1）编译器

编译器是一种将高级语言转换成机器语言前先对其进行分析的软件。由高级语言编写的程序指令叫作源代码。编译器的主要作用是将用高级语言编写的源代码转换成机器语言的形式，这种情况下生成的代码叫作目标代码。目标代码的优点在于它比直译（见下文）出来的代码运行速度快很多，缺点则是不同的机器操作系统适用于不同的机器代码。总体来说，编译器有两个作用：

- 检查源代码的句法错误。
- 将源代码转换成目标代码。

C 和 C++就是运用编译器进行转换的高级语言。

2）直译器

在直译过程中，每一行所对应的程序被解码，并被一种我们熟知的称作直译器的特殊程序翻译。不同的直译器对应不同的语言（不同的计算机）。直译过程发生在每一行程序被执行的过程中，也就是每执行一次程序，直译器也相应地工作一次，这使得计算机工作效率变低、运行相对缓慢。

尽管如此，直译器对源程序的反复检查使得直译器相对于编译器使用起来更加灵活，同时，它也提供了更加快速和简单的方式来测试小程序和程序片断。此外，对错误的检测也更加容易，因为直译器工作时直接接触源代码，错误的地方能精确地定位到每一行（程序被编译时则达不到这一点）。另外，还有一个优势，那些没有被执行的程序不需要被直译。

5. 第四代语言：超高级编程语言

相对于第三代计算机语言而言，第四代更加面向用户，使程序员能用更简洁的指令开发程序，尽管它要求更高的计算机水平。第四代计算机语言被称为非程序语言，因为程序员甚至用户只需要告诉计算机他们想要做什么就可以来编写程序，而不需要明确指定步骤。第四代语言由报告文件处理机、查询语言、应用程序生成器和交互式数据库管理系统语言构成。

- 报告文件处理机也被称作报告记录器，是一种为终端用户所使用来输出报告的程序，也是当今查询语言的前身。
- 查询语言是一种使用方便的语言，它能够很容易地从数据库管理系统检索所需数据。
- 应用程序生成器是程序员使用的一种工具，它通过对问题的描述来生成应用程序，而不是通过传统的编程方法。这样的优势在于程序员不需要明确指出数据得到何种处理。

第四代计算机语言尚未完全取代第三代语言，这是由于它们一般用来处理具体的程序，因此选择性较少。尽管如此，第四代语言还是提高了编程的效率，毕竟它促使编写程序更加容易。

6. 第五代语言：人工智能语言

第五代计算机语言是人工智能的，它使得计算机像人类一样工作。这个名词是由麻省理工学院的约翰麦卡锡在 1956 年提出来的。

课文 2：C

C 语言的开发始于 20 世纪 70 年代初期，它是由 Dennis Ritchie 为研制 UNIX 而开发出来的一种系统编程语言，现在已经成为非常流行的语言。C 语言也许最合适被称为“中级语言”。同真正的高级语言一样，一个 C 语句对应多个编译后的机器语言指令。然而，与大多数高级语言不同，C 语言使操作者很容易地做汇编语言能执行的工作（如位与指针操作）。因此，C 语言是开发操作系统（如 UNIX 操作系统）或其他系统软件的特别好的工具。

1. 什么是 C 语言

C 程序设计语言是一种用于编写计算机程序的广受欢迎而且被广泛使用的程序设计语言。它是当前正在使用的成千上万种程序设计语言中的一种。C 语言已经有几十年的历史了，而且得到了广泛的认可，因为 C 语言给了程序员极大的控制权而且效率高。如果你是一名程序员，或者希望成为一名程序员的话，那么学习 C 语言会使你受益匪浅：

- 你将可以在许多平台上阅读和编写程序。小到微控制器，大到最先进的科研系统，到处都是 C 语言的用武之地。许多现代操作系统也是用 C 语言编写的。
- 学习过 C 语言后，再进阶的面向对象 C++语言就容易多了。C++是对 C 的扩展。不先掌握 C 语言而直接学习 C++是非常困难的。

从字面上看，C 语言似乎比其他语言更加难懂。实际上，C 还是一种容易学习的语言。和其他一些语言相比，C 的风格稍有点古怪，但你会很快适应的。例如，在 C 语言里经常使用括号以避免使用其他关键字。然而，标识符中允许有下画线字符，这使得标识符易于理解。

有许多一元和二元运算符，有些具有意想不到的优先级。括号可能会被编译器所忽略，有时会出现令人惊讶的运算结果。由于移位运算的原因，整型数据算术运算上溢出现象可能会被忽略。有一些复合符号具有特殊的意义，例如“&&”是指“而且”，“||”是指“或者”。

C 语言中有几种大小不一样的整型类型，还有浮点型，指针（C 语言称为指向），数组和结构。C 语言不做类型检测，例如，有些编译器对数组下标不做即时运行检测等。类型的转换是允许的。地址运算可以通过指针得以实现，空（Null）的值为 0。

C 语言有过程和函数。除了过程和函数外，C 语言几乎没有其他特征来支持模块化，但分开的（严格地说是独立的）编译是可以进行的。

C 是所谓的编译型语言，这意味着一旦写了 C 程序，就必须使用 C 编译器来编译它，使它变成可以在计算机上运行（执行）的可执行文件。C 程序是供人阅读的，而编译器产生的可执行程序是供计算机阅读和执行的，也就是说，要编写并运行一个 C 程序，你必须使用 C 编译器。如果你使用 UNIX 系统（例如，在 UNIX 主机上用 C 编写 CGI 脚本，或者你是学生，在实验室的 UNIX 机器上工作），C 编译器是免费提供的。它的名字是 cc 或者 gcc，可以通过命令行调用。如果你是学生，那么学校很可能会向你提供编译器。你只要查一下学校提供什么样的编译器就可以学习使用。如果你是在家使用 Windows 操作系统的计算机，则需要下载免费的 C 编译器或是购买商业的 C 编译器。微软公司的 Visual C++是一套广泛使用的商业编译器（可以编译 C 和 C++程序），可惜这套软件要花费几百美元。如果不想花几百美元购买商业编译器，那么你可以在网上挑选一款免费的编译器。

我们会从一个极其简单的 C 语言程序为例开始，循序渐进。在这个例子中，假定你使用的是 UNIX 命令行，并以 gcc 作为开发环境。如果使用的不是 UNIX 系统，只需要去理解并使用任何一款你所拥有的 C 编译器，那么所有的代码依然可以正常运行。

2．最简单的C语言程序

开始C的最好方法是编写、编译和执行一个简单程序。现在，让我们从一个最简单的C程序开始，了解一下C的基本要素和编译过程。将下面的程序录入一个标准的文本编辑器，然后以samp.c的文件名保存。文件名如果没有.c结尾，编译的时候就很可能会出现一些错误，所以要确保文件名以.c结尾。另外，应确保你的编辑器不会自动在文件名末尾添加后缀（如.txt）。下面是我们的第一个程序：

```
#include<stdio.h>
int main()
{
    printf("This is output from my first program!\n");
    return 0;
}
```

执行完这段程序以后，程序会命令计算机输出："This is output from my first program!"，接着，程序便终止了，没有比这段程序更简单的了。

书写位置：录入这个程序时，要录入#include，这样，"#"被放在第一列（最左边）。此外，字符间的间距和缩进可以根据个人喜好随意控制。上例程序代码的间距和缩进风格可作为格式的范例。

编译这段代码时遵循以下步骤：

① 在UNIX系统上，键入gcc samp.c -o samp（若gcc不行就用cc）。此命令行的作用是调用名为gcc的C编译器来编译samp.c，并将生成的可执行文件命名为samp。要运行编译完的程序，请键入samp。

② 在使用DJGPP的DOS或Windows电脑上，请在MS-DOS命令中键入gcc samp.c -o samp.exe。这条命令会调用名叫gcc的C编译器，使它编译samp.c，并生成一个可执行文件samp.exe。要运行程序，只需键入samp。

③ 如果你使用的是其他编译器或其他开发系统，阅读并遵循该编译器的指南。

当运行这个程序后，应该会看到"This is output from my first program!"。如果没有正确地录入程序，那么它既不会通过编译也不能运行，你应该重新编辑，并检查输入的时候哪里有误，修正后再次运行。

让我们逐行讲解一下这个程序，看看每一行都做了什么。

- 这个C程序以#include<stdio.h>开始。这一行将"标准输入输出库（Standard I/O library）"包含到程序之中。使用标准输入输出库以后，你可以读入从键盘（称为"标准输入设备"）上输入的数据、向屏幕（称为"标准输出设备"）写出数据、处理磁盘上的文本文件等。这是一个极有用的函数库。C拥有大量像stdio一样的函数库，包括字符串、时间和数学函数库等。函数库就是一个别人写好供我们调用的代码包，这使编程变得更加容易。
- int main()这行声明了主函数。所有的C程序都必须包含一个名为main的函数。运行程序时，程序从主函数的第一行开始运行。
- C语言使用大括号（{ 和 }）标识一个代码块的开始和结束。本例中，构成主函数的代码块包括两行。

- C 语言中，printf 语句将结果输出至标准输出设备（对我们而言就是屏幕）。引号中的部分叫作格式字符串，用于表示数据输出时的格式。格式字符串可以包含字符串、回车换行符（\n）和标识变量值操作符等。如果使用的是 UNIX 系统，那么可以键入 man 3 printf 获得 printf 函数的完整说明文档。如果不是，那么可以通过阅读该编译器相关的文档来了解 printf 函数的细节。
- “return 0;”这行语句使主函数向运行它的命令行解释器返回错误代码 0（表示没有错误）。

3．变量

作为程序员，你常常需要你的程序“记住”某个数值。比如你的程序要求用户输入一个数，或者它计算出了一个结果，你需要把它保存在某处供以后使用。程序是靠使用变量来保存结果的。例如：

```
int b;
```

上面这行的意思是说：“我想创建一个可以容纳一个整数的存储空间，它的名字叫作 b。”每个变量都有名字（本例中是 b）和类型（本例中是 int，表示整型）。你可以用如下语句在 b 中存储一个值：

```
b=5;
```

你可以用如下语句使用 b 中保存的值：

```
printf("%d", b);
```

C 语言提供了几种标准的变量类型：

- int——整型
- float——浮点型
- char——字符型（如“m”或“Z”）

4．输出

printf 语句将结果送往标准输出设备。对我们而言，标准输出设备通常是屏幕，下面这个程序将会帮助你进一步了解 printf 语句：

```
#include<stdio.h>
int main()
{
    int a,b,c;
    a=5;
    b=7;
    c=a+b;
    printf("%d+%d=%d\n",a,b,c);
    return 0;
}
```

将这个程序录入到一个文件并保存为 add.c，使用 gcc add.c -o add 命令编译这个程序，再通过键入 add（或者/add）来运行它，你将会看到这样的输出结果“5+7=12”。

下面来解释一下程序中的每一句意思：

- int a,b,c 声明 3 个整型变量，分别为 a、b、c。
- 接下来一句将变量 a 赋值为 5。
- 下一行给 b 赋值为 7。
- 下一行是 a 加 b，并将结果赋给 c。计算机分别将 a、b 赋值 5 和 7，通过运算得出 12 这个结果，然后它赋予变量 c，c 便被指定为 12 这个数值，所以在这里“=”被称作“赋值运算”。
- printf 语句接着便打印“5+7=12.”，在 printf 语句中%d 是位置标志符，它的作用是作为值的位置标志符。在这个程序中有 3 个位置标识符，并且在最后的输出语句中 3 个变量都有各自的命名：a、b 和 c。a 与第一个位置标识符相匹配，其数值为 5。b 与第二个相匹配，数值为 7，c 则与第三个位置标识符匹配，数值为 12。

最后便会在屏幕上打印出完整的结果：5+7=12。“+”“=”以及空格都是语句的一部分，并且程序会指定将它们自动嵌入%d 运算符中。

课文 3：C++

C++是一种通用的编程语言，具有高级和低级处理的能力。它是一种静态输入、格式自由、多范例的编译语言，它支持过程编程、数据抽象、面向对象编程及类编程。

1. C++语言的由来

C++程序语言可以被看作 C 程序语言的增强版（且新增了现代特征）。C 程序语言是由美国电话电报公司贝尔实验室的丹尼斯·里奇于 1970 年开发的。它最初用来编写和维护 UNIX 操作系统（直到那时，UNIX 系统程序都还是由汇编语言或是一种叫由 UNIX 的创始人肯恩·汤普逊发明的名为 B 语言编写而成），C 是一种通用的语言，能用来编写任何类型的程序，但是它的这些荣誉与 UNIX 操作系统是密不可分的。如果你想要修整你的 UNIX 系统，你需要使用 C 语言。它们之间结合得如此之好，以至于不仅仅是系统程序，几乎所有在 UNIX 系统下运行的商业软件也都是由 C 语言编写的。C 变得如此之流行，促使多种版本被开发出来用于其他流行的操作系统。对它的使用因此就不限于使用 UNIX 的计算机。然而尽管其声名远扬，它也不是无懈可击的。

C 语言是一种奇怪的语言，因为它本身是一种高级语言，却拥有许多低级语言的特性。C 在某种程度上是介于极端高级语言和极端低级语言之间的一种语言，相比较于这两种语言 C 同时存在优点和缺陷。和（低级）汇编语言一样，C 语言程序能直接操纵计算机的内存。另一方面，它又包含高级语言的一些特性，使得它比汇编语言更容易读取或编写。这成就了 C 语言在编写系统程序时独一无二的地位。但是相对于其他程序（在某种意义上甚至是一些系统程序），C 语言不像其他语言那样简单易懂，并且它不像其他一些高级语言那样能够进行自动检测。

为了克服 C 语言的这些和其他的一些缺点，美国电话电报公司贝尔实验室的比亚内·斯特劳施特鲁普于 20 世纪 80 年代初期研发了 C++语言。斯特劳施特鲁普将 C++设计为 C 语言的升级版，C 语言的大部分是 C++的子集，并且大多数的 C 程序也就是 C++程序（反过来就不一定了，很多的 C++程序绝对不是 C 程序）。C 和 C++的基本语法和语义是相同的。如果熟悉 C 语言，那么可以很快地学会 C++编程。C++拥有在 C 中定义的相同的数据类型、操作符及其他工具，能直接适用于计算机体系结构。不同于 C 的是，C++含有类这一工具，所以它能用于面向对象编程。

2．C++与面向对象编程

面向对象的程序设计是一种程序设计技术，使得你能把一些概念看作各种各样的对象。通过使用对象，你可以描述要被执行的任务、它们之间的相互作用以及任何给定的必须被注意到的条件。一种数据结构往往形成一个对象的基础，因此，在 C 或 C++中，结构类型能形成某种基本对象，与对象的通信能通过使用消息来完成。消息的使用类似于在面向过程的程序中对函数的调用。当某对象收到一个消息时，包含在该对象内的一些方法做出响应。方法类似于面向过程程序设计的函数。然而，方法是对象的一部分。

面向对象程序设计（OOP）的主要特点是具有封装性、继承性和多样性。封装性是指信息隐藏和抽象化的一种形式。继承性则是指可以用可重用代码进行编写，多样性指一个名称在继承性的背景下可能会有多重含义。C++为面向对象编程提供了栖息之地，因为它含有类，类是一种将数据与算法相结合的数据类型。C++并不是一些专家所称的“纯面向对象编程语言”，它调和了面向对象编程的一些功能以增强其效率，也有人称之为“实用性语言”。这些结合使得 C++成了风靡至今的面向对象编程语言，虽然并不是所有的使用都严格遵循面向对象编程的准则。

3．C++的特点

C++允许程序员创建类，这有些类似于 C 中的结构。在 C++中，有各种原型的相关方法、函数，能够在类中访问和操作，类似于 C 函数操作支持的处理指针。C++类是 C 语言结构的扩展。由于结构与类的唯一区别在于结构成员的默认访问权限是公共的，而类成员的默认访问权限是私有的，因此可以使用关键字类或结构来定义相同的类。

C++的类是对 C 和 C++结构类型的扩充，并且形成了面向对象程序设计所需要的抽象数据类型。类能包含紧密相关的一些项，它们共享一些属性。更正式地说，一个对象只是类的一个实例。

最终，应该出现包含很多对象类型的类库，你能使用这些对象类型的实例去组织程序代码。

通常，对一个对象的描述是 C++类的一部分，包括该对象内部结构的描述、该对象如何与其他对象相关以及把该对象的功能细节和该类的外部相隔离的某种形式的保护。C++类结构做到了所有的这些。

在一个 C++类中，使用私有的、公共的或受保护的描述符来控制对象的功能细节。在面向对象的程序设计中，公共部分一般用于接口信息（方法），使得该类可在各应用中重复使用。如果数据或方法被包含在公共部分，那么它们在该类外部也可用。类的私有部分把数据或方法的可用性局限于该类本身。包含数据或方法的受保护部分被局限于该类和任何派生子类。

由于 C++与 C 之间的联系，使得它相对于其他新生的面向对象语言看起来更加传统，然而却比它们拥有更为强大的抽象机制。C++拥有一个模板工具，能全面直接地运行抽象算法。C++模板允许你使用参数类型编写代码。在最新的 C++标准以及大多数的 C++编译器中，允许使用多命名空间以容纳更多类和函数名的复用。在 C++中装卸设备这一概念与你在其他程序设计语言中发现的一样。C++中的内存管理也与 C 相似。程序员必须对其内存进行分配并处理碎片。因为 C 本质上是 C++的一个子类别，所以在 C++环境中大多数的编译器允许你做 C 类型的内存管理。然而 C++对在内存管理上也具有其独特的句法，并且也建议大家在用 C++进行编程时尽量使用 C++类型的内存管理办法。

面向对象程序设计中的继承性使得一个类能继承某个对象类的一些特性。父类用做派生类的模式，且能以几种方式被改变。如果某个对象从多个父类继承其属性，便称为多继承。继承是一个重要概念，因为它使得无须对代码做大的改变就能重用类定义。继承鼓励重用代码，因为子类是对父类的扩充。

与类层次结构相关的另一个重要的面向对象的概念是公共消息能被发送到各个父类对象和所有派生子类对象。按正式的术语，这称为多态性。

多态性使每个子类对象能以一种对其定义来说适当的方式对消息格式做出响应。假设一个收集数据的类层次结构，父类可能是负责收集姓名、社会安全号、职业和雇佣年数，那么你能使用子类来决定根据职业将添加什么附加信息。一种情况是管理职位会包括年薪，而另一种情况是销售员职位会包括小时工资和回扣信息。因此，父类收集一切子类公共的通用信息，而子类收集与特定工作描述相关的附加信息。多态性使得公共的数据收集消息能被发送到每个类。父类和子类两者都对该消息以恰当的方式做出响应。

多态性赋予对象这种当对象的精确类型还未知时响应来自例行程序的消息的能力。在 C++中这种能力是迟绑定的结果。使用迟绑定，地址在运行时刻动态地确定，而不是如同传统的编译型语言在编译时刻静态地确定。静态的方法往往称为早绑定。函数名被替换为存储地址。使用虚函数来完成迟绑定。一个父类，当随后的各派生类通过重定义一个函数的实现而重载该函数时，便在其中定义了虚函数。

虚函数利用了地址信息表，该表在运行时刻通过使用构造符初始化。每当创建它的类的一个对象时调用一个构造符。这里构造符的工作是把虚函数与地址信息表链接，在编译进行期间虚函数的地址是未知的；相反，给出的是地址表中将包含该函数地址的位置。

4．C++语言程序

C++中的所有程序类实体都被称作函数。在其他语言中可能会称作进程、方法、函数或子程序，但在 C++中统一称作函数。C++程序本质上就是一个主函数，当你运行程序时，运行时系统会自动向主函数发出请求。C++其他的一些术语大体上与其他大多数编程语言类似。下面是 C++程序。

```
1  #include<iostream>
2  using namespace std;
3  int main()
4  {
5    int numberOfLanguage;
6    cout<<"Hello reader.\n"
7        <<"Welcome to C++.\n";
8    cout<<"How many programming languages have you used?";
9    cin>>numberOfLanguages;
10    if(numberOfLanguages<1)
11      cout<<"Read the preface. You may prefer\n"
12         <<"a more elementary book by the same author.\n";
13    else
```

```
14    cout<<"Enjoy the book.\n";
15  return 0;
16  }
```


一个C++程序其实就是主函数的函数定义。当程序运行时，主函数得到请求。主函数的内容被大括号{}包围。当运行程序时，大括号内的声明得到执行。下面是当用户运行程序时可能会出现的两个屏幕显示。

对话 1

Hello reader. Welcome to C++. How many programming languages have you used? 0← User types in 0 on the keyboard. Read the preface. You may prefer a more elementary book by the same author.

对话 2

Hello reader. Welcome to C++. How many programming languages have you used? 1← *User types in 1 on the keyboard.* Enjoy the book

C++中的变量声明和其他程序语言类似。第 5 行声明了变量 numberOfLanguages。int 类型是 C++类型之一，用来表示整数。

如果你之前没有使用过 C++，那么输入输出控制台中的 cin 和 cout 使用对你来说可能有些陌生。但是其大意能在这个简单的 C++程序中得到体现。例如，考虑第 8 行和第 9 行：第 8 行在屏幕上输出引号中的文本。第 9 行读取用户用键盘输入的数字，并对 numberOfLanguages 变量赋予这个数值。

第 11 行和第 12 行输出两个字符串而不是一个。\n 是新起一行的符号，用来指示计算机另起一行进行输出。

课文 4：Java

Java 语言已经变得极为普遍。Java 的快速普及和广为接受可归因于其设计与编程的特点，特别是它承诺程序可一次编写，并可在任何地方运行。正如 Sun 公司的 Java 语言白皮书所说，Java 是一种简单的、面向对象的、分布式的、解释型的、健壮的、安全的、体系结构中立的、可移植的、高性能的、多线程的和动态的语言。

什么是 Java

Java 是超文本链接标识语言的主要分支语言，大部分网页都是 Java 编写的。和标识语言相比，譬如超文本链接标识语言和可扩展标记语言，Java 是一种面向对象的，具有网络亲和性的高级语言，它可以使程序员构建几乎能在任何操作系统上运行的应用程序。Java 的出现可以使一些庞杂的应用程序分解成简洁的小应用程序，它们可以直接从网站上下载并能在任何计算机上运行。此外，Java 能让网页传递富视觉内容和一些小的应用程序，这样用户在下载时能与更多的网页进行互动。

一些微型计算机包括特殊的 Java 微型计算机，可以直接运行 Java 软件，但缺点是 Java 与很多微处理器不兼容，如英特尔和摩托罗拉公司生产的微处理器。基于这个原因，这些用户需要使用一个称作 Java 虚拟机的小“直译器”程序，将 Java 程序转换成任何计算机和操作系统都能识别的语言。为了能浏览到 Java 网页上的一些特殊效果，用户还需要一个带有识别 Java 功能的浏览器。

每个程序员都可以利用 Java 开发程序，除此之外，Java 软件包——比如 ActionLine、Activator Pro、AppletAce 和 Mojo，非程序员用户可以通过这些软件，生成一些小程序，给具有 Java 浏览器功能的网页添加多媒体效果。对于懂得多媒体格式以及对菜单选项有兴趣的用户都可以运用这些软件包。

Java 是面向对象语言中一颗最耀眼的新星。面向对象程序设计语言的一个特别强大的功能是被称为继承的特性。继承允许一个对象获得与其功能上具有联系的其他对象的属性与功能。程序员把对象归为不同的类，并把类分成层次，以此将对象联系起来。这些类与层次使程序员可以定义对象的属性与功能，而不必重复源代码。因此，使用面向对象程序设计语言，可以大大缩短程序员编写应用程序所需的时间，也可以缩短程序长度。面向对象程序设计语言灵活而且适应性强，所以程序或程序的某些部分可用于不止一项任务。使用面向对象程序设计语言编写的程序，比使用非面向对象程序设计语言编写的程序，一般来说，长度要短且包含的错误也少。

Java 通过自身与互联网及网页浏览器的相关联性迅速占领了软件领域。它被设计成一种使用起来简单方便的语言，通过网页浏览器它能在任何连接网络的计算机上运行。就这点而言，使 Java 成了标准互联网和内部网程序设计语言。

因为 Java 具有 C++语言的语法，所以，它学起来是否容易（或困难）取决于你的经验。但在一些重要的方面 Java 对 C++做了改善。首先 Java 没有指针，而指针是低等的程序结构，会导致编程时更易出错。Java 还拥有垃圾收集功能，这个特征可以让程序员自由地进行明确分配和对内存进行取消配置。在虚拟计算机上它也能运行，因为软件内置了网页浏览器，允许标准的 Java 编制字节码得到执行，而不用考虑计算机的类型。

Java 的开发工具调用起来相当迅速，并且能从主要的软件公司如 IBM、微软和赛门铁克获得。

Java 的特点

1．Java 是一种简单的语言

语言没有简单的，但 Java 比 C++这种流行的面向对象程序设计语言要容易些。在 Java 之前，C++曾是占支配地位的软件开发语言。Java 部分仿照了 C++，但大大简化和改进了。例如，指针和多重继承常常使编程变得复杂。Java 用一种称为接口的简单语言结构取代了 C++中的多重继承，并取消了指针。

Java 采用自动的存储器分配和垃圾收集，而 C++要求程序员分配存储器和收集垃圾。此外，就这样一种功能强大的语言而言，语言结构的数量算小的。简洁的句法使 Java 程序易于编写和读取。有些人称 Java 为“C++--”，因为它像 C++，但与之相比功能增多了，缺点也得到了改进。

2. Java 是一种面向对象的语言

Java 是一种天生的面向对象语言。尽管不少面向对象的语言开始时完全是过程语言，但 Java 从一开始就是要面向对象的。面向对象编程是一种流行的编程方法，它正在取代传统的过程编程技术。

使用过程编程语言开发的软件系统基于过程范式。面向对象编程用对象来对真实世界进行建模。世界上的任何东西都可作为一个对象来建模。一个圆是一个对象，一个人是一个对象，一个 Windows 图标也是一个对象，甚至一笔贷款也可视为一个对象。Java 程序是面向对象的，因为用 Java 进行的编程是围绕着创建对象、操纵对象以及使对象一起工作的。

软件开发的一个中心问题是如何重复使用代码。面向对象编程凭借封装、继承和多态性提供了很大的灵活性、模块性、明确性和可复用性。多年来，面向对象技术被认为是精英技术，要求在培训和基础设备方面进行可观的投入。Java 帮助面向对象技术进入了计算机领域的主流。其简洁明了的句法使程序易于编写和读取。就设计和开发应用程序而言，Java 程序具有相当的表现力。

3. Java 是一种分布式的语言

分布式计算涉及数台计算机在网络上一起工作。Java 旨在使分布式计算变得容易。由于联网能力一开始即被结合进 Java，所以编写网络程序宛如向一个文件发送或从一个文件接收数据。

4. Java 是一种解释型的语言

运行 Java 程序需要解释器。程序被编译成称作字节码的 Java 虚拟机代码。字节码独立于机器，可在有 Java 解释器的任何机器上运行，而 Java 解释器是 Java 虚拟机的组成部分。

大多数编译器，包括 C++编译器，将高级语言程序翻译成机器码。这种代码只能在本机上运行，如果在其他机器上运行程序，则需要在本机上重新编译程序。例如，如果在 Windows 中编译 C++程序，由编译器生成的可执行代码只能在 Windows 平台上运行。就 Java 而言，一次编译源代码，由 Java 编译器生成的字节码可在任何有 Java 解释器的平台上运行。Java 解释器将字节码翻译成目标机的机器语言。

5. Java 是一种健壮的语言

健壮意指可靠。程序设计语言没有可确保完全可靠的。Java 非常重视对可能存在的错误进行早期检查，Java 编译器可检测出许多在其他语言中执行时才会首次暴露出来的问题。Java 取消了在其他语言中发现的某些类型的易出错编程结构。例如，它不支持指针，从而消除了改写内存和破坏数据的可能性。

Java 拥有运行期异常处理的功能，用于为健壮性提供编程支持。Java 迫使程序员编写用于处理异常的代码。Java 可捕捉异常情况并对其做出反应，以便在发生运行期错误时，程序能够继续其正常运行，并从容终止。

6. Java 是一种安全的语言

作为一种因特网程序设计语言，Java 用于联网的分布式环境。如果下载一个 Java 小程序（一种特殊的程序）并在计算机上运行，那么它不会破坏你的系统，因为 Java 实施了几种安全机制来保护你的系统免于杂乱程序造成的危害。安全基于什么也不应相信的前提。

7．Java 是一种体系结构中立的语言

Java 是一种解释型语言。这个特征使 Java 能够做到体系结构中立，或者换个说法，独立于平台。凭借 Java 虚拟机，你可以编写能在任何平台上运行的程序。

Java 最初的成功源自它的 Web 编程能力。你可以从一个 Web 浏览器运行 Java 小程序，但 Java 不仅是用于编写 Web 小程序。你还可以使用 Java 解释器从操作系统直接运行独立的 Java 应用程序。今天，软件供应商通常会将同一种产品开发成多个版本，以便在不同的平台上运行。开发者使用 Java，则只需编写一个版本即可，而该版本可在每个平台上运行。

8．Java 是一种可移植的语言

由于 Java 是体系结构中立的，所以 Java 程序是可移植的。它们可在任何平台上运行而不需重新编译。而且，Java 语言不存在平台特有的任何特征。在有些语言中，如 Ada 语言，最大的整数随着平台的不同而不同。但是，在 Java 中，整数类型的范围无论在哪个平台上都是一致的，正如其算术意义上的特性。固定的数字范围使程序具有可移植性。

Java 环境可移植到新的硬件和操作系统中。事实上，Java 编译器本身就是用 Java 编写的。

9．Java 的性能

Java 的性能有时受到批评。字节码的执行从来不如在 C++等编译型语言的环境下快。因为 Java 属于解释型语言，字节码不是由系统直接执行而是通过解释器来运行的。然而，其速度完全能够满足大多数交互式应用程序的需要。在这些应用程序的情况下，中央处理器经常空闲，等待着输入或来自其他来源的数据。

中央处理器的速度在过去几年间显著提高，而这种趋势将继续下去。改进性能有许多方法。使用过 Sun 公司出品的较早期 Java 虚拟机的人，无疑会注意到 Java 的速度慢。然而，新的 Java 虚拟机速度明显加快了。新的 Java 虚拟机采用了称作即时编译的技术。它将字节码编译成本机码，把本机码存储起来，并在执行其字节码时重新调用本机码。Sun 公司最近开发出 Java HotSpot Performance Engine，它包括一个优化常用代码的编译器。HotSpot Performance Engine 可插入 Java 虚拟机中，以显著提高其性能。

10．Java 是一种多线程的语言

多线程操作是一个能同时执行多项任务的程序。多线程编程被平稳地结合进 Java 语言，而在其他语言中，你要调用操作系统特有的进程才能实现多线程操作。多线程操作在图形用户界面和网络编程中特别有用。在图形用户界面编程中，有许多工作在同时进行。一个用户可在浏览网页的同时听录音。在网络编程中，一台服务器可同时为多个客户提供服务。多线程操作在多媒体与网络编程中必不可少。

11．Java 是一种动态的语言

Java 在设计上可适应不断演变的环境。新的类可在不重新编译的情况下快速装入。开发者没有必要创建，用户也没有必要安装重要的新软件版本。新的特征可以根据需要透明地被融合进来。

Java 技术可以做什么

用 Java 编程语言编写的大多数程序都是模块程序和应用程序。如果你在 Web 上冲浪过，你或许已经熟悉了这种模块程序。模块程序是符合一定约定的程序，这些约定使它可以运行于可用Java的浏览器内。在片头显示用Java技术制作的向你招手的吉祥物（Duke）动画。

然而，Java编程语言不仅仅用于编写Web上娇小可爱的娱乐性模块程序。通用的、高级Java编程语言也是一个功能强大的软件平台。使用丰富的API，你可以编写多种程序。

一个应用程序是可以直接运行于Java平台的独立程序。一种特殊的应用程序叫作服务器（server），它对网络客户提供服务和支持。例如，Web服务器、代理服务器、邮件服务器和打印服务器。另一种特殊的程序叫作“小服务程序”，它几乎可以被看成是运行在服务器端的模块程序。Java Servlet是建立交互网络应用的流行选择，用来替代CGI脚本。小服务程序类似于模块程序，在应用程序运行时扩展。虽然它不是工作在浏览器内，但运行在Java Web服务器内，配置或定制服务器。

API是怎样支持各种程序的呢？它用一个提供各种功能的软件包来做到这一点。每个Java平台都可以完整地提供如下功能：

- 基本要素——对象、串、线程、数字、输入和输出、数据结构、系统性能、日期和时间等。
- 模块程序——模块程序所用的惯例集。
- 网络——统一资源定位符、TCP（传输控制协议）、UDP（用户数据报协议）套接字及IP（网际协议）地址。
- 国际化——使所写程序可以让世界各地的用户使用。程序可以自动适应特定的场所并以适当的语言显示。
- 安全——有低级和高级两种，包括电子签名、公共和私有钥匙管理、访问权限控制和授权证书。
- 软件部件——叫作JavaBeans，可以插入到现有部件的体系结构中。
- 对象连续化——允许轻微持续和通过远程调用进行通信。
- Java数据库连接（JDBC）——对各种相关数据库提供统一的访问。

Java平台也有可以用于二维和三维图形、可访问性、服务器、协作、电话技术、语音、动画等的API。

注意：Java 2 SDK标准版本1.3版。Java 2的运行环境（JRE）由以下几部分组成：虚拟机、Java平台核心类及支持文件。Java 2 SDK包括JRE和像编译器及调试器这样的开发工具。

课文5：ActionScript基础

ActionScript代码要放在哪儿

当你有一个新的ActionScript工程，你知道程序代码该放在哪才能使其正确执行吗？答案是把ActionScript代码放在类的构造方法和其他方法中。

在ActionScript 1.0版本和2.0版本中，代码要放置的位置有很多种选择：时间轴、按钮和影片剪辑，影片剪辑的时间轴放在外部as文件内（由#include引用）或者作为外部类文件。ActionScript 3.0完全以类为基础，所以，所有代码都必须放在工程类的方法中。

当你创建新的ActionScript工程时，主类会自动创建，在Code视图中打开，看起来应该就像这样：

```
package {
    import flash.display.Sprite;
    public class ExampleApplication extends Sprite
    {
        public function ExampleApplication()
        {
        }
    }
}
```

即使你熟悉 ActionScript 2.0 的类，也还是有些新的东西要了解。

你会注意到的第一件事是代码顶端的单词 package（包）。包被用于将功能相关的类群集起来。在 ActionScript 2.0 中，包是通过保存类文件的目录结构而体现的。然而，在 ActionScript 3.0 中，你必须明确指明包。例如，你可以有一个 utility 工具类的包，而声明方式就像这样：

```
package com.as3cb.utils {
}
```

如果你没有指明包名称，那么类就会创建在默认顶层包中。但是你还是得引入 package 关键字和大括号。

接着，放置任何 import 语句。导入一个类，可以让该类在文件的代码中可用，同时建立快捷方式，以便每次你想引用那个类时，都不用再输入完整的包名称。例如，你可以使用下列 import 语句：

```
import com.as3cb.utils.StringUtils;
```

接下来，你可以直接引用 StringUtils 类，而无须输入路径其余的部分。如前例所述，你得从 flash.display 包中导入 Sprite 类，因为默认类都继承于 Sprite 类。

接着是主类 ExampleApplication。你可能会注意在类定义前面的关键字 public。虽然在包中不能有私有（private）类，你还是应该特别将该类标示为公开（public）类。请注意，主类继承于 Sprite。此外，.swf 文件本身就是一种影片 Sprite 或影片剪辑，这也就是为什么你可以把一个.swf 文件载入另一个.swf 文件中，而且多半能将其视为只是另一个嵌套的 Sprite 或影片剪辑的原因所在。主类代表整个.swf，所以应该继承自 Sprite 类或任何继承自 Sprite 类的类（例如 MovieClip 类）。

最后，有个公开函数（以类术语而言，是指方法）和类的名称相同。这就是构造函数，类的构造函数在类的实例创建时就会自动执行。既然这样，当.swf 载入到 Flash 播放器时，构造方法就会立刻执行。所以，你要把代码放在哪里使之执行？一般而言，先把一些代码放在构造函数的方法中。以下是非常简单的范例，在屏幕上随机画出一些线条：

```
package {
    import flash.display.Sprite;
```

```
    public class ExampleApplication extends Sprite {
        public function ExampleApplication() {
            graphics.lineStyle(1, 0, 1);
            for(var i:int=0;i<100;i++) {
                graphics.lineTo(Math.random() * 400, Math.random() * 400);
            }
        }
    }
}
```

保存并执行此应用程序。你的浏览器应该会打开所得的 HTML 文件，显示此.swf，而里面有 100 条随机画出的线条。如你所见，当此文件载入播放器时，构造函数就会立刻执行。

实际上，你通常会把放在构造方法的代码减到最少。理想情况是构造方法只会有一道方法调用，由那个方法对应用程序进行初始化。

就初学者而言，应当已经知道该在何处输入代码，以下是相关术语的快速入门。这些定义都很简洁，目的是针对那些没写过代码的人。

- 变量（variable）

变量是代码中为数据方便使用的占位符（placeholder），而且你可以用任何喜欢的名称予以命名，但是名称不能是 ActionScript 保留字，而且名称要以字母、下画线或美元符号开头（不能是数字）。随 Flex Builder 2 安装的帮助文件包含了保留字清单。变量可用于保存临时信息，非常方便，例如数字的和、变量或者指向某物，例如，文本域或影片 Sprite。变量首次在脚本（script）中使用时，是以 var 关键字声明的。可以用等号（=，也称为赋值运算符）指定值给变量。如果变量是在类的方法之外声明的，就是类变量。类变量或属性可以有访问修饰字：public、private、protected 或 internal。私有变量（private）只能在类本身的内部访问，而公开（public）变量可以让另一个类的对象访问。保护（protected）变量可以从该类的实例或任何子类的实例访问，而内部（internal）变量则可以由任何位于相同包内的类访问。如果没有指明任何访问修饰字，就默认为 internal。

- 函数（function）

函数是可做某种事的代码块。可以通过函数名称进行调用（call）或启用（invoke）（也就是执行）。当函数是类的一部分时，就称为该类的方法。方法可以使用的修饰字与属性（properties）所使用的完全相同。

- 作用域（scope）

变量的作用域是说明影片中的代码在何时何地可以对该变量进行操作。作用域定义变量的生命跨度（life span）以及脚本中其他代码块是否可以访问。作用域会决定变量可以存在多久以及你可以在代码中的何处设置或取得该变量的值。函数的作用域会决定其他代码块在何时何地可以访问该函数。

- 事件处理器（event handler）

处理器是一种函数或方法，它在执行时是为了响应某种事件，例如鼠标点击、按键或者时间轴中播放头（playhead）的移动。

- 对象和类（object and class）

对象是你可以在 ActionScript 里以代码操作的东西，例如影片 Sprite。对象还有很多类型。例如那些用于操作颜色、日期以及文本域的对象。对象是类的实例，也就是说，类是创建对象的范本，而对象是该类的特定实例（instance）。如果你还不清楚，那就从生物学角度考虑：把自己想象成对象（实例），属于一个通用类，名为人类。

- 方法（method）

方法是和对象关联的函数，而该函数会作用在对象身上。例如，文本域对象的 replaceSelectedText()方法可用于取代域中被选取的文字。

- 属性（property）

属性是指对象的特性（attribute），可以被读取和（或）设置。例如，影片 Sprite 的水平位置是由其 x 属性表示，而对其进行测试和设置。另一方面，文本域的 length 属性（表示域中有多少字符）只能做测试，而不能直接设置（然而，可以间接予以影响，也就是从域中新增或移除文字）。

- 语句（statement）

ActionScript 命令（command）被输入成一系列的语句。一条语句可能会告诉播放头跳到特定帧，或者也可能是修改影片 Sprite 的尺寸。多数 ActionScript 语句都以分号终结。

- 注释（comment）

注释是代码中的附注，专门给其他人看，Flash 会予以忽略。在 ActionScript 中，单行注释以//开头，在该行行尾自动终止。多行注释以/*开头，以*/终止。

- 解释器（interpreter）

ActionScript 解释器是 Flash Player 的一部分，用于检查、了解并执行你的代码。遵循 ActionScript 的严格语法规则可以确保解释器能轻易理解你的代码，如果解释器碰到错误，通常会悄悄地忽略，只会拒绝执行代码，而非生成明确的错误消息。

- 处理事件

如果你想让某些程序重复执行，给 enterFrame 事件添加一个事件侦听器，然后指定一个方法作为处理器。

在 ActionScript 2.0 中，处理 enterFrame 事件相当简单。你只需创建一个名为 onEnterFrame 的时间轴函数，而每次新帧开始时，此函数就会被自动调用。在 ActionScript 3.0 中，你对.swf 里的各种事件有更多控制权，但是，需要多做点工作才能实现这一点。

如果你熟悉 ActionScript 2.0 的 EventDispatcher 类，那么对于 ActionScript 3.0 处理事件的方式，你会感到得心应手。事实上，EventDispatcher 类已成功从外部定义类转变为所有交互对象（例如影片 Sprite）的基类（base class）。

要响应 enterFrame 事件，你得告诉应用程序去监听该事件，然后指定在事件发生时你想要调用什么方法，这就要使用 addEventListener 方法，其定义如下：

```
addEventListener(type:String, listener:Function)
```

type 参数是你想监听的事件类型。就此而言，就是字符串 enterFrame。然而，使用这种字符串文本会使得你的代码出现那些编译器无法捕捉的错误。例如，如果你不小心输入 enterFrame，那么应用程序就会监听 enterFrame 事件。为了防止这种情况，建议你使用 Event

类的静态属性。你应当先将 Event 类导入，才可以调用 addEventListener，方法如下：

```
addEventListener(Event.ENTER_FRAME, onEnterFrame);
```

现在，如果你不小心输入 Event.ENTER_FRANE，编译器会提示没有这个属性。

第二个参数 onEnterFrame 指的是类中的另一个方法。请注意，在 ActionScript 3.0 中，这个方法没有必要命名为 onEnterFrame。然而，对于事件处理方法，常用 on 加事件名称来命名。这个方法在调用时会传进一个 Event 类的实例。因此，你必须导入该类，定义该方法，使其接收一个事件对象：

```
import flash.events.Event;
private function onEnterFrame(event:Event) {
}
```

事件对象包含的信息和该事件有关，在处理事件时会有所帮助。即使你没有使用该事件对象，还是应该将处理器设置成准备接收该事件对象。如果你熟悉 EventDispatcher 的 ActionScript 2.0 版本，就会发现这里的实现方式有所不同。在早期版本中，存在这样一个问题，即处理事件的函数的作用域（scope）通常都需要使用 Delegate 类予以更正。在 ActionScript 3.0 中，处理方法的作用域就是其所在的类，因此，没有必要使用 Delegate 来更正作用域问题。

以下是一个关于绘制连续的随机线条的简单应用。

```
package {
   import flash.display.Sprite;
   import flash.events.Event;
      public class ExampleApplication extends Sprite {
            public function ExampleApplication() {
         graphics.lineStyle(1, 0, 1);
         addEventListener(Event.ENTER_FRAME, onEnterFrame);
      }
      private function onEnterFrame(event:Event):void {
         graphics.lineTo(Math.random() * 400, Math.random() * 400);
      }
   }
}
```

第六部分　计算机网络

课文 1：关于计算机网络

在计算机科学中，网络是指由通信设备连接在一起的一组计算机及其相关设备。网络可采用电缆等永久性连接，或者采用通过电话或其他通信链路而实现的临时性连接。网络既包括由少数计算机、打印机以及其他设备构成的局域网，也可以是由分布在广大地理区域的许多小型计算机与大型计算机组成的互联网。

计算机网络是信息时代的核心组成部分。无论大小，其存在的目的均是为计算机用户提供以电子方式进行信息通信与传送的方法。

网络基础

网络是一套通过媒介连接而互相联系起来的装置（通常称为节点）。一个节点可以是一台计算机、一台打印机或是其他任何能够传输或接收网络中其他节点生成的数据的装置。这种设备之间相互联系的纽带通常叫作通信渠道。网络需要依赖网络操作系统来管理网络资源。它可能是一个完全不需要依靠外界设备的操作系统，例如 NetWare，也有可能是需要现有的操作系统的支持才得以工作的操作系统，例如 Windows NT。下面的段落阐述了使用网络的优势。

- 促进交流：通过网络，人们可以利用电子邮件、即时信息、聊天室、电话、视频电话以及电视会议进行更加高效方便的沟通交流。有时这种类型的交流方式发生在商务网络里，有时则通过互联网遍布全球。
- 硬件共享：在同一个网络环境中的每台计算机都可以连接并使用这个网络环境中的硬件设施。假定有好几个用户需要使用一台激光打印机，如果他们的计算机和这台激光打印机都在同一个网络环境中，那么他们就可以共享它。商务及家庭用户通过这种方式共享硬件设备主要是用来节省费用，如果给每个用户都配备同样的一种硬件，例如打印机，那样花费就太大了。
- 数据和信息的共享：在同一网络环境里，任何经授权的计算机用户都能访问存储在同一网络中其他计算机上的数据和信息。这种在共享存储设备之间获取并存储信息和数据的能力是许多网络的重要特征。
- 软件共享：在同一网络里的用户都能获取这个网络内的软件。为了支持多用户类型也能够访问软件，大多数的软件供应商出售能够兼容他们生产的软件的网络版本。

计算机网络的类型

计算机网络可用于多种服务，既为公司也为个人。对公司而言，由使用共享服务器的个人计算机组成的网络提供了灵活的方式和很好的性价比。对个人而言，网络提供访问各种信息和娱乐资源的接口。

从传输技术的角度看，网络分为局域网（LAN）、广域网（WAN）和城域网（MAN）。局域网通常位于一座单独的建筑物或校园，处理办公室间的交流。广域网覆盖一个大的地理区域，它连接着多个城市和国家。城域网用来连接一个城市范围内的办公大楼。

1. 局域网

LAN 是 Local Area Network 的首字母缩写。它是计算机和其他设备集团分散在一个相对有限的区域并和通信链路连接。一个局域网可以在一个建筑内连接所有的工作站、外围设备、终端，或者其他设备。局域网利用计算机技术来有效地共享文件和打印机，并使电子邮件这样的通信方式成为可能。它们把数据、通信、计算机和文件服务器连接在一起。

局域网的设计可以做到以下几点：

- 在一个有限的地理区域内运作；
- 允许用户连接高带宽媒介；

- 为本地设备提供全天候的连接；
- 连接邻近的实质设备。

2．城域网

MAN 是英文 Metropolitan Area Network 的首字母缩写。它是一种高速网络，向大于75km 距离传输语音、数据和图像时的速度可以高达 200Mbps。如果距离短，传输速度会加快。城域网比一个广域网规模要小，但一般以较高的速度运行。范围是由最多相距 1mile（1609.344m）的办公室或者建筑物所构成。蜂窝电话系统扩大了汽车电话和手提电话连接城域网的灵活性。

所有这些网络都包含了各种计算机、存储设备和通信设备。

3．广域网

WAN 是英文 Wide Area Network 的首字母缩写，是地域广泛的网络，它依赖于连接多样的网络分支的通信能力。广域网可以是一个大的网络，也可以由一系列相互关联的局域网构成。

由于部门和机构的计算机使用的增加，很快会发现单有局域网是不够的。在一个局域网系统中，每个部门或企业类似一个电子岛屿。而为此所需要的是传送信息的渠道，有效地从一个企业传送到另一个，因此，才有了广域网的发展壮大。

20 世纪 80 年代个人计算机的广泛使用以及局域网的普及实现了访问遥远基地的信息，从海外网络上下载申请表，发送信息给一个不同国家的朋友，与同事分享文件——都来自个人计算机。

允许所有的这一切轻松变成现实的网络是复杂精密的组织。它们依赖于许多相互合作的部件来实现效率。世界性的计算机网络的设计和实施可以被视作近几十年来重大的科技奇迹。

网络拓扑

网络拓扑是在通信网络中各种设备的配置或实质安排。它可以以不同的方式布局。5 个基本拓扑是：总线、星形、环形、树形和混合。

1．总线拓扑

总线拓扑结构使用一个单一的骨干段（电缆长度）以使所有的主机直接连接。总线拓扑的运作像是一个巴士系统在高峰时期管理巴士停在不同的巴士区接送乘客。在一个总线网络，所有的设备都连接到一个共同的电缆通信，称为总线，使用 co-ax.STP 或者 UTP。在总线网络中没有中央计算机或服务器，数据传输为双向的，以大约 1～10Mbps 的速率传输。每个通信设备都可以传输电子信息到其他设备。如果这些信息出现冲突，设备等待并试图重新发送。

总线拓扑的优点是安装相对便宜。缺点是如果总线出现问题，那么整个网络就不能运行了。

2．星形网络

在星形网络，所有的计算机和其他通信设备通常都通过双绞线连接到中央集线器，如文件服务器或主机计算机。电子信息发送通过中央枢纽以 1～100Mbps 的速率前往目的地。中央枢纽监控流通情况。

星形网络的优势在于如果在任意通信设备和中央集线器之间的链接被打断，网络上其余设备将继续运作。主要缺点是中央集线器的失败是灾难性的。

3．环形网络

在环形网络，所有的微型计算机和其他通信设备构成一个循环。电子信息在循环中以一个方向传输，每个节点充当一个中继器，直至达到正确的目的地，之间没有中央主机或服务器。环路一般为 UTR、STR，或者光导纤维电缆。

代表性的、象征性的环网，其中的位模式（称为“标记”）决定哪个用户在网络上，并向其发送信息。环网的优点是只在一个方向传递信息流，因此，不存在碰撞危险。缺点是信息流速度的限制和相对较高的成本。

4．树形网络

树形网络结合了总线拓扑和星形拓扑结构的特点。它由星形网络配置的工作站连接到线性总线主干电缆。树形拓扑结构允许对现有网络进行扩容。

5．混合网络

混合网络是星形网络和总线网络的混合产物。例如，一所规模较小的学校可能使用一个总线网络来连接各个建筑，同时在建筑内部使用星形网络和环形网络。

企业内部互联网

由于许多组织认识到了互联网的力量和高效，他们将 Internet 和 Web 技术应用到自己的内部网络。企业内部互联网是使用互联网技术的内部网络。内部互联网一般使员工容易获得企业内部信息并且使团体协作更加便利。简单的内部网应用包括组织材料的电子版，如电话簿、企业工作日历、程序手册、员工福利信息、工作职位等。典型的内部网还包括与互联网的连接。更为复杂的内部网使用包括群件应用程序，如项目管理、聊天室、新闻组、团体调度、视频会议。

企业内部互联网本质上是存在于组织内部的互联网的缩小版。它使用 TCP/IP 技术，有一个 Web 服务器，支持在 HTML 编码的多媒体网页，并通过如 Microsoft Internet Explorer 或 Netscape Navigator Web 浏览器访问。用户可以通过创建和发布一个网页来发布和更新内联网，使用方法类似于互联网。

有时，一个公司使用外联网使客户或供应商来访问公司的内部网的一部分。例如，联邦快递使客户能够访问他们的内部网来打印传播账单，转播地点，甚至追踪运送包裹直至它们到达目的地。

网络操作系统

与 DOS 和 Windows 这样针对单个用户设计的一台计算机来控制的操作系统不同，网络操作系统（NOS）要协调跨多个计算机网络的活动。网络操作系统扮作主管来保持网络顺利运行。

两种类型的网络操作系统分别为：

1．点对点操作系统

点对点网络操作系统允许用户共享资源，共享其计算机上的文件和存取他们计算机上的资源。然而，他们没有一个文件服务器或集中管理源。在一个点对点网络中，所有计算机都视为平等的，它们有相同的权利去使用网络内的可用资源。点对点网络的设计主要是针对中小型局域网。AppleShare 和 Windows for Workgroups 的一些程序可作为点对点网络操作系统的例子。

点对点操作系统的优点如下：

- 很少的启动资金，无需专用的服务器。
- 安装方便——一个已经就绪的操作系统（如 Windows XP）可能只需要对点对点操作重新配置。

点对点操作系统的缺点如下：

- 较分散——没有主要的文件和应用软件的存放处。
- 安全性——不为客户机/主机提供安全计划。

2. 客户机/服务器

客户机/服务器网络操作系统允许网络集中功能和应用在一个或多个专用的文件服务器上。文件服务器成为系统的核心，提供对资源的访问和安全保障。（客户）个人工作站可以访问文件服务器上的可用资源。该网络操作系统提供整合所有的网络组件的机制，并允许多个用户同时共享同一资源而与物理位置无关。Novell 的 NetWare 和 Windows 2000 Server 是客户机/服务器网络操作系统的例子。

客户机/服务器网络的优点如下：

- 集中——资源和数据安全是通过服务器控制的。
- 可扩展性——任何或所有元素都可以随需求增加而被单个替换。
- 灵活性——新技术可以很容易地集成到系统。
- 互操作性——所有组件（客户端/网络/服务器）一起工作。
- 可达性——服务器可远程访问或通过多个平台访问。

客户/服务器网络的缺点如下：

- 费用——需要专用服务器的初始投资。
- 维护——大型网络需要工作人员以确保有效运作。
- 依赖性——当服务器出现故障时，操作将停止在网络上。

课文 2：计算机网络应用

计算机网络在短期内已经成为企业、各行各业以及娱乐不可缺少的一部分。下面介绍一些在不同领域的网络应用。

1. 市场营销及销售

计算机网络广泛应用于营销和销售组织。市场营销专业人士用它们来收集、交换和分析与客户需求和产品开发周期相关的数据。销售应用包括远程购物，这需要使用订单输入计算机或连接到订单处理网络的电话，还包括网上预订酒店、航班服务等。

2. 制造业

现在的计算机网络已应用于许多方面的制造，包括制造过程本身。利用网络来提供必要的服务的两个应用程序为计算机辅助设计（CAD）和计算机辅助制造（CAM），它们都可以实现多个用户同时在一个项目上进行工作。

3．目录查找

如果想知道一个特定类别的信息，而在头脑中只有一个大体概念的话，那么通过目录查找通常是很好的选择。目录查找也使用数据库，由于已经被人工编辑筛选过，所以这种目录通常很小，却更精确。举例来说，搜索引擎会把 computer chips（计算机芯片）的页面和 potato chips（薯片）的信息归入 chips 目录下，人工编辑却不会这样做，最大的目录之一——开放目录工程——声称已经有超过 30 000 名志愿编辑把超过 200 万个网页编入索引。

在一个搜索引擎上使用目录查找，通常选定的是目录类别而不是关键词。在选择了某个大的类别，下面就会显示相关的详细的子目录。最后，选了某个目录或子目录后，就会出现相关的网页。

4．网上支付

首先，需要一个有效的账号或一张国际信用卡（如 VISA 或 MasterCard）以及一个电子邮件的地址。要做的第一件事就是在网上注册账号到提供此类服务的公司，如 PayPal、Billpoint 等。当从参与公司或个人那里选择需要的货物以后，输入用户名和密码（在安全网页上）授权付款服务公司支付所需金额。提供网上支付的公司将向信用卡收费或者从指定付款的账户中扣钱；用户通常会通过电子邮件收到一份交易确认清单。

5．在线电视/电影

网上多媒体还处在发展的早期阶段，所以跟未来我们可以得到的在线影视相比，现在可以看到的只能说是很少很少。现在最常见的在线电视电影就是新电影的一个片段、电影预告片、音乐视频以及事先录制好的采访或者短的视频。这些剪辑片段可以在特定的提供影视服务的网站找到。视频剪辑链接也可以在新闻娱乐网站上找到。

尽管一些视频文件是完整下载以后再播放的，因为文件太大（相应下载时间长），所以流媒体技术十分常见。流媒体技术将视频（音频）文件分成一小块一小块的下载，并将其放入一定的缓冲区内，然后一边下载一边播放。此类文件通常使用 RealPlayer、QuickTime、Window Media Player 播放。如果计算机上网速度很慢，那么有时就需要停下来重新进行缓冲，此刻计算机播放的内容就会出现停顿的现象。

除了上述提到的几种视频之外，有些网站提供经典影视免费点播服务，也有一些网站提供经典电视节目点播服务以及在线电视直播服务等，虽然这样的网站现在还很少。通过因特网链接收看到的电视或电影称为网络电视，现在仍处于发展初期。但是，随着宽带因特网的广泛使用，网络电视最后一定会成为大众普及的因特网应用。可能会实现现场直播、电影点播服务（对下载的视频进行付费而并非去影碟店租取）和互动电视等一些服务。事实上，互动电视——电视和在线活动相结合——在过去的几年中已经得到了有限的实现。

6．邮件

可能最广泛的网络应用就数电子邮件了。下面我们将介绍如何使用电子邮件。

1）概述

电子邮件（E-mail）是通过计算机网络收发消息及文件的一种电子系统。它是因特网最早的应用之一，现在仍被广泛地使用。不过，与电话或纸质文档相比，电子邮件在工作中仍属于比较新的事物。

连接上因特网后，可以向任何拥有电子邮件地址的人发送邮件，而不必考虑当时收信人是否在线。电子邮件由发信人的计算机发送到他的因特网服务供应商，再通过因特网到达收信人的因特网服务供应商。当收信人登录上网，发出请求信息时，邮件就被发到他的计算机。若标明应该获取哪种电子邮件的话，应在浏览器的选项或者系统配置屏幕中设定出正确的电子邮件地址。一些浏览器允许同时设置多个电子邮件账号（如个人和学校账号并存)。另一些则只支持一个，所以在需要的时候查收不同邮件账号时必须更改设置。

基于网页的电子邮件，如 HotMail 和 Yahoo! Mail，它们的操作有些不同。使用该类邮件时，不需要设定浏览器里的邮件设置，而是在邮件服务器网站的指定框中输入用户名和密码。这种特性使网页电子邮件使用更灵活，因为用户从任何计算机都可以收取邮件，而不必更改浏览器设置。网页电子邮件也不是发送到用户的计算机，一般保存在邮件服务器上，只通过服务器的网站才能查看。旅行者、学生和其他经常使用不同计算机的用户广泛使用这种邮件。

这里有几条使用电子邮件时需要记住的指导原则：

- 电子邮件越来越多地用于专业目的——不久以前，电子邮件还被视为不重要的交流形式。而现在，电子邮件在大多数工作场所已成为一种主要的交流形式。因此，人们期待电子邮件具有专业性，而不肤浅琐屑。
- 电子邮件是一种公开的交流形式——你的读者可能把你的电子邮件有意发给或误发给数不清的其他人。所以，你不应该用电子邮件说那些你不愿公开说给主管、同事或客户听的事。
- 电子邮件越来越正式——过去，读者会原谅电子邮件中的打字错误、拼写错误以及过失，特别是在电子邮件刚出现并难以使用的时候。现在，读者期待电子邮件更为正式，能够显示出对其他交流形式所期待的质量。
- 电子邮件标准和常规还在形成中——电子邮件在工作场所应该如何使用，这一点还在探索中。关于电子邮件的适当（以及不适当）使用，人们持有相差甚远的看法。因此，你需要密切关注在你的公司和你的读者的公司中电子邮件是如何使用的。

你还应该记住，法律约束对在工作场所如何使用电子邮件具有决定作用。像任何其他书面文档一样，电子邮件受版权法保护。因此，你得小心，不要以可能触犯版权法的任何方式使用电子邮件。例如，如果你收到一位客户的电子邮件，在没有得到该客户允许的情况下，你不能将电子邮件马上发布到你公司的网站上。

还有，律师和法庭把电子邮件视为书面通信，将它等同于备忘录或信函。例如，20 世纪 90 年代末针对微软的反垄断案在很大程度上就是以之前的电子邮件为根据的；在这些电子邮件中，比尔·盖茨和其他经营主管人员非正式地聊到与其他公司展开大胆有力的竞争。

从法律上讲，你通过雇主的计算机网络发送的任何电子邮件都属于雇主。因此，你的雇主有权阅读你的电子邮件而不需要你知道或得到你的允许。另外，删除的电子邮件可从公司的服务器上得到恢复，而且它们可用于法律案件。

2）电子邮件的基本特征

（1）题头

题头有用于填写收件人和主题的行。通常，还会有允许你扩展邮件功能的其他行，如抄送、密送、附件等。

收件人行——这是你键入将接收你发送的电子邮件的人的地址的地方。你可以在该行输入多个地址，允许你向多人发送邮件。

抄送和密送行——这些行用于复制邮件给次要读者，如你的主管或者对你的交谈可能感兴趣的其他人。抄送行向收件人表明，有其他人在接收邮件的副本。密送行允许你在别人不知晓的情况下将邮件复制给他人。

主题行——该行通常使用简短的语句标明邮件的话题。如果该邮件是对以前邮件的回复，则电子邮件程序通常会在主题行自动插入“回复:”。如果转发邮件，则在主题行插入“转发:”。

附件行——该行标明是否有任何其他文件、图片或程序附加于电子邮件之中。附加的文档会保持其原有格式，并可直接下载到读者的计算机中。

（2）信息区

信息区在题头下面，是你可键入致读者的话语的地方。它应该有清楚明白的引言、正文和结尾。

引言——引言应尽可能简短地表明主题，说明用意并阐述要点。还有，如果你想让读者做某事，你应该在最前面提出，而不是在电子邮件的结尾。

正文——正文应提供证明或支持你的电子邮件用意所需的信息。

结尾——结尾应重申要点并展望未来。大多数电子邮件读者从来不看结尾，因此你应该在邮件中提早告诉他们任何行动项目，然后在结尾部分予以重申。

（3）签名

电子邮件程序通常允许你创建一个签名文件，该文件可在邮件结尾自动加上签名。签名文件可繁可简。它们允许对邮件进行个性化处理，并添加额外联系信息。通过创建签名文件，你可以避免在你写的每封邮件的结尾处键入姓名、头衔、电话号码等。

（4）附件

附件是读者可以下载到自己计算机上的文件、图片或程序。

发送附件——你如果想为电子邮件添加附件，那就点击你的电子邮件软件程序中的“添加文档”或“附件”按钮。大多数程序随即会打开一个文本框，允许你查找并选择想要附加的文件。

接收附件——如果某人发给你一个附件，你的电子邮件程序会使用一个图标来表示有文件附加于电子邮件。点击该图标，大多数电子邮件程序随即会让你将该文档保存到硬盘上。你可以再从那里打开该文件。

3）管理邮件

邮件发送后，该邮件被复制保存在发送（发送或其他类似名称）文件夹中，以备阅读或重发。这些邮件和用户收到的邮件（保存在 Inbox 文件夹中）一直保留除非被删除或被移动到其他文件夹（用户可以通过邮件程序的文件菜单建立新文件夹，把邮件拖入该文件夹）。邮件被删除后，它转移到为删除项目建立的特殊文件夹（Netscape Mail 中为垃圾邮件，Outlook Express 中为删除项目）。用户应该在定期清空这些文件夹释放占用的硬盘空间操作时，选择 Netscape“文件”菜单中的“清空垃圾邮件”或 Outlook Express“编辑”菜单中的清空“删除项目”。

课文 3：网络安全

安全是一个广泛的话题，它涵盖了许多的犯罪。大多数安全问题是那些心怀恶意的人故意制造的，他们希望从中获取利益或是为了伤害某些人。

网络安全问题大体上可以分成 4 个紧密联系的方面：隐私、验证、不可否认性和完整性控制。隐私，也称保密性，指避免信息落入未授权者手中，这也是人们一提到网络安全就会想到的。验证能确认在透漏敏感信息或进行商业交易前与你交谈的人的身份。不可否认性处理签名问题：顾客电子定购的 1000 万个为左撇子设计的装置一开始每个是 89 美分，而随后却又称每个 69 美分，你如何证明确有其事呢？再或者他可能声称他根本没有定购。最后，你如何确认收到的信息确属某人所发，而非经过竞争对手中途修改或捏造？

所有的这些问题（隐私、验证、不可否认性和完整性控制）在传统的系统中都有发生过，但也有着显著的差异。通过使用注册邮件，锁定文件可以实现完整性和隐私。

在致力于找到解决的途径之前，值得我们花一些时间来弄明白这些协议栈网络安全问题源于何处。这恐怕不是出于一个地方，而是许多原因造成的。

计算机安全年表

有几个重大的关键事件促成了计算机安全这一概念的发展。接下来将介绍其中几个意义重大的事件，它们让人们开始关注计算机和信息安全，也成就了如今计算机安全的重要地位。

1．20 世纪 60 年代

麻省理工学院的几名学生成立了技术模型铁路俱乐部，并开始入侵学校的 PDP-1 主计算机系统并对其进行编程。最后他们将现在大家所熟知的“黑客”这一名词作为其代名词。尽管“黑客”一词最初的意义表示对那些热心于计算机的人的一种称赞，而今却成了非法闯入计算机系统的人的贬义词。一些黑客对计算机进行攻击纯粹是为了享受挑战，另一些则为了盗取计算机资源或是破坏某些数据。

2．20 世纪 70 年代

吉姆埃利斯和汤姆特拉斯科特创立了用户网，它是一种公告板系统，能满足完全不同用户间的电子通讯。随后便一跃成为用户计算、联网，当然还有黑客等交换意见的最受欢迎的平台。

3．20 世纪 80 年代

末日军团和混乱电脑俱乐部是黑客集团的两大先驱，它们首次开始利用计算机和电子数据网的漏洞发起攻击。

根据计算机行骗和滥用法，法庭宣判一名毕业生罗伯特莫里斯有罪，原因是他将莫里斯蠕虫病毒传播到了超过 6000 台网络计算机上。

4．20 世纪 90 年代

图形化网页浏览器诞生了，用户对公共互联网访问的需求立即成倍地暴增。

弗拉基米尔·莱文及其同伙通过入侵花旗银行中央数据库，非法转移了其中的 1000 万美元资金到各个账户。莱文后来被国际刑事警察组织逮捕，被盗资金也几乎全部收回。在所有黑客入侵中最有预见性的恐怕要数凯文米特尼克了，他闯入了好几家公司的系统，盗取了名人的私人信息、超过2万张信用卡密码以及专利软件的源代码。随后他被逮捕并以电信欺诈的罪名被判入狱5年。

5. 现今的安全问题

2000 年 2 月，分布式拒绝服务攻击在一些大流量网站上全面展开。这次进攻波及yahoo.com、cnn.com、amazon.com、fbi.gov等各大网站，而一般用户对于这些网站都是无法接近的，因为它利用大字节ICMP数据包的调动将路由器进行捆绑至好几个小时，也被称作试通洪流。这次攻击中，不明身份的袭击者采用了一种特别设计且使用广泛的程序，它能扫描到网络服务器的漏洞，并在服务器内安装名为木马的客户端应用程序。每台被病毒感染的服务器大面积扩散到受害的网站使得进攻次数成倍增加，并使得这些网站无法被访问。许多指责的声音指向一个基本的缺陷，就是路由器和协议的结构方式致使它们接受任何输入的数据，而并不理会这些数据来自哪里，有何种目的。

常见的进攻方式

机密的信息以两种方式存在于网络中。一种是存储在物理存储介质里，例如硬盘和内存；也能存在于以数据包的形式经过有线物理网络进行传输的线路里。信息的这两种形态为黑客提供了多种机会，它们既可以对内部网络的用户发起进攻也能对互联网上的用户进行攻击。我们通常担忧的是在第二种状态下的信息，它涉及网络安全问题。攻击总是试图绕过计算机采取的安全控制措施，通常它会篡改、泄露或是删除数据。一次黑客攻击的成败取决于计算机系统的漏洞及现存对策的效率。有五种常见的进攻方式会威胁到网络上的信息安全：网络数据包嗅探器进行IP欺骗、拒绝服务攻击、口令攻击、对外部信源散布高度机密内部信息以及中间人进攻。

安全和保护技术

1. 传统密码系统

直到计算机的出现，对密码系统的主要制约因素之一是编码人员执行必要的转变的能力，就像战场上没有装备的士兵一样。另外一个制约因素是很快从一个加密方法切换到另一个，这非常困难，因为需要培训一大批人。然而，如果需要的话，一个代码员被敌人抓获的危险使他必须能够改变即时加密方法。

要加密的信息，即众所周知的明码文本，是由一个关键参数化功能转化。加密过程的输出称为密文，之后通常以专人或无线电来传送。我们假设敌人或入侵者，听到并且准确地将完整的密文复制下来。然而，与预期的收件人相比，他不知道解密密钥，因此无法轻易解密密文内容。有时候，入侵者既能听通信信道（被动入侵者），还可以记录信息，并且稍后破解它，注入他自己的信息，或修改合法信息在它到达接受者之前（有源入侵者）。打破密码艺术被称为密码分析。在制订密码（加密）之后破解的艺术（密码分析）共同被称为密码学。

真正的秘密在密码本中，它的长度是一个重要的设计议题。考虑一个简单的组合锁。一般原则是，你输入序列数字。大家都知道这一点，但答案是秘密的。一个答案的长度是两个数字意味着有100种可能性。一个有3个数字键的长度有1000种可能性，而6位密钥长度意味着100万种可能性。密钥越长，越多的关键因素需要解密专家处理。通过不断地搜索密码空间来打破密钥空间的工作基数是密码本的几何倍数。保密来源于强大的（但公开的）算法和一个长的密码。为了防止你顽皮的弟弟阅读电子邮件，64位密钥就可以了。

为了保持在海湾各主要国家政府系统的安全，至少有256位的密钥是必要的。

即使我们将学习许多不同的密码系统，有两种基于它们的根本原则的理解是非常重要的。

第一个原则是，所有加密的消息肯定包含一些冗余，也就是与了解密文无关的信息。一个测试会告诉我们为什么这是必要的。拿一个有60 000件产品的邮购公司TCP来说。想象他们非常有效率，TCP的程序员决定订购信息应该包括16B的客户名称并伴随着3B的数据字段（产品数量占1B而产品数目占用2B）。最后3B被加密成一个非常长的只有客户和TCP知道的密码。

起初，这似乎是安全的，并在某种意义上是因为被动入侵者无法解密消息。不幸的是，它也有一个致命的缺陷，使得它失去作用。假设最近被解雇员工要惩罚TCP解雇她的行为。临走时，她带走了（部分）她的客户名单。她连夜编写一个用真的用户名发送假的订单程序。由于她没有密码的列表，她只需放3个随机数字代替最后3B，然后对TCP发送出数百的订单。

当这些信息抵达时，TCP的计算机用客户的名称来定位密码并且解密信息。对TCP来说不幸的是，几乎各种3B的信息都符合逻辑，所以计算机开始打印运输说明。即使对于一个客户订购137套秋千或者沙盒看起来很奇怪，而计算机认为客户可能计划开一个有特许权的连锁游乐场。这样一个活跃的闯入者（即那个前雇员）可以引起巨大的麻烦，即使她不懂计算机生成的信息。

这个问题可以通过对所有信息加入冗余来解决。然而，加入冗余也使密码专家破解信息更加简单。

因此，加密的原则之一是所有的号码都必须使包含的信息冗余，以防止活跃的入侵者欺骗接收者并奉行一些错误的信息。然而，同样的冗余使得它更容易被被动入侵者破坏系统，所以这里有一些需要注意的地方。此外，那些冗余绝不能把一些数字0的字符串加在信息的开头或者结尾，因为通过一些加密算法，为运行这些信息提供了更多的可预见的结果，使得解密专家的工作更容易。一个英语单词的随机字符串将是一个用于冗余更好的选择。

第二个原则是加密必须采取一些措施，以防止旧消息回收给积极入侵者。如果不采取这些措施，我们的前雇员可以窃听TCP的电话线路，并不断重复以前发送的有效信息。

2．秘密密钥算法

现代密码学采用和传统密码学一样换位和替换的基本思路，但其重点是不同的。传统上，密码学家已经使用简单的算法加以很长的密码来确保安全。现在的情况正好相反：对象使用如此错综复杂的算法以至于即使密码分析员获得了他自己选择的巨大加密文本，他自己也无法理解。换位和替换可以实现简单的循环。

1）DES

1977年1月，美国政府通过了一项由IBM开发的产品密码作为非机密信息的官方标准。这个密码——DES（数据加密标准），由业界广泛采用在安全产品上。它不再是原来形式的安全，而且修改的形式仍然是有用的。

2）IDEA

IDEA（国际数据加密算法）是由两位在瑞士的研究人员设计的。它使用128位密钥，这将使它对外部力量免疫，也可承受差分分析。目前认为已知的技术或机器没有可以突破IDEA的。

3．公用密钥算法

历史上，密钥分配问题一直是密码系统中最薄弱的环节。无论多么强大的密码体制，如果一个入侵者可以窃取密钥，该系统则形同虚设。由于所有的密码学家总是理所当然地认为加密密钥和解密密钥是相同的（或容易从一个导出另一个），并且密钥要分发到系统的所有用户，它好像有一个内在的内置问题：按键不仅需要防止盗窃，同时它们也有要分发到用户，所以它们不能只是被锁在银行的金库了。

公用密钥密码系统要求每个用户有两个密码，一个是公用密码，被全世界用于加密信息发送到用户，一个是私用密码，为用户解密信息所需要。我们将一如既往地分别提到公用密钥和私用密钥，并通过用在传统的密码系统（也称为对称密钥）的密钥来加密和解密以区分它们。

RSA算法由最初开发它的3个人的英文名字（Rivest、Shamir、Adleman）的首字母缩写构成。虽然它被广泛使用，并以数论的一些原则为基础，但它并不是唯一的公共密钥算法。第一次公开的密钥算法是背包算法（梅克尔和赫尔曼，1978）。其他公共密钥计划是基于计算离散对数的困难（瑞彬，1979）。使用这种原理的算法是由厄尔贾迈勒（1985年）和石诺（1991年）发明的。还有几个其他的计划，如基于（梅内塞斯和范斯通，1993年）椭圆曲线的。基于需要因式分解的巨大数字，计算离散对数，并从其重量确定背包中的内容，这三大类问题被视作是真正难以解决的，因为数学家已经努力了多年，仍然没有什么大的突破。

4．认证协议

身份验证是通过一种技术的过程来检验它的合作伙伴应该是谁并且不是骗子。验证试图远程登录的邪恶活跃的入侵者的身份是出人意料地艰难，并且需要烦琐的基于密码学的协议。

5．数字签名

许多法律、金融和其他文件的真实性，是由存在或授权的情况下手写签名所决定的，复印件不算。随着电脑化的信息系统取代纸张和油墨物理传输文件，必须找到一个解决办法来解决这些问题。

数字签名的方法之一称为秘密密钥签名，要有一个中央机关知道一切并被大家信任，即所谓的“大哥”（BB）。然后每个用户选择一个秘密的钥匙，由专人携带到BB的办公室里。因此，只有Alice和BB或者鬼魂知道Alice的秘密。

一个与使用秘密密钥加密的数字签名的结构性问题是每个人都必须同意信任“大哥”。此外，“大哥”需要阅读所有签名的邮件。运行的“大哥”服务器的最合乎逻辑的候选人是各国政府、银行或律师。然而这些组织并没有激发全体公民的信心。因此，如果签署文件没有要求有一个值得信赖的权力机构是非常好的。幸运的是，公共密钥加密可以在这里做出重要贡献。它是公共密钥签名。

有一些社会问题，如网络安全涉及的个人隐私和社会的专利。一定范围内网络安全是政治化的；一些其他的科技议题也大抵如此，因为它涉及一个民主国家和专制国家在数字化时代的差异。

6．防火墙

防火墙是为了保护某个组织网络以抵抗外部攻击的安全系统。一个有局域网并允许员工通过局域网访问广域网的企业，安装防火墙以阻止外部用户使用它自己的私有数据资源并控制自己用户访问外部的资源。防火墙的主要工作是与路由器程序一起，检查每一个网络数据包以决定是否将其转发到目的地址。

防火墙有 4 种普遍认可的类型，它们都用于因特网的连接上：帧过滤防火墙、包过滤防火墙、电线网关防火墙和代理服务器防火墙。下面对每种类型进行简要介绍。

- 帧过滤防火墙：帧过滤防火墙能过滤布局中的位级别和局域网帧的内容。通过这种方式的过滤，那些不属于被信任部分的帧在到达任何存在价值的地方甚至是防火墙自身之前就被拒绝了。
- 包过滤防火墙：包过滤防火墙可以查看数据包的 IP 地址（网路层）和连接的类型（传输层），通过它们的信息来进行过滤。一个包过滤防火墙可以是一个独立的路由设备或者是一台含有两个网络接口卡的计算机（双宿主系统）。路由器连接两个网络并且进行包过滤来控制网络间的非法交易。程序员通过一套能够定义包过滤如何才算工作完成的规则来编写设计设备。包过滤能使端口也被封锁。如果一家公司的安全政策只允许网络浏览器（超文本传输协议）的使用，而不允许相对危险的文件传输协议，包过滤可以通过实施适当规则来做到。
- 电路级网关：电路级网关是一种代理服务系统，它提供了一种在内部系统和外部系统之间可控的网络连接。虚拟的“电路”存在于内部客户端和代理服务系统之间。互联网请求通过这个电路到达代理服务系统。在更改 IP 地址后，代理服务系统将这些请求传达给互联网。外部使用者只能看到代理服务系统的 IP 地址。答复随后被代理服务器所接受并通过电路回传给客户端。当流量被允许经过时，外部系统绝对察觉不到内部系统。
- 代理服务器：一个应用层的代理服务器拥有所有基本的代理器特征并且能进行广泛的数据包分析。当数据包从外部进入到网间连接器时，安全策略将会检查和评估它们是否被允许进入到内部网络。代理服务器不仅仅是检验 IP 地址，它还检查数据包内的数据是否安全，以防黑客在数据包里隐藏了恶意信息。如果任何一家公司的员工想要在互联网上访问某台服务器，这台计算机的请求将被传到代理服务器上。代理服务器随后使用它的地址作为源地址来与互联网上的这台服务器取得联系。信息从这台服务器经过代理服务器回传到请求得到数据的实际计算机上。通过这种方式，内部（公司）计算机的 IP 地址绝不会被外界的网络知道。代理服务器还会记录发出请求的计算机的信息和数据调用的细节，以便分析互联网接入的情况。

课文 4：HTML5 Canvas 简介

HTML5 是新一代的 HTML，即超文本标记语言。HTML 从 1993 年第一次标准化后，便奠定了万维网的基础。HTML 通过使用将标签用尖括号(<>)括起来的方式定义 Web 页面内容。

HTML5 Canvas 是屏幕上的一个由 JavaScript 控制的即时模式位图区域。即时模式是指在画布上呈现像素的方式，HTML Canvas 通过 JavaScript 调用 Canvas API，在每一帧中完全重绘屏幕上的位图。作为一名程序员，所要做的就是在每一帧渲染之前设置屏幕的显示内容，这样才能显示正确的像素。

这使得 HTML5 Canvas 与在保留模式下运行的 Flash、Silverlight 或 SVG 有很大的区别。在保留模式下，对象显示列表由图形渲染器保存，通过在代码中设置属性（例如，x 坐标、y 坐标和对象的 alpha 透明度）控制展示在屏幕上的对象。这使得程序员可以远离底层操作，但是它弱化了对位图屏幕最终渲染效果的控制。

基本的 HTML5 Canvas API 包括一个 2D 环境，允许程序员绘制各种图形和渲染文本，并且将图像直接显示在浏览器窗口定义的区域。读者可以对画布上放置的图形、文本和图像应用颜色、旋转、渐变色填充、alpha 透明度、像素处理等，并且可以使用各种直线、曲线、边框、底纹来增强其效果。

就其本身而言，HTML5 Canvas 2D 环境是一个用来在位图区域渲染图形显示的 API，但人们很少使用该技术在这个环境中创建应用程序。通过跨浏览器兼容的 JavaScript 语言可以调用键盘鼠标输入、定时器间隔、事件、对象、类、声音、数学函数等功能，希望读者能够学会并使用 HTML5 Canvas 创建优质的动画、应用程序和游戏。

本书将深入解读 Canvas API。在此过程中，本书将展示如何使用 Canvas API 来创建应用程序。本书中的很多技术已经被成功应用于其他平台，现在，本书要将它们应用到 HTML5 Canvas 这个令人兴奋的新技术上来。

支持 HTML5 Canvas 的浏览器

除了 IE 8 以外，很多新版本的浏览器都支持 HTML5 Canvas。几乎每天都会支持新的特性。支持最好的应该是 Google Chrome，紧接着是 Safari、Internet Explorer 10、Firefox、Opera。本书将利用名为 modernizr.js 的 JavaScript 库来帮助判断各个浏览器支持哪些 Canvas 特性。

什么是 HTML5

最近 HTML5 的定义已经发生了转变，当作者在 2010 年编写本书第一版的时候，W3C 的 HTML5 规范是一个独特的单元，它涵盖了有限的功能集合，其中包括了诸如新的 HTML 标签（<video>、<audio>和<canvas>）之类的东西。然而，在过去的一年中，这一定义已经发生了改变。

那么，究竟什么是 HTML5?在 W3C HTML5 的常见问题中，关于 HTML5 是这样说明的：HTML5 是一个开放的平台下开发的免费许可条款。

术语 HTML5 会被人们使用在以下两个方面。

- 指一组共同构成了未来开放式网络平台的技术。这些技术包括 HTML5 规范、CSS3、

SVG、MATHML，地理位置、XmlHttpRequest、Context 2D、Web 字体以及其他技术。这一套技术的边界是非正式的，且随时间变化的。

- 指 HTML5 规范，当然也是开放式网络平台的一部分。

在过去的几个月里，我们通过交谈和项目工作了解到的是：普通人（或者说那些急着要完成项目的客户）谁也不会严格遵守上述定义，这些都是 HTML5。因此，当有人说起“HTML5”的时候，他们实际上指的是“开放式网络平台”。

当人们提及“开放式网络平台”时，有一件可以确定的事是，这份邀请名单中一定不能漏掉 Adobe Flash。

HTML5 是什么？总之，它不是 Flash（也不是其他类似的技术）。HTML5 Canvas 是最有能力在网络和移动互联网上取代 Flash 功能的最好的技术。这本书将带领读者学习如何开始使用 HTML5 Canvas。

基础的 HTML5 页面

在开始讲解 Canvas 前，需要谈论一下 HTML5 的相关标准——这里将使用 HTML5 来创建 Web 页面。

HTML 是用于在互联网上构建页面的标准语言。本书不会将很多时间花费在讲解上，但 HTML 是<canvas>的基础，所以不能完全跳过它。

一个基本的 HTML 页面分成几个部分，通常有<head>和<body>，新的 HTML5 规范增加了一些新的部分，例如<nav>、<article>、<header>和<footer>。

<head>标签通常包含与使用<body>标签来创建 HTML 页面相关的信息。将 JavaScript 函数放在<head>中是约定俗成的，稍后讨论<canvas>标签时也会这样做。虽然有理由把 JavaScript 函数放在<body>中，但是简单起见，最好把 JavaScript 函数放在<head>中。

基本的 HTML 页面如例 1 所示。

例 1 简单的 HTML 页面。

```
<!doctype html>
<html lang="en">
<head>
<meta charset="UTF-8">
<title>CH1EX1: Basic Hello World HTML Page</title>
</head>
<body>
Hello World!
</body>
</html>
```

1. <!doctype html>

这个标签说明 Web 浏览器将在标准模式下呈现页面。根据 W3C 定义的 HTML5 规范，这是 HTML5 文档所必需的。这个标签简化了长期以来在不同的浏览器呈现 HTML 页面时出现的奇怪差异。它通常为文档中的第一行。

2. <html lang="en">

这是包含语言说明的<html>标签：例如，“en”为英语。下面是一些常见的语言值：

中文：lang = "zh"

法语：lang = "fr"

德语：lang = "de"

意大利语：lang = "it"

日语：lang = "ja"

韩语：lang = "ko"

波兰语：lang = "pl"

俄语：lang = "ru"

西班牙语：lang = "es"

3. <meta charset= "UTF-8">

这个标签说明 Web 浏览器使用的字符编码模式。如果没有需要特别设置的选项，一般没必要改变它。这也是 HTML5 页面需要的元素。

4. <title>…</title>

这个标签说明在浏览器窗口展示的 HTML 的标题。这是一个很重要的标记，它是搜索引擎用来在 HTML 页面上收录内容的主要信息之一。

5. 一个简单的 HTML5 页面

现在，在浏览器中看看这个页面（这是一个伟大的时刻，可以准备好工具开始开发代码了）。打开所选择的文本编辑器以及 Web 浏览器——Safari、FireFox、Opera、Chrome 或 IE。

（1）在文本编辑器中，输入例 1 中的代码。

（2）选择路径，保存为 CH1EX1.html。

（3）在 Chrome、Safari 或 Firefox 的 File 菜单中，找到 Open File 命令，单击它，将看到一个能够打开文件的对话框（在 Windows 下用 Chrome 时，也可以按 Ctrl+O 键来打开文件）。

（4）找到刚刚创建的 CH1EX1.html。

（5）单击“打开”按钮。

可以看到如图 47 所示的结果。

本书使用的基础 HTML 页面

可以使用多个 HTML 标签来创建 HTML 页面，在 HTML 以前的版本中，需明确指示 Web 浏览器如何渲染 HTML 页面的标签（例如<font>和<center>）。然而，在过去 10 年中，浏览器的标准越来越严格，这类标签就被束之高阁了。CSS（层叠样式表）成为定义 HTML 内容样式的主要方式。因为本书不是关于如何创建 HTML 页面的（页面中不包含 Canvas），因此这里不打算讨论 CSS 的内部工作原理。

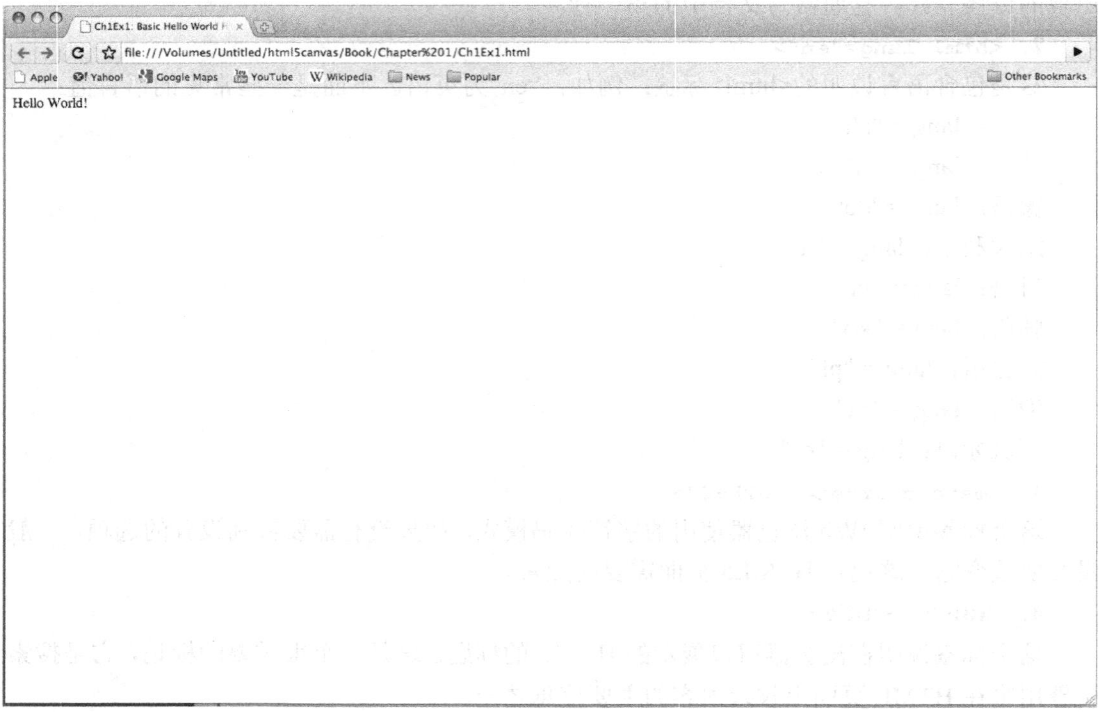

图 47 “Hello World!”页面

本节将只关注两个最基本的 HTML 标签：<div>和<canvas>。

1．<div>

<div>是本书主要使用的一个 HTML 标签，用来定位<canvas>在 HTML 页面的位置。例 2 使用<div>标签定义了“Hello World!”在屏幕上的位置，如图 48 所示。

例 2 HTML5 中的“Hello World!”。

```
<!doctype html>
<html lang="en">
<head>
<meta charset="UTF-8">
<title>CH1EX2: Hello World HTML Page With A DIV </title>
</head>
<body>
<div style="position: absolute; top: 50px; left: 50px;">
Hello World!
</div>
</body>
</html>
```

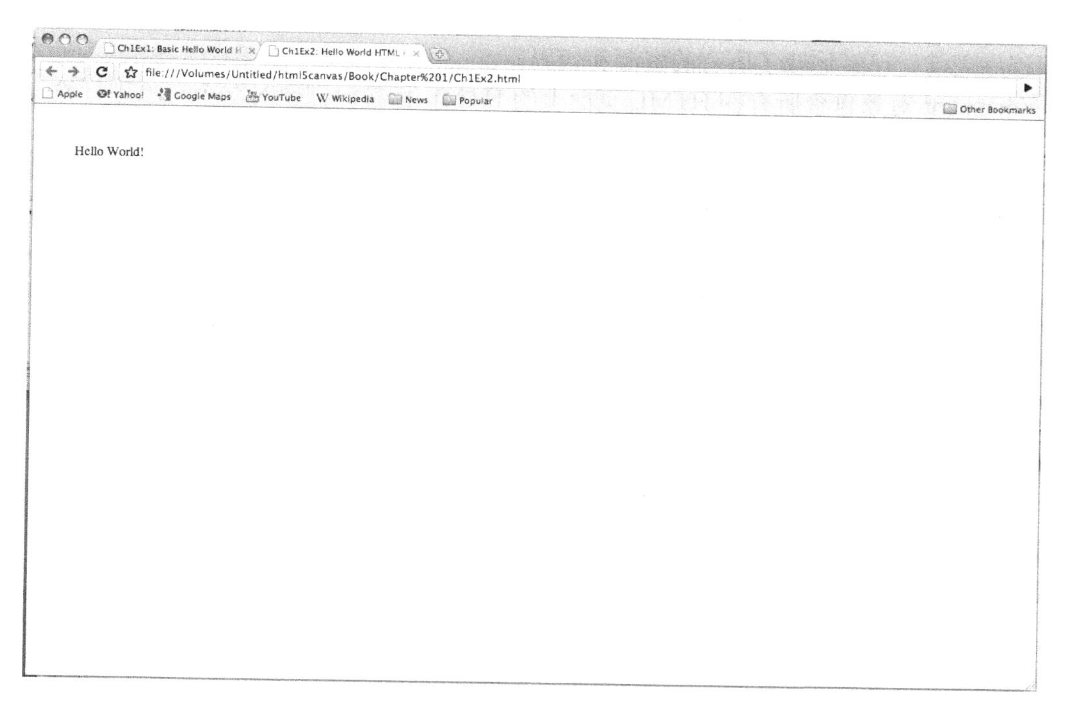

图 48　使用<div>的 HTML5 中的“Hello World！”

style="position:absolute;top:50px;left:50px;"——这段代码是在 HTML 页面中使用内联 CSS 的例子。它告诉浏览器呈现内容的绝对位置为：距离页面顶端 50 像素，并且距离页面左端 50 像素。

警告：这个<div>可以在浏览器中定位画布，但是对试图在画布上捕捉鼠标点击时则没有任何帮助。在第 5 章中，本书将讨论一种既能定位画布又能捕获正确的鼠标点击位置的方法。

2．<canvas>

利用对<div>进行绝对定位，有助于更好地使用<canvas>。把<canvas>放在<div>内，<div>可以帮读者获取信息。例如，当鼠标划过画布时，可以获取定义指针的相对位置。

文档对象模型（DOM）和 Canvas

文档对象模型代表了在 HTML 页面上的所有对象。它是语言中立且平台中立的。它允许页面的内容和样式被 Web 浏览器渲染之后再次更新。用户可以通过 JavaScript 访问 DOM。从 20 世纪 90 年代末以来，文档对象模型已经成为 JavaScript、DHTML 和 CSS 开发最重要的一部分。

画布元素本身可以通过 DOM，在 Web 浏览器中经由 Canvas 2D 环境访问。但是，在 Canvas 中创建的单个图形元素是不能通过 DOM 访问的。正如本章前面讲到的，画布工作在即时模式，它并不保存自己的对象，只是说明在每个单个帧里绘制什么。

例 2 在 HTML5 页面上使用 DOM 定位<canvas>标签，这也可以用 JavaScript 来操作。在开始使用<canvas>前，首先需要了解两个特定的 DOM 对象：windows 和 document。

window 对象是 DOM 的最高一级，需要对这个对象进行检测来确保开始使用 Canvas

应用程序之前，已经加载了所有的资源和代码。

document 对象包含所有在 HTML 页面上的 HTML 标签。需要对这个对象进行检索来找出用 JavaScript 操纵<canvas>的实例。

JavaScript 和 Canvas

JavaScript 是用来创建 Canvas 应用程序的一种程序设计语言，能在现有的任何 Web 浏览器中运行。如果需要重温 JavaScript 的相关内容，请关注 Douglas Crockford 的书 *JavaScript:The Good Parts*（O'Reilly）。这本书很流行，并且有很高的参考价值。

JavaScript 放置的位置及其理由

使用 JavaScript 为 Canvas 编程会产生一个问题：在创建的页面中，从哪里启动 JavaScript 程序？

把 JavaScript 放进 HTML 页面的<head>标签中是个不错的主意，这样做的好处是很容易找到它。但是，把 JavaScript 程序放在这里就意味着整个 HTML 页面要加载完 JavaScript 才能配合 HTML 运行，这段 JavaScript 代码也会在整个页面加载前就开始执行了。结果就是，运行 JavaScript 程序之前必须检查 HTML 页面是否已经加载完毕。

最近有一个趋势是将 JavaScript 放在 HTML 文档结尾处的</body>标签里，这样就可以确保在 JavaScript 运行时整个页面已经加载完毕。然而，由于在运行<canvas>程序前需要使用 JavaScript 测试页面是否加载，因此最好还是将 JavaScript 放在<head>中。如果读者不喜欢这样，也可以采用适合自己的代码习惯。

代码放在哪儿都行——可以放在 HTML 页面代码行内，也可以加载一个外部.js 文件。加载外部 JavaScript 文件的代码大致如下：

```
<script type="text/javascript" src="canvasapp.js"></script>
```

简单起见，这里将把代码写在 HTML 页面行内。不过，如果读者有把握，把它放在一个外部文件再加载运行也未尝不可。

提示：HTML5 不需要再指定脚本类型。

附录B 词汇表

A

abusive *adj.* 辱骂的，说人坏话的；滥用的，妄用的；虐待的
accelerate *v.* 加速，促进
accelerator *n.* 加速者，加速器
access *n.* 通路，访问，入门
access time 存取时间；访问时间
accumulate *vt.&vi* 累加，堆积
accurate *adj.* 准确的，精确的
additional *adj.* 增加的，添加的；附加的，追加的，外加的，额外的
adrenaline *n.* 肾上腺素
adventure *n.* 冒险，冒险的经历
algorithm *n.* 算法，运算法则
alley *n.* 胡同，小巷
allocation *n.* 分配，配置，安置
amortize *v.* 摊销，摊还，分期偿付
amplitude *n.* 振幅，丰富，广阔
analog *n.* 模拟
animation *n.* 活泼，生气，激励，卡通片绘制
anode *n.* 阳极
anomaly *n.* 异常，异常现象，反常现象
antenna *n.* 天线
antitrust *adj.* 反垄断的
approximation *n.* 近似，近似值
arc *n.* 弧，弧线
architecture *n.* 建筑学 建筑式样，建筑风格
arrange *vt.* 整理，布置，安排 *vi.* 作安排，作准备
arrangement *n.* 装置，设备，安装，布置
arrogate *v.* 非法霸占
artichoke *n.* 朝鲜蓟
artwork *n.* （书，杂志等的）插图；艺术作品；美术作品
ascend *v.* 攀登，上升
aspect ratio *n.* 屏幕宽高比

assembly language　汇编语言
assembly　*n.* 集合，集会；立法会议
asset　*n.* 资产，有用的东西
asynchronous　*adj.* 不同时的，异步的
asynchronous transfer mode (ATM)　异步传输模式
attenuation　*n.* 变薄，稀薄化，变细，衰减
audible　*adj.* 听得见的
authentication　*n.* 证明；鉴定；证实
authorization　*n.* 授权，委托，所授之权
authorized　*adj.* 经认可的；经授权的
authorize　*v.* 授权
automatically　*adv.* 自动地；机械地
axis　*n.* 轴，坐标轴

B

backbone　*n.* 脊椎，中枢，骨干，支柱，意志力，勇气，毅力，决心
backstage　*adv.* 在后台
bandwidth　*n.* 带宽
bar　*n.* 条，棒（常用作栅栏，扣栓物）
baud　*n.* 波特
beam　*n.*（光线的）束，柱，电波，梁
binary　*adj.* 双重的，双的，二进制的“数学”二元的
blanking pulse　消隐脉冲
blooper　*n.* 大挫折
bombard　*vt.* 轰击
boot　*n.* 引导；自引；启动；引导程序
booth　*n.* 货摊，售货亭，棚
border on　邻近，接界；近乎
bottleneck　*n.* 瓶颈
breakthrough　*n.* 突破
brightness　*n.* 亮度
broadband　*adj.* 多频率的
broadcast　*n.* 无线电；电视节目
brochure　*n.* 小册子
browser　*n.* 浏览者
bucket　*n.* 水桶
budge　*vt.&vi.*（使）稍微移动
buffer　*n.* 缓冲器

burst *v.* 爆炸，爆裂
buzzword *n.*（报刊等的）时髦术语，流行行话
bytecode *n.* 字节码，字节代码；位元码

C

cable *n.* 电缆；电报
cache *n.* [计算机]高速缓冲内存，高速缓冲存储器
camera tube 摄像管
capability *n.* 能力
caricature *n.* 漫画；夸张的描述或模仿
cartoon *n.* 卡通画，漫画，动画片
catalog *v.* 编目录
catastrophic *n.* 大灾难
catch a glimpse of 瞥见
category *n.* 种类，类型
cathode-ray tube 阴极射线管
community antenna TV（CATV） 共用天线电视
cellular *adj.* 细胞的
census *n.* 户口普查 *v.* 实施统计调查
central processing unit（CPU） 中央处理器，中央处理装置
chandelier *n.* 枝形吊灯
channel *n.* 信道、通道
characteristic *adj.* 表示特性的，典型的，特有的
checksum 检验[校验]和，核对和
chip *n.* 芯片
choppy *adj.* 波涛汹涌的
chore *n.* 零星工作（尤指家常杂务）
chroma *n.*（色彩的）浓度，色度
cinematographer *n.* 电影摄影技师，放映技师
ciphertext *n.* [计算机]密文
cladding *n.* 包层，镀层
classify *v.* 把……分类
client *n.* 委托人
clip *n.* 剪辑，从电影胶片或录像带剪出的片断 *vt.* 限幅，消波
coarse *adj.* 粗糙的，未精炼的
coat *n.& vt.* 涂，涂层
cocky *adj.* 狂妄自信的
codec *n.* 编码解码器

collapse *vi.* 倒坍，崩溃，瓦解
collide *v.* 冲突
collision *n.* 碰撞，冲突
communicate *vt. &vi.* 传递，传播
compatible *adj.* 兼容的
compiler *n.* 编辑者，汇编者，编译器，编译程序
complicated *adj.* 复杂的，难解的
complication *n.* 复杂化（使复杂化的）因素
composition *n.* 成分；合成；作品，著作，结构；构图；布置，布局
compress *vt.* 压，压缩；归纳，精简
compression *n.* 浓缩，压缩
comprise *v.* 包括，包含，构成
concept *n.* 概念，观念，思想
conceptualize *v.* 概念化
concurrent *adj.* 同时发生的，同时存在的
concurrently *adv.* 一致地，合作地，协力地
conductive *adj.* 传导的
conference *n.* 会议
connotation *n.* 内涵；含蓄；暗示，隐含意义；储蓄的东西（词、语等）
conservative *n.* 保守的人
considerable *adj.* 相当大的，相当多的
constraint *n.* 约束；强制
contact lenses *n.* 隐形眼镜
contain *vt.* 包含，容纳；相当于
context *n.* 背景，环境；上下文，语境
continuous *adj.* 继续的，连续的，持续的，延伸的
continuum *n.* 连续统一体
contour *n.* 轮廓，周线，等高线
convergence *n.* 集中，收敛
converter *n.* 转换器
convex *adj.* 凸的，凸面的
convey *v.* 运送，运输
convolution *n.* 卷积
coordinate *n.* 坐标系，坐标
coordinating *adj.* 协调的 *v.* 协调（coordinate 的 ing 形式）；整合
crisp *adj.* 脆的，鲜脆的；新奇的，整洁的；清新的，干冷的
cryptanalyst *n.* 密码专家；密码破译者
cryptography *n.* 密码使用法；密码学
cuff *n.* 袖口

curvature *n.* 弯曲，曲率
customer *n.* 顾客；[口]家伙
cyan *n.* 蓝绿色，青色
cycle *n.* 周期，循环
cylinder *n.* 圆柱体

D

deafening *adj.* 震耳欲聋的
debug *vt.* 驱除（某处的）害虫；排除程序等中的错误
decrease *v.* 减少，变少，降低
dedicate *v.* 奉献；贡献
dedicated *adj.*[计算机]专用的
deduct *v.* 扣除；减去
default *n.* 默认（值），缺省（值）
deflect *vt.&vi.* 偏斜
demonstration *n.* 示范
departure from 违反，违背
derivation *n.* 推导
derogatory *adj.* 贬损的
descend *v.* 下来，下降
description *n.* 描述，描写，说明书，类型
desktop search 桌面搜索
destination *n.* 目的地，终点
differential pulse code modulation 差分脉码调制
digital signal processor 数字信号处理
dimension *n.* 尺度，维度
discrete *adj.* 离散的，分立的，不连续的
dismay *vt.* 使灰心，使沮丧；使惊愕
disparate *adj.* 全异的
disperse *v.*（使）散开；（使）分散
disseminate *v.* 传播；散布；撒播
dissimilar *adj.* 不同的，相异的
distinction *n.* 区别；差别；不同之处；特征；特性；个性
distort *v.* 扭曲
distortion *n.* 扭曲，变形，曲解，失真
distraction *n.* 使人分心的事或人；娱乐，消遣
distribute *v.* 分发；发行
dizziness *n.* 头昏眼花

dodge *vt.&vi.* 闪躲
dominant *adj.* 支配的，统治的，占优势的
double-jointed *adj.* 可屈曲之关节的
download *vt.* [计算机]下载
drag *v.* 拖，拖曳，缓慢而费力地行动
drama *n.* 戏剧
drawn *adj.* 疲惫的，憔悴的
drill *n.* 操练，训练，演习 *vt.* 练习
drum *n.* 鼓，鼓状物，鼓膜 *v.* 打鼓
DTS（Digital Theatre System） *abbr.* 数字化影院系统
duplex *adj.* [电信，计算机]双工的，双向的
duration *n.* 持续时间，为期
DVI（Digital Visual Interface） *n.* 数字视频接口

E

elaborately *adv.* 苦心经营地，精巧地
electromagnetic *adj.* 电磁的
eliminate *v.* 除去，排除，剔除，[计算机]消除
encapsulate *vt.* 装入胶囊；总结；扼要概括；囊括
enclose *v.* 装入，围绕
encompass *vt.* 围绕，包围
encrypt *v.* 加密，将……译成密码
encryption *n.* 加密；加密术
encyclopedia *n.* 百科全书；专科全书
enhance *v.* 美化，提高
enormously *adv.* 巨大地，庞大地，非常地，在极大程度上
ensemble *n.* 整体，总效果
equilibrium *n.* 平衡，平静，均衡
equivalent *adj.* 相等的，相当的，同意义的 *n.* 等价物，相等物
erase *vt.* 擦掉，抹去
erratic *adj.* 不稳定的，无规律的，漂泊的
error *n.* 错误，误差
escapism *n.* 逃避现实，空想
essential *adj.* 必需的，基本的
eternity *n.* 永恒
evaluated *adj.* 估价的
evaluation *n.* 求值，评估
exaggerate *vt.&vi.* 夸大，夸张

examined *adj.* 验讫；检查过的 *v.* 检查；调查（examine 的过去式）
exception *n.* 额外，例外
executable *adj.* 可执行的，可实行的，可以做成的
execute *vt.* 实行，实施，执行；完成，实现，履行
executive *adj.* 执行的
explicitly *adv.* 明白地，明确地
exploration *n.* 搜寻，考察，探究，探索，考查，调查
exponentially *adv.* 以指数方式
extension *n.* 伸展，扩大，延长，延期，电话分机
external *adj.* 外部的；表面的；外面的；[药]外用的；外国的 *n.* 外部；外面；外观
extract *vt.* 提取
extraterrestrial *adj.* 地球外的，地球大气圈外的 *n.* 外星人
eyedropper *n.* 滴眼药器
eyelid *n.* 眼睑，眼皮

F

facilitate *n.* 便利，设备 *v.* 使便利
facility *n.* 设备
fade *vt.&vi.* 逐渐消失
federal *adj.* 联邦的
fiber-optic *n.* 视觉纤维
field *n.* 场地；领域；行业
film *n.* 薄膜
fine tuning 细调
firewall *n.* 防火墙 *vt.* 用作防火墙
flexible *adj.* 灵活的
flicker *vt.* 闪烁
flip *vt.&vi.* 翻转，倒转
floppy *adj.* 易掉落的，松软的
flyback *n.* 回扫，回描，逆程扫描
forecast *vt.* 预测；预报（天气） *n.* 预测；预报
format *n.* 格式，形式，板式 *vt.* 格式化，安排……的格局
Fourier transform 傅里叶变换
frame *n.* 帧，画面，框架
frequency *n.* 频率，周期，发生次数
frivolous *adj.* 轻薄的；轻浮的
front panel 前面板
fumble *vt.&vi.*（笨拙地）摸索或处理（某事物）

fundamental *adj.* 基本的，根本的 *n.* 基本原理；基本原则
fungus *n.* 菌类，蘑菇
fuzzy *adj.* 模糊的，失真的

G

garbage *n.* 垃圾，废物
gating signal 选通信号
geometric *adj.* 几何的，几何学的
geosynchronous *adj.* 与地球的相对位置不变的，相对地球是静止的
gigabit *n.* 吉（咖）比特
gigahertz *n.* 千兆赫
given *adj.* 特定的，假设的
gradation *n.* 分等级，顺序，阶级
graphical *adj.* 书写的，绘画的；印刷的，雕刻的
gravity *n.* [物]重力，（地心）引力；重量；认真，严肃，庄重
gray-scale *n.* 灰度
grid *n.* 栅极
GSM(Global System For Mobile Communication) *abbr.* 全球移动通信系统
Guaranteed *n.* 有保证的，被担保的

H

halogen *n.* 卤素
hand-held *adj.* 手持式的
hangover *n.* 宿醉（酒后醒来的头痛和不舒服）
harmonic *n.& adj.* 谐波
HDMI(High-Definition Multimedia Interface) *n.* 高清晰度多媒体接口
HE-AAC(High-Efficiency Advanced Audio Coding) *abbr.* 高效能高级音频编码，一种有损数据压缩技术
Hefner Candlepower *n.* 赫夫纳烛光
helix *n.* 螺旋结构
Hertz *n.* 赫，赫兹（频率单位：周/秒）
heterogeneous *adj.* 不同种类的，杂散的
hexagon *n.* 六边形；六角形
hierarchy *n.* 层级，等级制度
horizontal *adj.* 地平线的，水平的
hourly *adv.* 频繁地，随时，每小时地 *adj.* 每小时的，以钟点计算的，频繁的
hub *n.* 中心；枢纽

hue *n.* 色调
human-computer interface 人机接口
hybrid *n.* （动植物）杂种；混合型
hydrogen *n.* [化学]氢
hypermedia *n.* 超媒体

I

icon *n.* 图标，肖像，偶像
identification *n.* 认明，识别，鉴定
identifier *n.* 标志（标识，识别）符
idle *adj.* 空闲的，闲着的；懒散的，无所事事的
illuminate *v.* 照亮
illustrative *adj.* 说明性的
immediate *adj.* 直接的；最接近的；即时的，立即的
immerse *vt.* 使浸入
impedance *n.* 阻抗
implementation *n.* 实施，实现，安装启用
inadvertent *adj.* 疏忽的；漫不经心的；非有意的；因疏忽所致的
inadvertently *adv.* 不注意地
incarnation *n.* 具体化，化身，体现
inch *n.* 寸，英寸
incident *adj.* 入射的
incorporate *vt.* （使）合并，并入，合编，组成公司，具体表现，使混合，使加入
incorporated *adj.* 组成公司的，合成一体的
incur *v.* 招致，遭受
indent *vt.* 切割……使呈锯齿状，缩进排版
index *v.* 编入索引中，指出
indicate *v.* 表示
indigo *n.* 靛，靛青色
indispensable *adj.* 必须的
infancy *n.* 婴儿期，幼年时代，未成年，初期，幼年期
inherent *adj.* 固有的；内在的；与生俱来的，遗传的
inherit *vt.* 继承，遗传而得 *vi.* 成为继承人
inheritance *n.* 阶层，层级，分层，分类，遗产，继承，遗传
initial *adj.* 最初的
innately *adv.* 天赋地，天生就有地
instance *n.* 实例，情况
instant messaging （网上的）即时通信（服务）

instantaneous *adj.* 瞬间的，即刻的，即时的
insulator *n.* 绝缘体，绝热器
intact *adj.* 完整无缺的
integer *n.* 整数；完整的东西
integrated circuits 集成电路
intensity *n.* 强度
interaction *n.* 交互作用，相互作用
interactive *adj.* 相互作用的，交互式的
interactive video 交互视频
interference *n.* 冲突，干涉
interim *adj.* 暂时的，临时的，间歇的 *n.* 过渡时期
interior *n.* 内部
interlace *v.* 隔行，交错
intermediate *adj.* 中间的，居中的
intermittent *adj.* 间歇的，断断续续的
internationalization *n.* 国际化
interoffice *adj.* （同一组织的）各部门间的；各办公室之间的
interoperability *n.* 可由双方共同操作
interphone *n.* 对讲机
interrelated *adj.* 相互关联的
intersection *n.* 横断；交叉
interwine *v.* 缠绕
intimidate *vt.* 恐吓，威胁
intranet *n.* 内联网
introspection *n.* 内省，反省，自省
invariably *adv.* 不变地，总是
invariance *n.* 不变性
investment *n.* 投资
invoke *vt.* 援引，援用；行使（权利等）
irreparable *adj.* 不能挽回的；不能修补的
irrespective *adj.* （与 of 连用）不顾……的，不考虑……的，不论……的
isochronous *adj.* 等时的
item *n.* 条款，项目

J

jerky *adj.* 跳动的，不平稳的
jitter *n.* 抖动，跳动
juggling *v.* 欺骗，杂耍；*adj.* 欺骗的，欺诈的，变戏法（似）的

K

Kelvin *n.* 绝对温标，开氏温标，简写为 K
kernel *n.*（果实的）核，仁；内核，核心，要点
kilobit *n.* 千（二进制）位，千比特

L

landscape *n.* 风景，景色
laptop *n.* 膝上计算机
latency *n.* 等待时间
leverage *n.* 杠杆作用，势力
linear *adj.* 线的；直线的
Linux *n.* 一个自由操作系统，一种可免费使用的类 UNIX 操作系统
loop *n*. 循环
loop *n.* 环形；圈
lossless *adj.* 无损的
lossy *adj.* 有损的
luminosity *n.* 发光度
lumped *adj.* 集总的
Lux *n.* 勒克斯（照明单位）

M

magenta *n.* 红紫色，洋红
magnifier *n.* 放大器，放大镜
magnify *vt.* 放大，扩大 *vi.* 放大，扩大；有放大能力
mainframe computer 大型计算机
mainframe *n.* 主（计算）机，大型机
maintenance *n.* 维持；维护；保养；维修
malfunction *vi.* 失灵，发生故障
manipulate *v.* 操纵，利用，假造；[计算机]操作
manual *adj.* 手工的；人力的
manufacturing *n.* 制造业
marine *adj.* 海的，海产的，航海的，船舶的，海运的
marketplace *n.* 集会场所，市场；商场
mass *n.* 大量，主题，（物体的）质量
match *n.&vt.&vi.* 匹配

mathematician *n.* 数学家
matrix *n.* 矩阵
matte *adj.* 不光滑的
maximum *n.& adj.* 最大量（的），最大值（的）
mechanics *n.* 力学；机械学
megabit *n.* 兆位，百万位
memorabilia *n.* 大事记
merge *vt.* 使合并，使结合
message *v.* 通知，发信息
metaphor *n.* 隐喻，暗喻
methodology *n.* 一套方法，方法学，方法论
metropolitan *n.* 大城市；首都
microphone *n.* 麦克风，话筒，扩音器（也作 mike）
microprocessor *n.* 微处理器
mighty *adj.* 强有力的，强大的；*adv.* 非常，很
miniaturize *vt.* 使小型化
minicomputer *n.* 小型计算机；小型电脑
miscellaneous *adj.* 不同种类的，多种多样的；混杂的
mixer *n.* 混频器，混合器
modifier *n.* [语]修饰语
modularization *n.* 模块化
molecular *adj.* 分子的，摩尔的
monetary *adj.* 货币的，金钱的
monitor *n.* 班长，监听器，监视器，监控器
monster *n.* 怪物，恶人，巨物
motivate *v.* 激发，促动
motivation *n.* 动机
MPEG(Moving Pictures Experts Group/Motion Pictures Experts Group) *abbr.* 运动图像专家组，一种压缩比率较大的活动图像和声音的压缩标准
multimedia file system 多媒体文件系统
multimode *n.* 多方式，多态，多型
multiple *n.& adj.* 倍数，多倍的，多数的
multiplex *adj.* 复合的，多重的
multipoint *adj.* 多点（式）的，多位置的
multithreaded *adj.* 多线程，多线程的，多重线串的
multithreading *n.* 多线程，多线索

N

narration *n.* 叙述

NASA（National Aeronautics and Space Administration） *n.* 美国国家航空和航天局

nausea *n.* 作呕；恶心；反胃

nested *adj.* 嵌套的

Netscape *n.* 网景公司

node *n.* 节点，中心点

non-interactive 非交互式的

nonrepudiation *n.* 认可；不可抵赖性；非否认性

novice *n.* 新手，初学者

NTSC（National Television Systems Committee） *n.* NTSC 制式，全国电视系统委员会制式

nudge *vt.* 轻推，推进 *n.* 轻推，推动

O

occupied *v.* 占领已占用的，使用中的；无空闲的

ongoing *adj.* 前进的，进行的 *n.* 前进，举止，行为

onscreen *adv.& adj.* 在银幕上（的）

opportunities *n.* 因素；机会；机遇

optimize *vt.* 有效地进行，使完美 *vi.* 持乐观态度

optimizing *n.* 优化，最佳化 *adj.* 最佳的

optimum *adj.* 最好的，最佳的，最有利的

organism *n.* 生物体；有机体；社会组织；机关

orthogonal *adj.* 正交的

outfit *n.* 全套装备，全套工具

outgoing *adj.* 往外去的，即将离任的，好交往的

outrageous *adj.* 骇人的；残暴的；无耻的；不道德的

overhead *n.* 管理费用，经常费用

overlap *vt. & vi.* 部分重叠 *n.* 重叠的部分

overpowering *adj.* 无法抵抗的，压倒性的

override *vt.* 超越，取代，不顾

P

PAL（Phase Alternating Line） *n.* PAL 制式，逐行倒相制式

palette *n.* 调色板，控制面板

pan *vt.* 摇动（镜头），使拍摄全景；*n.* 摇镜头，拍全景
paradigm *n.* 典范，范例，示例
parallel *adj.* 平行的，相同的，类似的，并联的；
parametric *adj.* 参（变）数的，参（变）量的
parity *n.* 同等，平等，[计算机]奇偶校验
partition *n.* 划分
pavement *n.* 人行道
PDA(Personal Digital Assistant) *abbr.* 个人数字助理
peculiarity *n.* 古怪；特性
penetrate *vt.& vi.* 穿透
pentagon *n.* 五边形；五角形
perceive *v.* 感知，感到，认识到
perception *n.* 感知（能力），觉察（力），认识，观念，看法
permanent *adj.* 永久的，不变的，耐久的；持久的，经久的
persistence *n.* 坚持，持续
pertinent *adj.* 恰当的
phosphor *n.* 黄磷
picture tube 显像管
pigment *n.* 色素，颜料
pitch *n.* 斜度，程度，倾斜 *vt.* 投，掷，定位于 *vi.* 投掷，坠落，倾斜
pivotal *adj.* 枢轴的，关键的
planar *adj.* 平面的，平坦的
platform *n.* 平台，讲台；计算机平台
polygon *n.* 多角形，多边形
polymorphism *n.* 多态性，多形性
PON(passive optical network) *abbr.* 无源光纤网络
portable *adj.* 轻便的，手提（式）的，便携式的
portion *n.* 一部分；一定数量；区；段
potential *adj.* 可能存在或出现的，可能的 *n.* 潜在性，可能性
precursor *n.* 先驱者，前导，先进者
predominant *adj.* 占主导地位的，显著的
preeminent *adj.* 卓越的，杰出的，出类拔萃的
preliminary *adj.* 初步的，预备的，开端的 *n.* 准备工作，初步行动
prevalent *adj.* 普遍的，流行的
prey *n.* 被掠食者，牺牲者 *vi.* 捕食
primary memory 主存储器
primitive *adj.* 原始的，早期的
priority *n.* 先，前，优先，优先权
prism *n.* 棱镜，棱柱

profound *adj.* 深刻的，意义深远的
progressive *adj.* 渐次的，逐渐的，渐进的
proliferate *n.* 增生增殖，扩散，激增 *vt.* 使激增
proliferation *n.* 增殖，扩散
propagate *v.* 繁殖，增殖
proportion *n.* 比例，比率
provoking *adj.* 激怒的
punctuation *n.* 标点；标点法；全部标点符号
pyramidal *adj.* 金字塔形的，锥体的

Q

quantum *n.* 量，额；定量，定额；份；总量
quartz *n.* 石英

R

randomness *n.* 随意，无安排
rate *n.* 比率，速度，等级，价格 *vt.* 估价，认定，鉴定等级 *vi.* 被评价
ratio *n.* 比率
real-time scheduling 实时调度
recalculate *v.* 重新计算
recipient *n.* 接受者
reduction *n.* 缩减；降低；减小
refraction *n.* 折光，折射
refractive *adj.* 折射的
regarding *prep.* （表示论及）关于；至于；就……而论
reinforcement *n.* 加强
reinstalling *n.* 重新安装
relatively *adv.* 相对地，比较地
remainder *n.* 剩余物，其余（的人）
render *vt.* 渲染；报答；归还；给予；呈递；提供；开出
repel *vt.* 排斥
repository *n.* 贮藏室，智囊团，知识库，仓库
represent *vt.* 表现，描绘，声称，扮演 *vi.* 提出异议
resemble *v.* 相似
reservation *n.* 预定；预留
reshape *vt.* 改造，再成形
resolution *n.* （监视器的）分辨率

retain *v.* 保留；保持
retrace *vt.* 回扫（描）
retrieval *n.* 检索
retrieve *vt.* 取回，挽回，弥补，恢复，补偿，回忆，检索
revolutionize *v.* 使革命化
ribbon *n.* 带，缎带，丝带
ripped *adj.* [美俚]喝醉的；受毒品麻醉的

S

sample *n.* 采样，标本，样品 *vt.* 采样，取样，抽取……的样品，尝试
saturation *n.* 饱和（状态），浸润，浸透
scalability *n.* 可量测性，可伸缩性
scalable *adj.* 可攀登的，可升级的
scanning rate 扫描速率
scene *n.* 情景，现场
sculpt *v.* 雕刻，造型
seamless *adj.* 无缝合线的，无伤痕的
SECAM(法文 Sequentiel Couleur A Memoire) *n.* SECAM 制式，顺序与存储彩色电视系统
secondary storage 辅助存储器，二级存储器
security *n.* 安全；抵押品；证券；保证 *adj.* 安全的；保密的；保安的
segment *n.* 部分，片段
semantics *n.* 语义学，语义论
semicolon *n.* 分号
serial *adj.* 连续的
set-top box *n.* 置顶盒
shatter *vt.* 砸碎，粉碎
shivering *adj.* 颤抖的
shuttle *n.* 往返汽车（列车、飞机），航天飞机，梭子，穿梭
signature *n.* 签名
silhouette *n.* 轮廓
simplex *n.* 单工
simplicity *n.* 简单，简易，朴素，朴实，单纯
simulator *n.* 模拟器，模拟装置
skeptic *n.* 无神论者
sketch *n.* 草图；素描；速写
slider *n.* 滑动器；滑子（块，板，座）
snap *vt.* 猛咬，突然折断

sniffer　n. 嗅探器；嗅探犬；以鼻吸毒者
snippet　n. （尤指讲话或文字的）小片，片段，零星的话
somewhat　adv. 有点儿，微微
sophisticate　adj. 使复杂，使老于世故；老于世故的人
sophistication　n. 复杂性；尖端性
specification　n. 详述，规格，说明书，规范
specify　v. 规定；详述
spectator　n. 观众
spectral　adj. 光谱的
speech processing　语音处理
sphere　n. 球（体）
splitter　n. 分路器，分裂机
spot　n. 光点
squiggly　adv. 弯弯曲曲地
squint　vi. 〈医〉斜视；眯着眼睛，斜着眼睛（看某物）；瞟；从小孔或缝隙里看
stage　n. 舞台，活动场景
stand-alone multimedia workstation　独立多媒体工作站
standardization　n. 标准化
statement　n. 语句
still　adj. 静止的，静寂的
still image　静止图像
storage　n. 贮存，保管，存储器
streamline　vt. 把……做成流线型，简化使效率更高
stylish　adj. 有风度的，有气派的，有格调的
stylized　adj. 程式化的，格式化的
stymied　adj. [美]被侵袭的
subclasses　n. 子类，亚类
submultiple　n. 约（因）数
subordinate　adj. 次要的，从属的
subscript　adj. 写在下面的　n. 添标，下标，下角数码
substitutions　n. 代替；代用；替换（substitution 的复数）
subvert　vt. 颠覆，破坏（政治制度、宗教信仰等）
successive　adj. 连续的，相继的
sufficient　adj. 充分的
superb　adj. 卓越的，杰出的，极好的
supercomputer　n. 超级计算机，巨型计算机
superfluous　adj. 多余的，过剩的，过量的
superimpose　v. 叠加
supervisor　n. 监督者

susceptible *adj.* （与 to 连用）易受影响的
S-Video(Separate Video) *n.*二分量视频接口
sweep *vt.& vi.* 扫描
sync *n.* 同时；同步
sync pulse 同步脉冲
synchronization *n.* 同步
synchronize *v.* 使同步
synchronized mode 同步模式
synchronous *adj.* 同时发生的，同步的
STM(synchronous transfer mode) 同步传输模式
syntax *n.* 句法
synthesis *n.* 综合，合成

T

tangible *adj.* 切实的，通过触摸可以感知的，确实的，真实的
TDMA(Time Division Multiple Access) *abbr.* 时分多址
TD-SCDMA(Time Division-Synchronous Code Division Multiple Access) *abbr.* 时分同步码分多址
tease *vt.&vi.* 取笑，戏弄
technique *n.* 方法，技术
tendril *n.* 卷须，蔓，卷须状之物
terminate *vt.&vi.* 结束；使终结
terminology *n.* 专门用语，术语
text-to-speech conversion 文本语音变换
texture *n.* 纹理
thereby *adv.* 因此，从而，由此
thickness *n.* 厚度，浓度，稠密
threshold *n.* 门槛，入口，开端，阈，[计算机]阈；阈值
thumbnail *adj.* 极小的，极短的
timeline *n.* 时间线
tint *n.* 带白的颜色，淡色
toggle *v.* 切换
token ring 令牌环网
torso *n.*（人体的）躯干；（人体的）躯干雕塑像
toxicity *n.* 毒性
trademark *n.* （注册）商标
trailer *n.* 预告片
trajectory *n.*（射体）轨道，弹道，流轨
transaction *n.* 事务；事项

transfer rate　传输速率
transfer　*v.* 转移，调转，调任，传递，转让，改变
transistor　*n.* 晶体管（收音机）
transition　*n.* 转变，转换，跃迁，过渡，变调
translator　*n.* 翻译者；[无线电]译码机；变换器；传送器；转发器
transmission　*n.* 传输，播送，传递，传动装置，变速器
transparent　*adj.* 明显的
transponder　*n.* 异频雷达收发机
traverse　*vt.& vi.* 经过，横过
triangulate　*vt.* 分成三角形，对……作三角测量
trigger　*n.* 触发器
trim　*v.* 整理，修整，装饰
tuner assembly　调谐器组件
turquoise　*n.* 绿宝石，绿松石色，青绿色
typo　*n.* 打字排印错误

U

ultrasound　*n.* 超声，超声波
ultraviolet　*adj.* 紫外的；紫外线的；紫外线辐射的
uncompress　*vt.* 未压缩
underlie　*vt.* 位于或存在于（某物）之下；构成……的基础（或起因）；引起
underpinning　*n.* 基础，支柱，支撑
underscore　*vt.* 在……下划线；强调
underway　*adj.* 起步的，进行中的，航行中的
unity　*adj.* 和谐，统一
unprecedented　*adj.* 空前的
unsheathe　*v.* 抽出鞘，拔出
upheaval　*n.* 动乱，剧变
utilization　*n.* 利用，被利用
utilize　*v.* 利用

V

vacuum　*n.* 真空（度）；真空状态；空虚；空白；空处
vacuum　tube 电子管
valid　*adj.* 有效的
validation　*n.* 证明正确，批准，确认
value　*n.* 色彩中的明暗关系
variable　*adj.* 易变的，不稳定的

variation *n.* 变量，变度，偏差
vast *adj.* 巨大的
vector *n.* 向量，矢量
vendor *n.* 摊贩；卖家
vernier trimmer 微调电容器
version *n.* 版本，形式
vertices *n.* 制高点，头顶
vestigial *adj.* 残余的
veteran *n.* 经验丰富的人；老兵
via *prep.* 经由，经过
vibrate *v.* （使）振动，（使）摇摆
vibration *n.* 振动，颤动，摇动，摆动
vicinity *n.* 邻近，附近，接近
viewer *n.* 阅读器
violate *v.* 侵犯；冒犯
violet *n.* 紫罗兰色
vision *n.* 视觉
VOD(video on demand) *abbr.* 视频点播
void *n.* 空间，空旷，空虚
voltage *n.* 电压，伏特数
volume *n.* 音量，卷，册，体积
vulnerability *n.* 弱点；易损性
vulnerable *adj.* 易受伤的，脆弱的，敏感的

W

waveform *n.* 波形
wavelength *n.* （无线电）波长
wavelet *n.* 子波，小波
WCDMA(Wideband Code Division Multiple Access) *abbr.* 宽带码分多址
webmaster *n.* 网管
whack *n.* 重打，重击
whenever *adv.* 随便什么时候
wipe off 揩去，擦掉
with a vengeance 猛烈地；极度地
workstation *n.* 工作站

Z

zoom *n.&v.* 缩放；突然扩大，急速上升，变焦

参 考 文 献

1. 王春艺，刘艺. 新编计算机英语[M]. 北京：机械工业出版社，2007.
2. 卜艳萍，周伟. 计算机专业英语[M]. 北京：清华大学出版社，2010.
3. 郭涛. 计算机专业英语[M]. 武汉：华中科技大学出版社，2007.
4. 张海波，何谦卫，黄亦军，等. 信息技术英语[M]. 北京：中国水利水电出版社，2004.
5. 张强华，司爱侠. 计算机专业英语[M]. 北京：科学出版社，2006.
6. 陈枫艳，陈志峰. 计算机英语[M]. 北京：清华大学出版社，北京交通大学出版社，2010.
7. 李大友，方娟，朱学斌. 计算机专业英语[M]. 北京：清华大学出版社，2004.
8. 程显林. 信息技术科技英语[M]. 北京：清华大学出版社，2004.
9. 司爱侠，张强华. 计算机英语教程[M]. 北京：电子工业出版社，2009.
10. 支丽平，武文婷，刘晓魁. 计算机专业英语[M]. 北京：中国水利水电出版社，2010.
11. 姜同强，王雯，孔凡航，等. 计算机英语（学生用书）[M]. 北京：清华大学出版社，2004.
12. 陈俊宇，辛燕清. 计算机专业英语[M]. 北京：冶金工业出版社，2006.
13. 赵萱，郑仰成. 科技英语翻译[M]. 北京：外语教学与研究出版社，2006.
14. 张彩霞，黄清清. 多媒体与艺术设计专业英语[M]. 北京：北京师范大学出版社，2008.
15. 谢小苑. 科技英语翻译技巧与实践[M]. 北京：国防工业出版社，2008.
16. 王卫平，潘丽蓉. 英语科技文献的语言特点与翻译. 上海：上海交通大学出版社，2009.
17. 马新英，等. 科技英语教程[M]. 合肥：中国科学技术大学出版社，1997.
18. 苏雪. 通信专业英语[M]. 武汉：华中科技大学出版社，2007.
19. Joseph Lowery. Dreamweaver CS3 Bible. Indianapolis, Indiana: Wiley Publishing, Inc, 2007.
20. Joey Lott, Darron Schall, Keith Peters. ActionScript 3.0 Cookbook. O'Reilly, 2006.
21. Rafael C. Gonzalez, Richard E. Woods. Digital Image Processing[M]. 2nd ed. Prentice Hall, 2002.
22. Mark Christiansen. Adobe After Effects 7.0 Studio Techniques[M]. Adobe Press, 2006.
23. David Austerberry. Digital Asset Management. Focal Press, 2006.
24. Kelly L. Murdock. 3ds Max 2009 Bible. Canada: Wiley Publishing, Inc., 2008.
25. Adele Droblas, Seth Greenberg. Adobe Premiere Pro Bible. Canada: Wiley Publishing, Inc., 2004.
26. Steve Fulton, Jeff Fulton. HTML5 Canvas: Native Interactivity and Animation for the Web. O'Reilly, 2011.

图书资源支持

感谢您一直以来对清华版图书的支持和爱护。为了配合本书的使用，本书提供配套的资源，有需求的读者请扫描下方的"书圈"微信公众号二维码，在图书专区下载，也可以拨打电话或发送电子邮件咨询。

如果您在使用本书的过程中遇到了什么问题，或者有相关图书出版计划，也请您发邮件告诉我们，以便我们更好地为您服务。

我们的联系方式：

地　　址：北京海淀区双清路学研大厦 A 座 707

邮　　编：100084

电　　话：010－62770175－4604

资源下载：http://www.tup.com.cn

电子邮件：weijj@tup.tsinghua.edu.cn

QQ：883604(请写明您的单位和姓名)

资源下载、样书申请

书圈

用微信扫一扫右边的二维码，即可关注清华大学出版社公众号"书圈"。